INSTRUCTOR'S MANUAL

to accompany

General Chemistry

CAROLE H. McQUARRIE

DONALD A. McQUARRIE
University of California, Davis

PETER A. ROCK
University of California, Davis

W. H. FREEMAN AND COMPANY
New York

Printed in the United States of America

Contents

Preface

This Instructor's Manual is meant to accompany the text *General Chemistry* by Donald A. McQuarrie and Peter A. Rock. It contains detailed solutions to all the even-numbered problems that appear at the end of each chapter and answers to all the questions that appear after each Interchapter. The solutions to the even-numbered problems for all twenty six chapters are presented first, followed by the answers to the questions of all fourteen (A through N) Interchapters. Detailed solutions to all the odd-numbered problems can be found in the *Study Guide/Solutions Manual* that accompanies the text.

We have used SI units almost exclusively and have paid careful attention to expressing numerical answers to the appropriate number of significant figures. The use of hand calculators, however, introduces a slight difficulty in presenting intermediate numerical results in the solutions to the problems. If we insist upon presenting intermediate numerical results to the correct number of significant figures and use those results to carry out subsequent calculations, then round-off errors can arise, and the final numerical answer may differ in the last digit from the numerical answer obtained by carrying out the calculation continuously on a hand calculator. To avoid this difficulty, we usually display intermediate results to one more digit than is significant, and then round off the final result to the appropriate number of significant figures. Even this procedure is not always foolproof, however, and so the answers to the numerical problems occasionally may be inaccurate to one digit in the last significant figure. We do not consider this to be of any importance, but we do alert you to this inevitable possibility.

The authors wish to thank Dr. Joseph Ledbetter for checking all the problems, Elaine Rock for typing the material, and Michael Rock for assembling the material in camera-ready form.

CAROLE H. MCQUARRIE
DONALD A. MCQUARRIE
PETER A. ROCK

January 26, 1984

CHAPTER 1

SOLUTIONS TO THE EVEN-NUMBERED PROBLEMS

1-2. a) V d) Mg g) Br j) As

b) Au e) Fe h) Kr

c) Zn f) Cs i) Sb

See the alphabetical list of the elements in the inside front cover.

1-4. a) platinum d) tungsten g) nickel j) carbon

b) strontium e) calcium h) tin

c) lead f) chromium i) sulfur

See the alphabetical list of the elements in the inside front cover.

1-6. The mass percentage of lanthanum is given by

$$\text{mass \% of La} = \frac{\text{mass of lanthanum}}{\text{mass of compound}} \times 100$$

$$= \frac{7.08 \text{ g}}{8.29 \text{ g}} \times 100 = 85.4\%$$

The mass percentage of oxygen is given by

$$\text{mass \% of O} = \frac{\text{mass of oxygen}}{\text{mass of compound}} \times 100$$

$$= \frac{1.21 \text{ g}}{8.29 \text{ g}} \times 100 = 14.6\%$$

1-8. The mass percentage of tin in the sample of stannous fluoride is given by

$$\text{mass \% of Sn} = \frac{\text{mass of Sn}}{\text{mass of stannous fluoride}} \times 100$$

$$= \frac{1.358 \text{ g}}{1.793 \text{ g}} \times 100 = 75.74\%$$

The mass of fluorine in the sample of stannous fluoride is

$$\text{mass of F} = \text{mass of stannous fluoride} - \text{mass of tin}$$

$$= 1.793 \text{ g} - 1.358 \text{ g} = 0.435 \text{ g}$$

The mass percentage of fluorine in the sample of stannous fluoride is given by

$$\text{mass \% of F} = \frac{\text{mass of F}}{\text{mass of stannous fluoride}} \times 100$$

$$= \frac{0.435\text{ g}}{1.793\text{ g}} \times 100 = 24.3\%$$

Note that the mass percentage of fluorine is given to only three significant figures.

1-10. The respective mass percentages are

$$\text{mass \% of C} = \frac{\text{mass of C}}{\text{mass of ethyl alcohol}} \times 100$$

$$= \frac{1.93\text{ g}}{3.70\text{ g}} \times 100 = 52.2\%$$

$$\text{mass \% of H} = \frac{\text{mass of H}}{\text{mass of ethyl alcohol}} \times 100$$

$$= \frac{0.49\text{ g}}{3.70\text{ g}} \times 100 = 13\%$$

$$\text{mass \% of O} = \frac{\text{mass of O}}{\text{mass of ethyl alcohol}} \times 100$$

$$= \frac{1.28\text{ g}}{3.70\text{ g}} \times 100 = 34.6\%$$

1-12. Using Table 1-6 and the list of the elements in the inside front cover, we have

a) barium fluoride
b) magnesium nitride
c) rubidium bromide
d) cesium chloride
e) calcium sulfide

1-14.
a) aluminum oxide
b) magnesium fluoride
c) aluminum nitride
d) magnesium selenide
e) lithium phosphide

1-16.
a) antimony trichloride and antimony pentachloride
b) iodine trichloride and iodine pentachloride
c) krypton difluoride and krypton tetrafluoride
d) selenium dioxide and selenium trioxide
e) carbon monosulfide and carbon disulfide

1-18. Refer to the inside front cover for atomic masses

a) The molecular mass of $C_5H_{10}O_2$ is

molecular mass = (5 x atomic mass of C)+(10 x atomic mass of H)+(2 x atomic mass of O)

= (5 x 12.01) +(10 x 1.008)+(2 x 16.00)

= 102.13

b) The molecular mass of $C_8H_{16}O_2$ is

molecular mass = (8 x atomic mass of C)+(16 x atomic mass of H)+(2 x atomic mass of O)

= (8 x 12.01) +(16 x 1.008)+(2 x 16.00)

= 144.21

c) The molecular mass of $C_6H_{12}O_2$ is

molecular mass = (6 x atomic mass of C)+(12 x atomic mass of H)+(2 x atomic mass of O)

= (6 x 12.01) +(12 x 1.008)+(2 x 16.00)

= 116.16

d) The molecular mass of $C_9H_{18}O_2$ is

molecular mass = (9 x atomic mass of C)+(18 x atomic mass of H)+(2 x atomic mass of C)

= (9 x 12.01) +(18 x 1.008)+(2 x 16.00)

= 158.23

1-20. a) molecular mass of $C_{20}H_{30}O$ = (20 x atomic mass of C)+(30 x atomic mass of H)+(atomic mass of O)

= (20 x 12.01) +(30 x 1.008)+(16.00)

= 286.44

b) molecular mass of $C_{12}H_{17}ClN_4OS$ = (12 x atomic mass of C)+(17 x atomic mass of H)+(atomic mass of Cl)+ (4 x atomic mass of N)+(atomic mass of O)+(atomic mass of S)

= (12 x 12.01) +(17 x 1.008)+(35.45)+ (4 x 14.01) +(16.00) +(32.06)

= 300.81

c) molecular mass of $C_{17}H_{20}N_4O_6$ = (17 x atomic mass of C)+(20 x atomic mass of H)+(4 x atomic mass of N)+ (6 x atomic mass of O)

= (17 x 12.01) +(20 x 1.008)+ (4 x 14.01) +(6 x 16.00)

= 376.37

d) molecular mass of $C_{56}H_{88}O_2$ = (56 x atomic mass of C)+(88 x atomic mass of H)+(2 x atomic mass of O)

= (56 x 12.01) +(88 x 1.008)+(2 x 16.00)

= 793.26

e) molecular mass of $C_6H_8O_6$ = (6 x atomic mass of C)+(8 x atomic mass of H)+(6 x atomic mass of O)

= (6 x 12.01) +(8 x 1.008)+(6 x 16.00)

= 176.12

1-22. molecular mass of NO = 14.01 + 16.00 = 30.01

$$\text{mass \% of N} = \frac{\text{atomic mass of N}}{\text{molecular mass of NO}} \times 100$$

$$= \frac{14.01}{30.01} \times 100 = 46.68\%$$

$$\text{mass \% of O} = \frac{\text{atomic mass of O}}{\text{molecular mass of NO}} \times 100$$

$$= \frac{16.00}{30.01} \times 100 = 53.32\%$$

1-24. molecular mass of SiO_2 = (28.09) + (2 x 16.00) = 60.09

$$\text{mass \% of Si} = \frac{\text{atomic mass of Si}}{\text{molecular mass of SiO}_2} \times 100$$

$$= \frac{28.09}{60.09} \times 100 = 46.75\%$$

$$\text{mass \% of O} = \frac{2 \times \text{atomic mass of O}}{\text{molecular mass of SiO}_2} \times 100$$

$$= \frac{2 \times 16.00}{60.09} \times 100 = 53.25\%$$

1-26. molecular mass of Na_3AlF_6 = 209.95

$$\text{mass \% of Na} = \frac{3 \times \text{atomic mass of Na}}{\text{molecular mass of Na}_3\text{AlF}_6} \times 100$$

$$= \frac{3 \times 22.99}{209.95} \times 100 = 32.85\%$$

$$\text{mass \% of Al} = \frac{\text{atomic mass of Al}}{\text{molecular mass of Na}_3\text{AlF}_6} \times 100$$

$$= \frac{26.98}{209.95} \times 100 = 12.85\%$$

$$\text{mass \% of F} = \frac{6 \times \text{atomic mass of F}}{\text{molecular mass of Na}_3\text{AlF}_6} \times 100$$

$$= \frac{6 \times 19.00}{209.95} \times 100 = 54.30\%$$

1-28. $$\text{mass \% of Li in Li}_2\text{CO}_3 = \frac{2 \times 6.941}{(2 \times 6.941) + (12.01) + (3 \times 16.00)} \times 100$$

$$= 18.79\%$$

$$\text{mass \% of Li in LiC}_2\text{H}_3\text{O}_2 = \frac{6.941}{(6.941) + (2 \times 12.01) + (3 \times 1.008) + (2 \times 16.00)} \times 100$$

$$= 10.52\%$$

$$\text{mass \% of Li in Li}_3\text{C}_6\text{H}_5\text{O}_7 = \frac{3 \times 6.941}{(3 \times 6.941) + (6 \times 12.01) + (5 \times 1.008) + (7 \times 16.00)} \times 1$$

$$= 9.92\%$$

$$\text{mass \% of Li in } Li_2SO_4 = \frac{2 \times 6.941}{(2 \times 6.941)+(32.06)+(4 \times 16.00)} \times 100$$

$$= 12.63\%$$

1-30. Using a 100 g sample, we can write

$$\text{compound I:} \quad \frac{\text{mass of F}}{\text{mass of Br}} = \frac{41.64 \text{ g F}}{58.36 \text{ g Br}} = \frac{0.7135 \text{ g F}}{1.000 \text{ g Br}}$$

$$\text{compound II:} \quad \frac{\text{mass of F}}{\text{mass of Br}} = \frac{54.32 \text{ g F}}{45.68 \text{ g Br}} = \frac{1.189 \text{ g F}}{1.000 \text{ g Br}}$$

The ratio of the mass of F per gram of Br in compound II to compound I is

$$\text{ratio} = \frac{1.189 \text{ g F}/1.000 \text{ g Br}}{0.7135 \text{ g F}/1.000 \text{ g Br}} = 1.666$$

which is the ratio of the small whole numbers 5 to 3. If we assume that each compound has one bromine atom, then compound I has three fluorine atoms and compound II has five fluorine atoms. The compounds are BrF_3 and BrF_5.

1-32. Using a 100 g sample, we can write

$$\text{compound I:} \quad \frac{\text{mass of Cl}}{\text{mass of Sn}} = \frac{37.40 \text{ g Cl}}{62.60 \text{ g Sn}} = \frac{0.5974 \text{ g Cl}}{1.000 \text{ g Sn}}$$

$$\text{compound II:} \quad \frac{\text{mass of Cl}}{\text{mass of Sn}} = \frac{54.43 \text{ g Cl}}{45.56 \text{ g Sn}} = \frac{1.195 \text{ g Cl}}{1.000 \text{ g Sn}}$$

The ratio of the mass of Cl per gram of Sn in compound I to compound II is

$$\text{ratio} = \frac{1.195 \text{ g Cl}/1.000 \text{ g Sn}}{0.5974 \text{ g Cl}/1.000 \text{ g Sn}} = 2.000 = 2$$

which is a small, whole number. Compound II has twice as many chlorine atoms per tin atom than does compound II. The compounds are $SnCl_2$ and $SnCl_4$.

1-34. For compound I, the mass of Fe is 5.58 g Fe and the mass of Cl is (12.68 - 5.58) = 7.10 g Cl. For compound II, we have 5.58 g Fe and (16.23 - 5.58) = 10.65 g Cl.

compound I: $\dfrac{\text{mass of Cl}}{\text{mass of Fe}} = \dfrac{7.10\text{ g Cl}}{5.58\text{ g Fe}} = \dfrac{1.27\text{ g Cl}}{1.00\text{ g Fe}}$

compound II: $\dfrac{\text{mass of Cl}}{\text{mass of Fe}} = \dfrac{10.65\text{ g Cl}}{5.58\text{ g Fe}} = \dfrac{1.91\text{ g Cl}}{1.00\text{ g Fe}}$

The ratio of the mass of Cl per gram of Fe in compound II to compound I is

$$\text{ratio} = \frac{1.91\text{ g Cl}/1.00\text{ g Fe}}{1.27\text{ g Cl}/1.00\text{ g Fe}} = 1.5 \text{ or } \frac{3}{2}$$

If we assume each compound contains one Fe atom, then compound I has two Cl atoms and compound II has three Cl atoms. The compounds are $FeCl_2$ and $FeCl_3$.

1-36.

	protons	electrons	neutrons
$^{14}_{6}C$	6	6	14-6=8
$^{206}_{82}Pb$	82	82	206-82=124

1-38.

	protons	electrons	neutrons
a) P-30	15	15	15
b) Tc-97	43	43	54
c) Fe-55	26	26	29
d) Am-240	95	95	145

1-40.

symbol	atomic number	number of neutrons	mass number
$^{48}_{20}Ca$	20	28	48
$^{90}_{40}Zr$	40	50	90
$^{131}_{53}I$	53	78	131
$^{99}_{42}Mo$	42	57	99

1-42.

symbol	atomic number	number of neutrons	mass number
$^{39}_{19}K$	19	20	39
$^{56}_{26}Fe$	26	30	56
$^{84}_{36}Kr$	36	48	84
$^{120}_{50}Sn$	50	70	120

1-44. Using the data in Table 1-9, we have that

$$\text{atomic mass of Mg} = (23.9850)\left(\frac{78.99}{100}\right) + (24.9858)\left(\frac{10.00}{100}\right) + (25.9826)\left(\frac{11.01}{100}\right) = 24.31$$

1-46. $$\text{atomic mass of Si} = (27.977)\left(\frac{92.23}{100}\right) + (28.977)\left(\frac{4.67}{100}\right) + (29.974)\left(\frac{3.10}{100}\right) = 28.09$$

1-48. Let x be the percentage of boron-10 in naturally occurring boron. The percentage of boron-11 must be 100-x. Now set up the equation

$$\text{atomic mass of B} = 10.811 = 10.013\left(\frac{x}{100}\right) + (11.009)\left(\frac{100-x}{100}\right)$$

Multiply this equation through by 100 and collect terms to obtain

$$1081.1 = 10.013x + 1100.9 - 11.009x$$

Collecting terms, we get

$$0.996x = 19.8$$

$$x = 19.9\% = \%\ \text{boron} - 10$$

The percentage of boron-11 is

$$\%\ \text{boron-11} = 100-x = 80.1\%$$

1-50. Let x be the percentage of europium-151 in naturally occurring europium. The percentage of europium-153 must be 100-x. Now set up the equation

$$\text{atomic mass of Eu} = 151.96 = (150.9199)\left(\frac{x}{100}\right) + (152.9212)\left(\frac{100-x}{100}\right)$$

Multiply through by 100 to obtain

$$15196 = 150.9199x + 15292.12 - 152.9212x$$

Collecting terms, we get

$$2.0013x = 96$$

or

$$x = 48\% = \%\ \text{Eu-151}$$

The percentage of Eu-153 = 100-x = 52%.

1-52. a) $\text{atomic mass of Li} = (6.0151)\left(\frac{7.42}{100}\right) + (7.0160)\left(\frac{92.58}{100}\right)$

$$= 6.942$$

b) Let x be the percentage of lithium-6 in the lithium sample. The percentage of lithium-7 is 100-x. Now set up the equation

$$7.000 = (6.0151)\left(\frac{x}{100}\right) + (7.0160)\left(\frac{100-x}{100}\right)$$

Multiply through by 100 to obtain

$$700.0 = 6.0151x + 701.60 - 7.0160x$$

Collecting terms, we get

$$1.0009x = 1.6$$

or

$$x = 1.6\% = \%\ \text{Li-6}$$

1-54. The mass percentage of carbon in sucrose is

$$\text{mass \% of C} = \frac{12 \times \text{atomic mass of C}}{\text{molecular mass of sucrose}} \times 100$$

$$= \frac{12 \times 12.01}{342.30} \times 100 = 42.10\%$$

The mass of carbon in a metric ton of sucrose is

$$\text{mass of C} = (1000\text{ kg})(0.4210) = 421.0\text{ kg}$$

Using the data in Table 1-9, we see that the mass percentage of ^{13}C in naturally occurring carbon is 1.11%. Therefore, the mass of $^{13}_{6}C$ in one metric ton of sucrose is

$$\text{mass of } ^{13}_{6}C = (421.0\text{ kg})\left(\frac{1.11}{100}\right) = 4.67\text{ kg}$$

$$= 4670\text{ g}$$

1-56. The number of electrons = Z - ionic charge

a) 36
b) 18
c) 28
d) 46
e) 78

1-58. a) 54
b) 54
c) 54
d) 78
e) 74

1-60. a) N^{2-}, O^{-}, Ne^{+}

b) Li^{-}, B^{+}, C^{2+}

c) La^{+}, Ce^{2+}, Cs^{-}

d) N^{3-}, O^{2-}, F^{-}, Na^{+}, Mg^{2+}, Al^{3+}

e) C^{-}, O^{+}, F^{2+}

1-62. If the diameter of the nucleus were 3 cm, then the diameter of the atom would be

$$\text{diameter atom} = (3\ \text{cm})(10^4) = 3 \times 10^4\ \text{cm}$$

$$= (3 \times 10^4\ \text{cm})\left(\frac{1\ \text{m}}{100\ \text{cm}}\right) = 300\ \text{m}$$

1-64. a) 578 has three significant figures: 5, 7, 8, or 578 is an exact number and has no uncertainty associated with it.

b) 0.000578 has three significant figures: 5, 7, 8.

c) 1000 is an exact number and has no uncertainty associated with it.

d) 93,000,000 has two significant figures: 9, 3.

e) 400,000 has one significant figure: 4. 10,000 has one significant figure: 1.

1-66. a) The molecular mass of CH_4 = (12.011)+(4 x 1.0079) = 16.043

b) The molecular mass of CaF_2 = (40.08)+(2 x 18.998403) = 78.08

c) The molecular mass of $TiCl_4$ = (47.90)+(4 x 35.453) = 189.71

d) The molecular mass of $^{243}AmCl_3$ = (243)+(3 x 35.453) = 349

1-68. a) 33209. The result cannot be more accurate than zero digits past the decimal point.

b) 254

c) 3.4×10^{22} . The result cannot be expressed to more than two significant figures.

d) 1.43883×10^{-2}

e) -1.25×10^{-13}

1-70. a) To convert from feet to meters, we use the unit conversion factors 1.0936 yards/1 m and 3 ft/yard.

$$(325\ \text{ft})\left(\frac{1\ \text{yd}}{3\ \text{ft}}\right)\left(\frac{1\ \text{m}}{1.0936\ \text{yd}}\right) = 99.1\ \text{m}$$

b) To convert Ångstroms to picometers, we use the unit conversion factor 100 pm/1 Å.

$$(1.54\ \text{Å})\left(\frac{100\ \text{pm}}{1\ \text{Å}}\right) = 154\ \text{pm}$$

and

$$(1.54\ \text{Å})\left(\frac{1 \times 10^{-10}\ \text{m}}{1\ \text{Å}}\right)\left(\frac{1 \times 10^{9}\ \text{nm}}{1\ \text{m}}\right) = 0.154\ \text{nm}$$

c) To convert from pounds to kilograms, we use the unit conversion factor 0.45359 kg/1 lb.

$$(175\ \text{lb})\left(\frac{0.45359\ \text{kg}}{1\ \text{lb}}\right) = 79.4\ \text{kg}$$

1-72. The volume in cubic centimeters is

$$\text{volume} = (454\ \text{in}^3)\left(\frac{2.540\ \text{cm}}{1\ \text{in}}\right)^3 = 7.44 \times 10^3\ \text{cm}^3$$

The volume in liters is

$$\text{volume} = (7.44 \times 10^3\ \text{cm}^3)\left(\frac{1\ \text{mL}}{1\ \text{cm}^3}\right)\left(\frac{1\ \text{L}}{10^3\ \text{mL}}\right)$$

$$= 7.44\ \text{L}$$

1-74. The speed in meters per hour is

$$\text{speed} = (90\ \text{miles}\cdot\text{hr}^{-1})\left(\frac{1.61\ \text{km}}{1\ \text{mile}}\right)\left(\frac{10^3\ \text{m}}{1\ \text{km}}\right) = 1.45 \times 10^5\ \text{m}\cdot\text{hr}^{-1}$$

The speed in meters per second is

$$\text{speed} = (1.45 \times 10^5\ \text{m}\cdot\text{hr}^{-1})\left(\frac{1\ \text{hr}}{60\ \text{min}}\right)\left(\frac{1\ \text{min}}{60\ \text{s}}\right) = 40\ \text{m}\cdot\text{s}^{-1}$$

The distance between the pitcher's mound and home plate in meters is

$$\text{distance} = (60.5\ \text{ft})\left(\frac{1\ \text{yd}}{3\ \text{ft}}\right)\left(\frac{1\ \text{m}}{1.0936\ \text{yd}}\right) = 18.4\ \text{m}$$

The time it takes for a 90 mph fastball to travel from the pitcher's mound to home plate is

$$\text{distance} = \text{speed} \times \text{time}$$

or

$$\text{time} = \frac{\text{distance}}{\text{speed}}$$

$$\text{time} = \frac{18.4\ \text{m}}{40\ \text{m}\cdot\text{s}^{-1}} = 0.46\ \text{s}$$

CHAPTER 2

SOLUTIONS TO THE EVEN-NUMBERED PROBLEMS

2-2. a) $Li_3N(s) + 3H_2O(l) \longrightarrow 3LiOH(s) + NH_3(g)$

b) $Al_4C_3(s) + 12HCl(aq) \longrightarrow 4AlCl_3(aq) + 3CH_4(g)$

c) $H_2S(g) + 2NaOH(aq) \longrightarrow Na_2S(aq) + 2H_2O(l)$

d) $2HCl(aq) + CaCO_3(s) \longrightarrow CaCl_2(aq) + CO_2(g) + H_2O(l)$

e) $4CoO(s) + O_2(g) \longrightarrow 2Co_2O_3(s)$

Note that details of the general balancing procedure are given in the solution to Problem 2-1 (Study Guide-Solutions Manual).

2-4. a) $3N_2H_4(g) \longrightarrow 4NH_3(g) + N_2(g)$

b) $2GeO_2(s) \longrightarrow 2GeO(g) + O_2(g)$

c) $2KHF_2(s) \longrightarrow 2KF(s) + H_2(g) + F_2(g)$

d) $2H_2O_2(l) \longrightarrow 2H_2O(l) + O_2(g)$

e) $2N_2O(g) \longrightarrow 2N_2(g) + O_2(g)$

2-6. a) $2AgNO_3(aq) + Cu(s) \longrightarrow Cu(NO_3)_2(aq) + 2Ag(s)$

b) $Zn(s) + 2HCl(aq) \longrightarrow ZnCl_2(aq) + H_2(g)$

c) $2KI(aq) + Br_2(l) \longrightarrow 2KBr(aq) + I_2(s)$

d) $2ZnS(s) + 3O_2(g) \longrightarrow 2ZnO(s) + 2SO_2(g)$

e) $2GaBr_3(aq) + 3Cl_2(g) \longrightarrow 2GaCl_3(aq) + 3Br_2(l)$

2-8. a) $Ca(s) + H_2(g) \longrightarrow CaH_2(s)$

b) $2S(s) + 3O_2(g) \longrightarrow 2SO_3(g)$

c) $PCl_5(s) + 4H_2O(l) \longrightarrow H_3PO_4(l) + 5HCl(g)$

d) $P_4O_{10}(s) + 6H_2O(l) \longrightarrow 4H_3PO_4(l)$

e) $2Sb(s) + 3Cl_2(g) \longrightarrow 2SbCl_3(s)$

2-10. a) $NaH(s) + H_2O(l) \longrightarrow NaOH(s) + H_2(g)$

sodium hydride + water $\longrightarrow$ sodium hydroxide + hydrogen

b) $Li_3N(s) + 3D_2O(l) \longrightarrow 3LiOD(s) + ND_3(g)$

lithium nitride + deuterium oxide $\longrightarrow$ lithium deuteroxide + nitrogen trideuteride

c) $2NaN_3(s) \longrightarrow 2Na(s) + 3N_2(g)$

sodium azide ⟶ sodium + nitrogen

d) $Li(s) + D_2O(l) \longrightarrow LiOD(s) + D_2(g)$

lithium deuteride + deuterium oxide ⟶ lithium deuteroxide + deuterium

e) $LiOH(s) + HCl(g) \longrightarrow LiCl(s) + H_2O(g)$

lithium hydroxide + hydrogen chloride ⟶ lithium chloride + water

The isotope of hydrogen with mass number = 2 has the special name deuterium. Thus LiOD is lithium deuteroxide.

2-12. Using Tables 2-1 and 2-2, we have that

a) $2K(s) + Cl_2(g) \longrightarrow 2KCl(s)$

b) $Sr(s) + S(s) \longrightarrow SrS(s)$

c) $Ba(s) + H_2O(g) \longrightarrow BaO(s) + H_2(g)$

d) $2Li(s) + H_2(g) \longrightarrow 2LiH(s)$

e) $2Na(s) + Br_2(l) \longrightarrow 2NaBr(s)$

2-14. Consulting Table 2-1 for the analogous representative reactions, we have

a) $2Li(s) + I_2(s) \longrightarrow 2LiI(s)$

b) $2Na(s) + S(s) \longrightarrow Na_2S(s)$

c) $2K(s) + H_2(g) \longrightarrow 2KH(s)$

d) $2Li(s) + F_2(g) \longrightarrow 2LiF(s)$

e) $2K(s) + Br_2(l) \longrightarrow 2KBr(s)$

2-16. Consulting Table 2-3 for the representative reactions, we have

a) $2P(s) + 3Cl_2(g) \longrightarrow 2PCl_3(l)$

b) $H_2(g) + F_2(g) \longrightarrow 2HF(g)$

c) $2Sb(s) + 3Cl_2(g) \longrightarrow 2SbCl_3(s)$

d) $2As(s) + 3Br_2(l) \longrightarrow 2AsBr_3(s)$

e) $2P(s) + 3I_2(s) \longrightarrow 2PI_3(s)$

The physical states of the products in (a), (c), (d) and (e) cannot be deduced from Table 2-3; they are given here for your information.

2-18. Argon is an unreactive gas as are helium and neon, and so argon must belong to the noble gas family. Potassium has chemical properties similar to those of sodium and lithium, therefore, potassium belongs to the alkali metal family.

2-20. By analogy with the properties of the other noble gases we predict the following:

a) colorless

b) odorless

c) Ra

d) no reaction

2-22. Te is a main group (6) nonmetal; P is a main group (5) nonmetal; Mn is a transition metal; Kr is a main group (8) nonmetal; W is a transition metal; Pb is a main group (4) metal; Ga is a main group (3) metal.

2-24. Sodium is a main group (1) metal; carbon is a main group (4) nonmetal; helium is a main group (8) nonmetal; iron is a transition metal; copper is a transition metal; zinc is a transition metal.

2-26. Based on the properties described and its lack of chemical reactivity, we conclude that krypton is a noble gas (Group 8).

2-28. Strontium has similar chemical properties to calcium which is a component of the human body.

2-30. For positive atomic ions, the number of electrons is equal to the atomic number Z minus the charge. For negative atomic ions, the number of electrons is equal to the atomic number plus the charge.

a) 23

b) 18

c) 54

d) 54

e) 54

2-32. We first determine the number of electrons in the ion and then compare the result with the Z values for the noble gases.

a) yes (Xe)

b) no

c) yes (Kr)

d) yes (Ar)

e) yes (Kr)

f) no

2-34. We determine the charges on the atomic ions in ionic compounds using Figure 2-18.

a) Li^{+}, O^{2-}

b) Ca^{2+}, S^{2-}

c) Mg^{2+}, N^{3-}

d) Al^{3+}, S^{2-}

e) K^{+}, I^{-}

2-36. To determine the chemical formula of an ionic compound, we use the procedure outlined in Example 2-12.

a) Al_2S_3

b) Na_2O

c) $GaBr_3$

d) BaF_2

e) KI

2-38. a) Cs_2O

b) Na_2Se

c) SrS

d) Li_2S

e) CaI_2

2-40. a) PtF_4

b) Au_2O_3

c) Fe_2Se_3

d) $BaAt_2$

e) Zn_3N_2

2-42. To work a problem like this you have to know how to write a chemical formula from the name. Once we have written down the formulas for the reactants and the products, we then proceed to balance the equation.

a) $2Na(s) + S(s) \longrightarrow Na_2S(s)$

b) $Ca(s) + Br_2(l) \longrightarrow CaBr_2(s)$

c) $2Ba(s) + O_2(g) \longrightarrow 2BaO(s)$

d) $2SO_2(g) + O_2(g) \longrightarrow 2SO_3(g)$

e) $3Mg(s) + N_2(g) \longrightarrow Mg_3N_2(s)$

2-44. a) $2CO(g) + O_2(g) \longrightarrow 2CO_2(g)$

b) $2Cs(s) + Br_2(l) \longrightarrow 2CsBr(s)$

c) $2NO(g) + O_2(g) \longrightarrow 2NO_2(g)$

d) $4NH_3(g) + 5O_2(g) \longrightarrow 4NO(g) + 6H_2O(l)$

e) $Ga(s) + As(s) \longrightarrow GaAs(s)$

2-46. a) Lithium atoms are neutral and so have no charge. Selenium atoms are neutral and so have no charge. In Li_2Se, the ionic charge of lithium is +1 and the ionic charge of selenium is -2. Thus lithium is oxidized and selenium is reduced.

b) Scandium atoms have no charge and iodine atoms in I_2 have no charge. In ScI_3, the ionic charges of scandium and iodine are +3 and -1, respectively. Thus scandium is oxidized and iodine is reduced.

c) Gallium is oxidized and phosphorus is reduced.

d) Potassium is oxidized and fluorine is reduced.

2-48. a) Each of the two lithium atoms loses one electron and each selenium atom gains two electrons. Thus a total of two electrons are transferred.

b) Each of the two scandium atoms loses three electrons and each of the six iodine atoms gains one electron. Thus a total of 2 x 3 or 6 x 1 = 6 electrons are transferred or three electrons per scandium iodide unit.

c) Three electrons.

d) One electron.

CHAPTER 3

SOLUTIONS TO THE EVEN-NUMBERED PROBLEMS

3-2. a) formula mass of aspirin = (9 x atomic mass of C)+(8 x atomic mass of H)+(4 x atomic mass of O)

= (9 x 12.01) +(8 x 1.008)+(4 x 16.00)

= 180.15

b) formula mass of vitamin D_1 = (56 x atomic mass of C)+(88 x atomic mass of H)+(2 x atomic mass of O)

= (56 x 12.01) +(88 x 1.008)+(2 x 16.00)

= 793.26

c) formula mass of bromoacetone = (3 x atomic mass of C)+(5 x atomic mass of H)+(atomic mass of Br)+(atomic mass of O)

= (3 x 12.01) +(5 x 1.008)+(79.90) + (16.00)

= 136.97

d) formula mass of choral hydrate = (2 x atomic mass of C)+(3 x atomic mass of H)+(3 x atomic mass of Cl)+(2 x atomic mass of O)

= (2 x 12.01) +(3 x 1.008)+(3 x 35.45) +(2 x 16.00)

= 165.39

3-4. a) the formula mass of malathion = 330.34

$$\text{number of moles} = (1000\text{ g})\left(\frac{1\text{ mol}}{330.34\text{ g}}\right) = 3.03\text{ mol}$$

b) the formula mass of parathion = 291.25

$$\text{number of moles} = (1000\text{ g})\left(\frac{1\text{ mol}}{291.25\text{ g}}\right) = 3.43\text{ mol}$$

c) the formula mass of methoxychlor = 345.6

$$\text{number of moles} = (1000\ \text{g})\left(\frac{1\ \text{mol}}{345.63\ \text{g}}\right) = 2.89\ \text{mol}$$

3-6. a) the formula mass of vitamin C = 176.12

$$\text{number of moles} = (60\ \text{mg})\left(\frac{1\ \text{g}}{10^3\ \text{mg}}\right)\left(\frac{1\ \text{mol}}{176.12\ \text{g}}\right) = 3.4 \times 10^{-4}\ \text{mol}$$

b) the formula mass of vitamin A = 286.44

$$\text{number of moles} = (1.5\ \text{mg})\left(\frac{1\ \text{g}}{10^3\ \text{mg}}\right)\left(\frac{1\ \text{mol}}{286.44\ \text{g}}\right) = 5.2 \times 10^{-6}\ \text{mol}$$

c) the formula mass of vitamin B_{12} = 1355.37

$$\text{number of moles} = (6.0\ \mu\text{g})\left(\frac{1\ \text{g}}{10^6\ \mu\text{g}}\right)\left(\frac{1\ \text{mol}}{1355.37\ \text{g}}\right) = 4.4 \times 10^{-9}\ \text{mol}$$

3-8. $$\text{dollars per person} = \frac{6.022 \times 10^{23}\ \text{dollars}}{2.30 \times 10^{8}\ \text{persons}}$$

$$= 2.62 \times 10^{15}\ \text{dollars}\cdot\text{person}^{-1}$$

3-10. See Problems 3-7 and 3-8 for examples.

3-12. a) formula mass of O_2 = 32.00

$$\text{mass of one } O_2 \text{ molecule} = \left(\frac{32.00\ \text{g}}{1\ \text{mol}}\right)\left(\frac{1\ \text{mol}}{6.022 \times 10^{23}\ \text{molecules}}\right)$$

$$= 5.314 \times 10^{-23}\ \text{g}$$

b) formula mass of $FeSO_4$ = 151.91

$$\text{mass of one } FeSO_4 \text{ formula unit} = \left(\frac{151.91\ \text{g}}{1\ \text{mol}}\right)\left(\frac{1\ \text{mol}}{6.022\times10^{23}\ \text{formula units}}\right)$$

$$= 2.523 \times 10^{-22}\ \text{g}$$

c) formula mass of $C_{12}H_{22}O_{11}$ = 342.30

$$\text{mass of one } C_{12}H_{22}O_{11} \text{ molecule} = \left(\frac{342.30\text{ g}}{1\text{ mol}}\right)\left(\frac{1\text{ mol}}{6.022\times10^{23}\text{molecules}}\right)$$

$$= 5.684 \times 10^{-22}\text{ g}$$

3-14. a) formula mass of $C_3H_5N_3O_9$ = 227.10

$$\text{mass of 100 } C_3H_5N_3O_9 \text{ molecules} = \left(\frac{227.10\text{ g}}{\text{mol}}\right)\left(\frac{1\text{ mol}}{6.022\times10^{23}\text{molecules}}\right)(100\text{ molecules})$$

$$= 3.771 \times 10^{-20}\text{ g}$$

b) formula mass of $C_7H_5N_3O_6$ = 227.14

$$\text{mass of 5000 } C_7H_5N_3O_6 \text{ molecules} = \left(\frac{227.14\text{ g}}{1\text{ mol}}\right)\left(\frac{1\text{ mol}}{6.022\times10^{23}\text{molecules}}\right)(5000\text{ molecules})$$

$$= 1.886 \times 10^{-18}\text{ g}$$

c) formula mass of C_8H_{18} = 114.22

$$\text{mass of } 10^{10}\ C_8H_{18} \text{ molecules} = \left(\frac{114.22\text{ g}}{\text{mol}}\right)\left(\frac{1\text{ mol}}{6.022\times10^{23}\text{molecules}}\right)(10^{10}\text{molecules})$$

$$= 1.897 \times 10^{-12}\text{ g}$$

d) formula mass of O_3 = 48.00

$$\text{mass of 10 } O_3 \text{ molecules} = \left(\frac{48.00\text{g}}{1\text{ mol}}\right)\left(\frac{1\text{ mol}}{6.022\times10^{23}\text{molecules}}\right)(10\text{ molecules})$$

$$= 7.971 \times 10^{-22}\text{ g}$$

3-16. 20.0 g CH_3OH corresponds to $(20.0\text{g})\left(\frac{1\text{ mol}}{32.04\text{ g}}\right) = 0.624\text{ mol}$

The number of CH_3OH molecules in 0.624 mol is

$$(0.624\text{ mol})\left(\frac{6.022\times10^{23}\text{molecules}}{1\text{ mol}}\right) = 3.76 \times 10^{23}\text{ molecules}$$

There are four hydrogen atoms in each molecule and so the number of H atoms is $(4\ \text{atoms}\cdot\text{molecule}^{-1})(3.76 \times 10^{23}\text{molecules}) = 1.50 \times 10^{24}$ atoms. There is one carbon atom in each molecule and so the number of C atoms is $(1\ \text{atom}\cdot\text{molecule}^{-1})(3.76 \times 10^{23}\ \text{molecules}) = 3.76 \times 10^{23}$ atoms. There is one oxygen atom in each molecule and so the number of O atoms is $(1\ \text{atom}\cdot\text{molecule}^{-1})$ $(3.76 \times 10^{23}\ \text{molecules}) = 3.76 \times 10^{23}$ atoms.

3-18. The number of nerve gas molecules per liter per breath is

$$\frac{5.24 \times 10^{8}\ \text{molecules}}{0.500\ \text{L}} = 1.05 \times 10^{9}\ \text{molecules}\cdot\text{L}^{-1}$$

The number of nerve gas molecules per liter of air at a concentration of 1.0 picogram per liter is

$$(1.0\ \text{pg}\cdot\text{L}^{-1})\left(\frac{1\ \text{g}}{10^{12}\text{pg}}\right)\left(\frac{1\ \text{mol}}{238\ \text{g}}\right)\left(\frac{6.022\text{x}10^{23}\text{molecules}}{\text{mol}}\right) = 2.5\text{x}10^{9}\text{molecules}\cdot\text{L}^{-1}$$

The nerve gas is present in harmless amounts.

3-20. Take a 100 gram sample and write

$$69.9\ \text{g Fe} \Leftrightarrow 30.1\ \text{g O}$$

Divide each quantity by its corresponding atomic mass to get

$$(69.9\ \text{g})\left(\frac{1\ \text{mol}}{55.85\ \text{g}}\right) = 1.25\ \text{mol Fe} \Leftrightarrow (30.1\ \text{g})\left(\frac{1\ \text{mol}}{16.00\ \text{g}}\right) = 1.88\ \text{mol O}$$

Divide by the smaller quantity (1.25) to obtain

$$1\ \text{mol Fe} \Leftrightarrow 1.50\ \text{mol O}$$

Recognize that 1.50 is 3/2 so that

$$2\ \text{mol Fe} \Leftrightarrow 3\ \text{mol O}$$

The empirical formula is Fe_2O_3.

3-22. The mass percentage of Fe in the compound is

$$\text{mass \% Fe} = \frac{3.78\text{ g}}{5.95\text{ g}} \times 100 = 63.5\%$$

Therefore, mass % S = 36.5%. Assume a 100 g sample and write

$$63.5\text{ g Fe} \Bumpeq 36.5\text{ g S}$$

Divide each quantity by the corresponding atomic mass to get

$$(63.5\text{ g})\left(\frac{1\text{ mol}}{55.85\text{ g}}\right) = 1.14\text{ mol Fe} \Bumpeq (36.5\text{ g})\left(\frac{1\text{ mol}}{32.06\text{ g}}\right) = 1.14\text{ mol S}$$

Divide both quantities by 1.14 to obtain

$$1\text{ mol Fe} \Bumpeq 1\text{ mol S}$$

The empirical formula is FeS.

3-24. The mass percentage of aluminum in the compound is

$$\text{mass \% of Al} = \frac{5.00\text{ g}}{9.45\text{ g}} \times 100 = 52.9\%$$

Therefore, mass % of O = 47.1%. Taking a 100 g sample, we have

$$52.9\text{ g Al} \Bumpeq 47.1\text{ g O}$$

Divide each quantity by its corresponding atomic mass to get

$$1.96\text{ mol Al} \Bumpeq 2.94\text{ mol O}$$

Divide by the smaller quantity to get

$$1\text{ mol Al} \Bumpeq 1.50\text{ mol O}$$

or

$$2\text{ mol Al} \Bumpeq 3\text{ mol O}$$

The empirical formula is Al_2O_3.

3-26. Take a 100 gram sample and write

$$68.3 \text{ g Pb} \Bumpeq 10.6 \text{ g S} \Bumpeq 21.1 \text{ g O}$$

Divide each quantity by its corresponding atomic mass to obtain

$$0.330 \text{ mol Pb} \Bumpeq 0.331 \text{ mol S} \Bumpeq 1.32 \text{ mol O}$$

Divide by the smallest quantity (0.330) to obtain

$$1 \text{ mol Pb} \Bumpeq 1.00 \text{ mol S} \Bumpeq 4.00 \text{ mol O}$$

The empirical formula is $PbSO_4$.

3-28. mass % of X = 100% - 75.0% = 25.0%

Assuming a 100 g sample, we have

$$75.0 \text{ g Cl} \Bumpeq 25.0 \text{ g X}$$

We do not know the atomic mass of M, but we can divide 75.0 g Cl

by 35.45 to obtain

$$2.116 \text{ mol Cl} \Bumpeq 25.0 \text{ g X}$$

We know from the given empirical formula that

$$4 \text{ mol Cl} \Bumpeq 1 \text{ mol X}$$

Divide both quantities by 4 and multiply by 2.116 to obtain

$$2.116 \text{ mol Cl} \Bumpeq 0.529 \text{ mol X}$$

We also have that

$$2.116 \text{ mol Cl} \Bumpeq 25.0 \text{ g X}$$

and so

$$25.0 \text{ g X} \Bumpeq 0.529 \text{ mol X}$$

Divide by 0.529 to get

$$47.3 \text{ g X} \approx 1.00 \text{ mol X}$$

The atomic mass of X is 47.3 which corresponds to titanium (Ti).

3-30. formula mass of $HXO_3 = \dfrac{1.123 \text{ g}}{0.0133 \text{ mol}}$

$$= \frac{84.44 \text{ g}}{1 \text{ mol}}$$

Also

$$\text{formula mass of } HXO_3 = (1 \times 1.008) + X + (3 \times 16.00)$$

$$= 49.01 + X$$

Thus

$$49.01 + X = 84.44$$

$$X = 35.4$$

The atomic mass of X is 35.4 which corresponds to chlorine (Cl).

3-32. Take a 100 g sample and write

$$40.0 \text{ g C} \approx 6.71 \text{ g H} \approx 53.3 \text{ g O}$$

Divide by the corresponding atomic masses

$$3.33 \text{ mol C} \approx 6.66 \text{ mol H} \approx 3.33 \text{ mol O}$$

Now divide by the smallest quantity (3.33) to get

$$1 \text{ mol C} \approx 2.00 \text{ mol H} \approx 1 \text{ mol O}$$

The simplest formula is CH_2O. The molecular mass corresponding to this simplest formula is 30.03 which divides into the observed molecular mass, 180.2, six times. Thus the molecular formula is $C_6H_{12}O_6$.

3-34. Assume a 100 g sample and write

$$0.373 \text{ g Fe} \approx 99.627 \text{ g hemoglobin}$$

Divide by the atomic mass of Fe to obtain

$$0.00668 \text{ mol Fe} \approx 99.627 \text{ g hemoglobin}$$

We also have that

$$4 \text{ mol Fe} \approx 1 \text{ mol hemoglobin}$$

Divide both quantities by four and multiply by 0.00668 to obtain

$$0.00668 \text{ mol Fe} \approx 0.00167 \text{ mol hemoglobin}$$

Thus

$$99.627 \text{ g hemoglobin} \approx 0.00167 \text{ mol hemoglobin}$$

Dividing by 0.00167, we have that

$$59{,}700 \text{ g hemoglobin} \approx 1 \text{ mol hemoglobin}$$

The molecular mass of hemoglobin is 59,700.

3-36. The reaction is

$$C_6H_6(l) + Cl_2(g) \xrightarrow{AlCl_3} C_6H_5Cl(l) + HCl(g)$$

150 g of benzene corresponds to

$$(150 \text{ g})\left(\frac{1 \text{ mol}}{78.11 \text{ g}}\right) = 1.92 \text{ mol } C_6H_6$$

We see from the reaction that one mole of C_6H_6 produces one mole of C_6H_5Cl and

$$\text{moles of } C_6H_5Cl = (1.92 \text{ mol } C_6H_6)\left(\frac{1 \text{ mol } C_6H_5Cl}{1 \text{ mol } C_6H_6}\right)$$

$$= 1.92 \text{ mol}$$

The mass of C_6H_5Cl produced is

$$\text{mass of } C_6H_5Cl = (1.92 \text{ mol})\left(\frac{112.55 \text{ g}}{1 \text{ mol}}\right)$$

$$= 216 \text{ g}$$

3-38. The reaction is

$$2Ca_3(PO_4)_2(s) + 6SiO_2(s) \longrightarrow 6CaSiO_3(s) + P_4O_{10}(g)$$

The quantity of pure SiO_2 involved is

$$\text{mass of } SiO_2 = (1.00 \text{ kg})(0.60) = 0.60 \text{ kg} = 600 \text{ g}$$

The number of moles of SiO_2 is

$$\text{moles of } SiO_2 = (600 \text{ g})\left(\frac{1 \text{ mol}}{60.09 \text{ g}}\right) = 9.985 \text{ mol}$$

According to the reaction, one mole of P_4O_{10} is produced from six moles of SiO_2. Therefore

$$\text{moles of } P_4O_{10} = (9.985 \text{ mol } SiO_2)\left(\frac{1 \text{ mol } P_4O_{10}}{6 \text{ mol } SiO_2}\right) = 1.664 \text{ mol}$$

The mass of P_4O_{10} produced is

$$\text{mass of } P_4O_{10} = (1.664 \text{ mol})\left(\frac{283.88 \text{ g}}{\text{mol}}\right) = 472 \text{ g}$$

3-40. The reaction is

$$C_6H_{12}O_6(aq) + 6O_2(g) \longrightarrow 6CO_2(g) + 6H_2O(l)$$

The number of moles of glucose is

$$\text{moles of } C_6H_{12}O_6 = (28 \text{ g})\left(\frac{1 \text{ mol}}{180.16 \text{ g}}\right) = 0.155 \text{ mol}$$

The number of moles of O_2 required is

$$\text{moles of } O_2 = (0.155 \text{ mol } C_6H_{12}O_6)\left(\frac{6 \text{ mol } O_2}{1 \text{ mol } C_6H_{12}O_6}\right) = 0.930 \text{ mol}$$

The mass of O_2 required is

$$\text{mass of } O_2 = (0.930 \text{ mol})\left(\frac{32.00 \text{ g}}{\text{mol}}\right) = 30 \text{ g}$$

The number of moles of CO_2 produced is

$$\text{moles of } CO_2 = (0.155 \text{ mol } C_6H_{12}O_6)\left(\frac{6 \text{ mol } CO_2}{1 \text{ mol } C_6H_{12}O_6}\right) = 0.930 \text{ mol}$$

The mass of CO_2 produced is

$$\text{mass of } CO_2 = (0.930 \text{ mol})\left(\frac{44.01 \text{ g}}{\text{mol}}\right) = 41 \text{ g}$$

3-42. The mass percentage of Ag in AgCl is

$$\text{mass \% of Ag in AgCl} = \frac{\text{atomic mass of Ag}}{\text{formula mass of AgCl}} \times 100$$

$$= \frac{107.9}{143.4} \times 100 = 75.24\%$$

The mass of Ag in the AgCl is

$$\text{mass of Ag} = (3.74 \text{ g})(0.7524) = 2.81 \text{ g}$$

The percentage of Ag in the ore is

$$\text{\% Ag} = \frac{\text{mass of Ag}}{\text{mass of ore}} \times 100 = \frac{2.81 \text{ g}}{24.31 \text{ g}} \times 100 = 11.6\%$$

3-44. The number of moles of TiO_2 is

$$\text{moles of } TiO_2 = (4.10 \times 10^3 \text{ kg})\left(\frac{10^3 \text{ g}}{1 \text{ kg}}\right)\left(\frac{1 \text{ mol}}{79.90 \text{ g}}\right)$$

$$= 5.13 \times 10^4 \text{ mol}$$

Note that one mole of Ti results from each mole of TiO_2.

$$\text{moles of Ti} = 5.13 \times 10^4 \text{ mol}$$

The mass of Ti produced is

$$\text{mass of Ti} = (5.13 \times 10^4 \text{ mol})\left(\frac{47.90 \text{ g}}{1 \text{ mol}}\right)$$

$$= 2.46 \times 10^6 \text{ g} = 2.46 \times 10^3 \text{ kg}$$

3-46. The number of moles of Sb_2S_3 is

$$\text{moles of } Sb_2S_3 = (500\text{ g})\left(\frac{1\text{ mol}}{339.8\text{ g}}\right) = 1.471\text{ mol}$$

From the two reactions, we see that four moles of Sb are produced from two moles of Sb_2S_3. Thus

$$\text{moles of Sb} = (1.471\text{ mol } Sb_2S_3)\left(\frac{4\text{ mol Sb}}{2\text{ mol } Sb_2S_3}\right) = 2.943\text{ mol}$$

The mass of Sb produced is

$$\text{mass of Sb} = (2.943\text{ mol})\left(\frac{121.8\text{ g}}{\text{mol}}\right) = 358\text{ g}$$

3-48. Because we are given the quantities of two reactants, we must check to see if one of them acts as a limiting reagent. The number of moles of P_4 and NaOH are

$$\text{moles of } P_4 = (0.610\text{ g})\left(\frac{1\text{ mol}}{123.88\text{ g}}\right) = 0.00492\text{ mol}$$

$$\text{moles of NaOH} = (0.250\text{ g})\left(\frac{1\text{ mol}}{40.00\text{ g}}\right) = 0.00625\text{ mol}$$

Each mole of P_4 reacts with three moles of NaOH and so we see that 0.00492 mol of P_4 requires 0.0148 mol of NaOH. Thus P_4 is in excess and NaOH is the limiting reagent. The mass of PH_3 produced is

$$\text{mass of } PH_3 = (0.00625\text{ mol NaOH})\left(\frac{1\text{ mol } PH_3}{3\text{ mol NaOH}}\right)\left(\frac{33.99\text{ g } PH_3}{1\text{ mol } PH_3}\right)$$

$$= 0.0708\text{ g}$$

3-50. We must determine which reactant is the limiting reagent. The number of moles of NaBr and Cl_2 are

$$\text{moles of NaBr} = (25.0\text{ g})\left(\frac{1\text{ mol}}{102.89\text{ g}}\right) = 0.243\text{ mol}$$

$$\text{moles of } Cl_2 = (25.0\text{ g})\left(\frac{1\text{ mol}}{70.90\text{ g}}\right) = 0.353\text{ mol}$$

Each mol of Cl_2 requires two moles of NaBr and so we see that 0.353 mol of Cl_2 requires 0.706 mol of NaBr. Thus Cl_2 is in excess and NaBr is the limiting reagent. The mass of Br_2 produced is

$$\text{mass of } Br_2 = (0.243 \text{ mol NaBr})\left(\frac{1 \text{ mol } Br_2}{2 \text{ mol NaBr}}\right)\left(\frac{159.80 \text{ g } Br_2}{1 \text{ mol } Br_2}\right)$$

$$= 19.4 \text{ g}$$

3-52. The number of moles of reactants are

$$\text{mol of } CS_2 = (1000 \text{ g})\left(\frac{1 \text{ mol}}{76.13 \text{ g}}\right) = 13.1$$

$$\text{mol of NaOH} = (1000 \text{ g})\left(\frac{1 \text{ mol}}{40.00 \text{ g}}\right) = 25.0 \text{ mol}$$

The CS_2 is in excess and NaOH is the limiting reagent. The mass of each product that is produced is

$$\text{mass of } Na_2CS_3 = (25.0 \text{ mol NaOH})\left(\frac{2 \text{ mol } Na_2CS_3}{6 \text{ mol NaOH}}\right)\left(\frac{154.17 \text{ g } Na_2CS_3}{1 \text{ mol } Na_2CS_3}\right)$$

$$= 1280 \text{ g}$$

$$\text{mass of } Na_2CO_3 = (25.0 \text{ mol NaOH})\left(\frac{1 \text{ mol } Na_2CO_3}{6 \text{ mol NaOH}}\right)\left(\frac{105.99 \text{ g } Na_2CO_3}{1 \text{ mol } Na_2CO_3}\right)$$

$$= 442 \text{ g}$$

$$\text{mass of } H_2O = (25.0 \text{ mol NaOH})\left(\frac{3 \text{ mol } H_2O}{6 \text{ mol NaOH}}\right)\left(\frac{18.02 \text{ g } H_2O}{1 \text{ mol } H_2O}\right)$$

$$= 225 \text{ g}$$

3-54. The number of moles of $NaHCO_3$ is

$$\text{moles of } NaHCO_3 = (69.0 \text{ g})\left(\frac{1 \text{ mol}}{84.01 \text{ g}}\right) = 0.821 \text{ mol}$$

Because the 0.821 mol are dissolved in one liter of solution, the molarity of the solution is 0.821 M.

3-56. The number of moles of caffeine is

$$\text{moles of } C_8H_{10}N_4O_2 = (300 \text{ mg})\left(\frac{1 \text{ g}}{10^3 \text{ mg}}\right)\left(\frac{1 \text{ mol}}{194.20 \text{ g}}\right)$$

$$= 0.001545 \text{ mol}$$

The molarity is calculated using Equation (3-6)

$$\text{molarity} = \frac{n}{V} = \left(\frac{0.001545\text{mol}}{1 \text{ cup}}\right)\left(\frac{4 \text{ cups}}{0.946 \text{ L}}\right) = 0.00653 \text{ M}$$

3-58. The number of moles of $CuSO_4$ in 50.0 mL of a 0.200 M solution is

$$\text{moles of } CuSO_4 = \text{molarity x V}$$

$$= (0.200 \text{ mol}\cdot\text{L}^{-1})(50.0 \text{ mL})\left(\frac{1 \text{ L}}{1000 \text{ mL}}\right)$$

$$= 0.0100 \text{ mol}$$

The number of moles of $CuSO_4 \cdot 5H_2O$ required is

$$\text{moles of } CuSO_4 \cdot 5H_2O = (0.0100 \text{ mol } CuSO_4)\left(\frac{1 \text{ mol } CuSO_4 \cdot 5H_2O}{1 \text{ mol } CuSO_4}\right)$$

$$= 0.0100 \text{ mol}$$

The mass of $CuSO_4 \cdot 5H_2O$ required is

$$\text{mass of } CuSO_4 \cdot 5H_2O = (0.100 \text{ mol})\left(\frac{249.69 \text{ g}}{1 \text{ mol}}\right)$$

$$= 2.50 \text{ g}$$

Dissolve 2.50 g of $CuSO_4 \cdot 5H_2O$ in about 40 mL of water and then dilute the solution to 50 mL using a volumetric flask.

3-60. We first determine the number of moles of NaBr. Using Equation (3-6), we have

$$\text{moles of NaBr} = MV = (4.00 \times 10^{-3}\ \text{mol}\cdot\text{L}^{-1})(1\text{m}^3)\left(\frac{10^3\ \text{L}}{1\ \text{m}^3}\right)$$

$$= 4.00\ \text{mol}$$

The number of moles of Br_2 produced is

$$\text{moles of Br}_2 = (4.00\ \text{mol NaBr})\left(\frac{1\ \text{mol Br}_2}{2\ \text{mol NaBr}}\right) = 2.00\ \text{mol}$$

The mass of Br_2 produced is

$$\text{mass of Br}_2 = (2.00\ \text{mol})\left(\frac{159.80\ \text{g}}{1\,\text{mol}}\right) = 320\ \text{g}$$

The number of moles of Cl_2 required is

$$\text{moles of Cl}_2 = (4.00\ \text{mol NaBr})\left(\frac{1\ \text{mol Cl}_2}{2\ \text{mol NaBr}}\right) = 2.00\ \text{mol}$$

The mass of Cl_2 required is

$$\text{mass of Cl}_2 = (2.00\ \text{mol})\left(\frac{70.90\ \text{g}}{1\,\text{mol}}\right) = 142\ \text{g}$$

3-62. The number of moles in the AgCl is

$$\text{moles of AgCl} = (0.231\ \text{g})\left(\frac{1\ \text{mol}}{143.4\ \text{g}}\right) = 0.00161\ \text{mol}$$

The number of moles of NH_3 required is

$$\text{moles of NH}_3 = (0.00161\ \text{mol AgCl})\left(\frac{2\ \text{mol NH}_3}{1\ \text{mol AgCl}}\right) = 0.00322\ \text{mol}$$

The volume of NH_3 solution required can be found using Equation (3-6)

$$V = \frac{n}{M} = \frac{0.00322\ \text{mol}}{0.100\ \text{mol}\cdot\text{L}^{-1}} = 0.0322\ \text{L}$$

3-64. The number of moles of NaOH that reacts is

$$\text{moles of NaOH} = (4.00\text{ g})\left(\frac{1\text{ mol}}{40.00\text{ g}}\right) = 0.100\text{ mol}$$

The number of moles of acetic acid is

$$\text{moles of } HC_2H_3O_2 = (0.100\text{ mol NaOH})\left(\frac{1\text{ mol } HC_2H_3O_2}{1\text{ mol NaOH}}\right) = 0.100\text{ mol}$$

The concentration of the solution can be found using Equation (3-6)

$$\text{molarity} = \frac{n}{V} = \frac{0.100\text{ mol}}{0.0752\text{ L}} = 1.33\text{ M}$$

3-66. The number of moles of oxalic acid is

$$\text{moles of } H_2C_2O_4 = (10.0\text{ g})\left(\frac{1\text{ mol}}{90.04\text{ g}}\right) = 0.111\text{ mol}$$

The number of moles of NaOH required is

$$\text{moles of NaOH} = (0.111\text{ mol } H_2C_2O_4)\left(\frac{2\text{ mol NaOH}}{1\text{ mol } H_2C_2O_4}\right) = 0.222\text{ mol}$$

The volume of the solution required can be found using Equation (3-6)

$$\text{volume} = \frac{n}{M} = \frac{0.222\text{ mol}}{0.50\text{ mol}\cdot\text{L}^{-1}} = 0.44\text{ L} = 440\text{ mL}$$

3-68. The mass of caffeine that must be administered is

$$\text{mass of } C_8H_{10}N_4O_2 = \left(\frac{200\text{ mg}}{1\text{ kg rat}}\right)(0.450\text{ kg rat})$$

$$= 90.0\text{ mg}$$

The number of moles of caffeine that must be administered is

$$\text{moles of } C_8H_{10}N_4O_2 = (90.0 \text{ mg})\left(\frac{1 \text{ g}}{10^3 \text{ mg}}\right)\left(\frac{1 \text{ mol}}{194.20 \text{ g}}\right)$$

$$= 0.000463 \text{ mol} = 4.63 \times 10^{-4} \text{ mol}$$

The volume of solution required can be found using Equation (3-6)

$$\text{volume} = \frac{n}{M} = \frac{0.000463 \text{ mol}}{0.11 \text{ mol}\cdot\text{L}^{-1}} = 0.0042 \text{ L} = 4.2 \text{ mL}$$

CHAPTER 4

SOLUTIONS TO THE EVEN-NUMBERED PROBLEMS

4-2. In order to name chemical compounds, it is necessary to know the names and symbols of the elements as well as the names and symbols of the common polyatomic ions (Table 4-1).

a) sodium acetate
b) calcium chlorate
c) ammonium carbonate
d) potassium hydroxide
e) barium nitrate

4-4.
a) ammonium thiosulfate
b) sodium carbonate
c) sodium sulfite
d) potassium carbonate
e) sodium thiosulfate

4-6. In going from the name of an ionic compound to the chemical formula, we first write down the formulas for the cation and the anion with the appropriate ionic charges. Then we use the fact that an ionic compound has no net charge to determine the relative numbers of cations and anions in the formula unit. Polyatomic ions are enclosed in parentheses if there are two or more polyatomic ions in the chemical formula.

a) $HC_2H_3O_2$
b) $HClO_3$
c) H_2CO_3
d) $HClO_4$
e) $HMnO_4$

4-8.
a) $NaClO_4$
b) $KMnO_4$
c) $AgC_2H_3O_2$
d) $CaSO_3$
e) $LiCN$

4-10. These compounds involve transition metal cations which have more than one possible charge. The value of the metal ion charge is indicated by a Roman numeral in parentheses.

a) chromium(II)sulfate

b) iron(III)oxide

c) cobalt(II)cyanide

d) tin(II)nitrate

e) copper(I)carbonate

4-12. a) $Hg_2(C_2H_3O_2)_2$

b) $Hg(CN)_2$

c) $Co(OH)_3$

d) $Fe(HCO_3)_2$

e) $CrSO_3$

4-14. The reactions are combination reactions and thus only a single reaction product is formed. To determine the reaction product we have to know the resulting charges on the positively charged metal ion and the negatively charged nonmetal ion.

a) $2Na(s) + Cl_2(g) \longrightarrow 2NaCl(s)$

b) $K(s) + O_2(g) \longrightarrow KO_2(s)$

c) $MgO(s) + CO_2(g) \longrightarrow MgCO_3(s)$

d) $2H_2(g) + O_2(g) \longrightarrow 2H_2O(l)$

e) $N_2(g) + 3H_2(g) \longrightarrow 2NH_3(g)$

4-16. In these reactions, water acts as a reactant and/or a solvent.

a) $SrO(s) + H_2O(l) \xrightarrow[H_2O(l)]{} Sr^{2+}(aq) + 2OH^-(aq)$

Strontium oxide is a basic anhydride.

b) $HNO_3(l) \xrightarrow[H_2O(l)]{} NO_3^-(aq) + H^+(aq)$

c) $Cs_2O + H_2O(l) \xrightarrow[H_2O(l)]{} 2Cs^+(aq) + 2OH^-(aq)$

Cesium oxide is a basic anhydride.

d) $HI(g) \xrightarrow[H_2O(l)]{} H^+(aq) + I^-(aq)$

4-18. Decomposition reactions involve the breakdown of a substance into smaller molecules. We predict the products of the reaction given by analogy with known cases.

a) $MgCO_3(s) \xrightarrow{\text{high T}} MgO(s) + CO_2(g)$

b) Recall that $KClO_3(s)$ decomposes on heating to $KCl(s) + O_2(g)$, (Example 4-5), thus, by analogy, we write

$2NaClO_3(s) \xrightarrow{\text{high T}} 2NaCl(s) + 3O_2(g)$

c) Sodium azide contains only two elements and thus the decomposition reaction must be

$2NaN_3(s) \xrightarrow{\text{high T}} 2Na(s) + 3N_2(g)$

d) $2Au_2O_3(s) \longrightarrow 4Au(s) + 3O_2(g)$

4-20. In order for a metal-metal replacement reaction to occur, the free metal must be a more reactive metal than the metal in the compound (Table 4-6). In order for a metal to replace hydrogen in an acid or in water, the metal must be a reactive metal (Table 4-6).

a) $Ba(s) + H_2O(g) \longrightarrow BaO(s) + H_3(g)$

b) $Fe(s) + H_2SO_4(aq) \longrightarrow FeSO_4(aq) + H_2(g)$

c) $Ca(s) + 2HBr(aq) \longrightarrow CaBr_2(aq) + H_2(g)$

d) $Pb(s) + 2HCl(aq) \longrightarrow PbCl_2(aq) + H_2(g)$

e) $Zn(s) + CaCl_2(aq) \longrightarrow$ N.R.

4-22. a) $2Na(l) + NiO(s) \longrightarrow Na_2O(s) + Ni(s)$

b) $Ca(s) + PbCl_2(s) \xrightarrow{\text{high T}} CaCl_2(s) + Pb(s)$

c) $4Mg(s) + Fe_3O_4(s) \xrightarrow{\text{high T}} 4MgO(s) + 3Fe(s)$

d) $Cu(s) + Al_2O_3(s) \longrightarrow$ N.R.

e) $Ag(s) + Zn(NO_3)_2(s) \longrightarrow$ N.R.

4-24. See Table 4-6. K, Zn, Ag, Pt.

4-26. $2CuO(s) + C(s) \xrightarrow{\text{high T}} 2Cu(s) + CO_2(g)$

$SnO_2(s) + C(s) \xrightarrow{\text{high T}} Sn(s) + CO_2(g)$

$2Fe_2O_3(s) + 3C(s) \xrightarrow{\text{high T}} 4Fe(s) + 3CO_2(g)$

4-28. To determine the net ionic equation that corresponds to the complete ionic equation, we determine the precipitate or other favored product formed. The ions that are used to form this species are the reactants in the net ionic equation.

a) $H^+(aq) + OH^-(aq) \longrightarrow H_2O(l)$

b) $Pb^{2+}(aq) + CO_3^{2-}(aq) \longrightarrow PbCO_3(s)$

c) $2Ag^+(aq) + SO_4^{2-}(aq) \longrightarrow Ag_2SO_4(s)$

d) $S^{2-}(aq) + Zn^{2+}(aq) \longrightarrow ZnS(s)$

e) $Hg_2^{2+}(aq) + 2Cl^-(aq) \longrightarrow Hg_2Cl_2(s)$

4-30. a) $2AgNO_3(aq) + Na_2S(aq) \longrightarrow Ag_2S(s) + 2NaNO_3(aq)$

$2Ag^+(aq) + S^{2-}(aq) \longrightarrow Ag_2S(s)$

b) $H_2SO_4(aq) + Pb(NO_3)_2(aq) \longrightarrow PbSO_4(s) + 2HNO_3(aq)$

$SO_4^{2-}(aq) + Pb^{2+}(aq) \longrightarrow PbSO_4(s)$

c) $Hg(NO_3)_2(aq) + 2NaI(aq) \longrightarrow HgI_2(s) + 2NaNO_3(aq)$

$Hg^{2+}(aq) + 2I^-(aq) \longrightarrow HgI_2(s)$

d) $CdCl_2(aq) + 2AgClO_4(aq) \longrightarrow 2AgCl(s) + Cd(ClO_4)_2(aq)$

$Cl^-(aq) + Ag^+(aq) \longrightarrow AgCl(s)$

e) $2LiBr(aq) + Pb(ClO_4)_2(aq) \longrightarrow PbBr_2(s) + 2LiClO_4(aq)$

$2Br^-(aq) + Pb^{2+}(aq) \longrightarrow PbBr_2(s)$

4-32. a) basic
b) acidic
c) acidic
d) basic
e) basic

4-34. $NH_3(aq) + \underline{H}CHO_2(aq) \longrightarrow NH_4CHO_2(aq)$ —— $NH_4^+(aq) + CHO_2^-(aq)$

4-36. a) double replacement

b) decomposition

c) double replacement

d) single replacement

4-38. a) $SrCO_3(s) \xrightarrow{heat} SrO(s) + CO_2(g)$ decomposition

b) $4Li(s) + O_2(g) \longrightarrow 2Li_2O(s)$ combination

c) $Zn(s) + H_2SO_4(aq) \longrightarrow ZnSO_4(aq) + H_2(g)$ single replacement

d) $Na_2O(s) + CO_2(g) \longrightarrow Na_2CO_3(s)$ combination

e) $XeF_4(s) \xrightarrow{heat} Xe(g) + 2F_2(g)$ decomposition

4-40. a) $ZnS(s) + 2HCl(aq) \longrightarrow H_2S(g) + ZnCl_2(aq)$

b) $3CaCl_2(aq) + 2H_3PO_4(aq) \longrightarrow Ca_3(PO_4)_2(s) + 6HCl(aq)$

c) $2HCl(aq) + Na_2CO_3(s) \longrightarrow CO_2(g) + H_2O(l) + 2NaCl(aq)$

d) $ZnCl_2(aq) + Na_2S(aq) \longrightarrow ZnS(s) + 2NaCl(aq)$

4-42. $HgS(s) + O_2(g) \xrightarrow{heat} Hg(g) + SO_2(g)$

cool

$Hg(l) + SO_2(g)$

4-44. The number of moles of H_2SO_4 required to neutralize the NaOH solution is

$$\text{moles of } H_2SO_4 = MV = (0.300 \text{ mol}\cdot\text{L}^{-1})(0.0246 \text{ L})$$

$$= 0.00738 \text{ mol}$$

We see from the neutralization reaction

$$H_2SO_4(aq) + 2NaOH(aq) \longrightarrow Na_2SO_4(aq) + 2H_2O(l)$$

that it required one mole of H_2SO_4 to neutralize two moles of NaOH. Thus we have

$$\text{moles of NaOH} = (0.00738 \text{ mol } H_2SO_4)\left(\frac{2 \text{ mol NaOH}}{1 \text{ mol } H_2SO_4}\right)$$

$$= 0.01476 \text{ mol}$$

The concentration of the NaOH solution is

$$\text{molarity} = \frac{n}{V} = \frac{0.01476 \text{ mol}}{0.0200 \text{ L}} = 0.738 \text{ M}$$

4-46. We first have to determine the number of moles of $Mg(OH)_2$ and $Al(OH)_3$ in 500 mg

$$\text{moles of } Mg(OH)_2 = (0.500 \text{ g})\left(\frac{1 \text{ mol}}{58.33 \text{ g}}\right) = 8.57 \times 10^{-3} \text{ mol}$$

$$\text{moles of } Al(OH)_3 = (0.500 \text{ g})\left(\frac{1 \text{ mol}}{78.00 \text{ g}}\right) = 6.41 \times 10^{-3} \text{ mol}$$

From the neutralization reaction

$$Mg(OH)_2(s) + 2HCl(aq) \longrightarrow MgCl_2(aq) + 2H_2O(l)$$

the number of moles of HCl neutralized by the $Mg(OH)_2$ is

$$\text{moles of HCl} = (8.57 \times 10^{-3} \text{ mol } Mg(OH)_2)\left(\frac{2 \text{ mol HCl}}{1 \text{ mol } Mg(OH)_2}\right)$$

$$= 1.71 \times 10^{-2} \text{ mol}$$

The volume of HCl neutralized by the $Mg(OH)_2$ is

$$V = \frac{n}{M} = \frac{1.71 \times 10^{-2} \text{ mol}}{0.10 \text{ mol} \cdot \text{L}^{-1}} = 0.170 \text{ L} = 170 \text{ mL}$$

From the neutralization reaction

$$Al(OH)_3(s) + 3HCl(aq) \longrightarrow AlCl_3(aq) + 3H_2O(l)$$

the number of moles of HCl neutralized by the $Al(OH)_3$ is

$$\text{moles of HCl} = (6.41 \times 10^{-3} \text{ mol } Al(OH)_3)\left(\frac{3 \text{ mol HCl}}{1 \text{ mol } Al(OH)_3}\right)$$

$$= 1.92 \times 10^{-2} \text{ mol}$$

The volume of HCl neutralized by the $Al(OH)_3$ is

$$V = \frac{n}{M} = \frac{1.92 \times 10^{-2} \text{ mol}}{0.10 \text{ mol}\cdot\text{L}^{-1}} = 0.19 \text{ L} = 190 \text{ mL}$$

4-48. The neutralization reaction is

$$NaOH(aq) + HBr(aq) \longrightarrow NaBr(aq) + H_2O(l)$$

The moles of NaOH added is

$$\text{moles of NaOH} = (0.500 \text{ L})(0.200 \text{ mol}\cdot\text{L}^{-1}) = 0.100 \text{ mol}$$

The moles of HBr initially present is

$$\text{moles of HBr} = (0.200 \text{ L})(0.100 \text{ mol}\cdot\text{L}^{-1}) = 0.0200 \text{ mol}$$

The total volume of the final solution is

$$\text{total volume} = 0.500 \text{ L} + 0.200 \text{ L} = 0.700 \text{ L}$$

The moles of NaOH added exceeds the moles of HBr initially present, thus all of the HBr is neutralized. The final concentration of HBr is thus zero. The final number of moles of NaOH equals the initial moles of NaOH minus the moles of NaOH used to neutralize the HBr. The final concentration of NaOH equals the final moles of NaOH divided by the total volume.

$$\text{molarity of NaOH} = \frac{0.100\text{ mol} - 0.0200\text{mol}}{0.700\text{ L}} = 0.11\text{ M}$$

The final concentration of NaBr equals the moles of HBr neutralized divided by the final volume

$$\text{molarity of NaBr} = \frac{0.0200\text{ mol}}{0.700\text{ L}} = 0.0286\text{ M}$$

4-50. The actual molarity, M, of the NaOH(aq) solution is computed from the neutralization condition

$$(25.0\text{ mL})M = (23.2\text{ mL})(0.100\text{ M})$$

$$M = \left(\frac{23.2\text{ mL}}{25.0\text{ mL}}\right)(0.100\text{ M})$$

$$M = 0.0928\text{ M}$$

The mass of NaOH in the 100 mL of NaOH solution is

$$\text{mass NaOH in 100 mL} = (0.0928\text{ mol}\cdot\text{L}^{-1})(0.100\text{ L})\left(\frac{40.00\text{ g}}{1\text{ mol}}\right)$$
$$= 0.371\text{ g}$$

The percent of NaOH in the sample is equal to the purity of the NaOH

$$\%\text{ purity} = \frac{0.371\text{ g NaOH}}{0.400\text{ g sample}} \times 100 = 92.8\%$$

We assumed that any impurities do not react with HCl.

4-52. The number of moles of NaOH required to neutralize the acid is

$$\text{moles of NaOH} = MV = (0.250\text{ mol}\cdot\text{L}^{-1})(0.0869\text{ L}) = 0.0217\text{ mol}$$

Therefore, the number of moles of acid present in the original 100 mL solution was 0.0217 mol. Thus we have that

$$1.00\text{ g acid} \approx 0.0217\text{ mol acid}$$

Dividing both sides by 0.0217, we get that

$$46.1\text{ g} \approx 1.00\text{ mol}$$

Thus the formula mass of the acid is 46.1.

4-54. The number of moles of base required to neutralize the acid is

$$\text{moles of NaOH} = MV = (0.250\ \text{mol}\cdot\text{L}^{-1})(0.0677\ \text{L}) = 0.0169\ \text{mol}$$

Because the acid has two acidic protons, it requires two moles of NaOH to neutralize one mole of the acid. Thus

$$\text{moles of acid} = (0.0169\ \text{mol NaOH})\left(\frac{1\ \text{mol acid}}{2\ \text{mol NaOH}}\right) = 0.00846\ \text{mol}$$

We have that

$$1.00\ \text{g acid} \approx 0.00846\ \text{mol}$$

$$118\ \text{g} \approx 1.00\ \text{mol}$$

The formula mass of the acid is 118.

CHAPTER 5

SOLUTIONS TO THE EVEN-NUMBERED PROBLEMS

5-2. Recall that for unit conversions, we multiply by the appropriate conversion factor; that is, the units in the numerator of the conversion factor are the units desired for the result and the units in the denominator are the units from which we want to change. See Table 5-1 for the relations between pressure units.

$$P = (24\ \text{torr})\left(\frac{1\ \text{atm}}{760\ \text{torr}}\right) = 0.032\ \text{atm}$$

$$P = (24\ \text{torr})\left(\frac{1\ \text{atm}}{760\ \text{torr}}\right)\left(\frac{1013\ \text{mbar}}{1\ \text{atm}}\right) = 32\ \text{mbar}$$

5-4. The pressure in atmospheres is

$$P = (990\ \text{mbar})\left(\frac{1\ \text{atm}}{1013\ \text{mbar}}\right) = 0.977\ \text{atm}$$

The pressure in torr is

$$P = (0.977\ \text{atm})\left(\frac{760\ \text{torr}}{1\ \text{atm}}\right) = 743\ \text{torr}$$

5-6. A pressure of 740 torr corresponds to a column of mercury of 740 mm high. The density of gallium is 6.095 $g\cdot mL^{-1}$; the density of mercury is 13.6 $g\cdot mL^{-1}$. The height of a gallium column at a gas pressure of 740 torr is

$$\text{height} = (740\ \text{mm})\left(\frac{13.6\ \text{g}\cdot\text{mL}^{-1}}{6.095\ \text{g}\cdot\text{mL}^{-1}}\right)$$

$$= 1650\ \text{mm} = 1.65 \times 10^3\ \text{mm} = 1.65\ \text{m}$$

5-8.

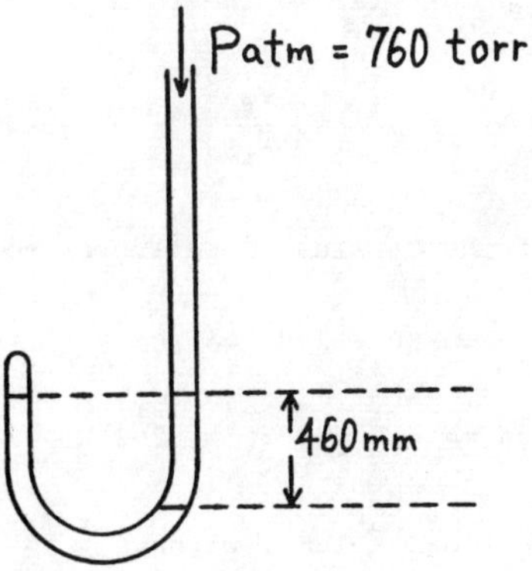

5-10. Boyle's law problems are worked out using Boyle's law in the form

$$P_iV_i = P_fV_f$$

where i stands for initial and f stands for final. Thus we have for the initial volume

$$V_i = \frac{P_fV_f}{P_i} = \frac{(0.10\ \text{atm})(100\ \text{m}^3)}{(100\ \text{atm})} = 0.10\ \text{m}^3$$

$$V_i = (0.10\ \text{m}^3)\left(\frac{100\ \text{cm}}{1\text{m}}\right)^3\left(\frac{1\ \text{mL}}{1\ \text{cm}^3}\right)\left(\frac{1\ \text{L}}{10^3\ \text{mL}}\right)$$

$$= 100\ \text{L}$$

We need two cylinders.

5-12. $P_iV_i = P_fV_f$ or $V_f = \frac{P_iV_i}{P_f}$

The total volume of air at 1.00 atm is

$$V_f = \frac{(200\ \text{atm})(50\ \text{L})}{(1.00\ \text{atm})} = 1.0 \times 10^4\text{L}$$

The number of breaths is

$$\text{no. of breaths} = (1.0 \times 10^4\ \text{L})\left(\frac{1\ \text{breath}}{0.50\ \text{L}}\right) = 2.0 \times 10^4\ \text{breaths}$$

5-14. We shall use Charles' law in the form

$$\frac{V_i}{T_i} = \frac{V_f}{T_f} \qquad (5\text{-}4)$$

We must convert the Celsius temperatures to Kelvin temperatures

$$T_i = 30 + 273 = 303 \text{ K}$$

$$T_f = -10 + 273 = 263 \text{ K}$$

Solving for the final volume, we have

$$V_f = \frac{T_f V_i}{T_i} = \frac{(263 \text{ K})(3.25 \text{ L})}{(303 \text{ K})} = 2.82 \text{ L at } -10^\circ\text{C}$$

5-16. The reaction involved is

$$2H_2(g) + O_2(g) \longrightarrow 2H_2O(g)$$

We see from Gay-Lussac's law that 1.0 L of O_2 and 2.0 L of H_2 produce 2.0 L of H_2O. The volume of oxygen needed is

$$V = \left(\frac{1}{2}\right)(0.55 \text{ L}) = 0.28 \text{ L}$$

The volume of water produced is

$$V = \text{volume of } H_2O = 0.55 \text{ L}$$

5-18. Solving the ideal-gas equation

$$PV = nRT$$

for n, we have

$$n = \frac{PV}{RT}$$

where

$$R = 0.0821 \text{ L·atm·mol}^{-1}\text{·K}^{-1}$$

$$T = 25 + 273 = 298 \text{ K}$$

Thus substituting in the values for P, V, R and T, we have

$$n = \frac{(9.6\ \text{atm})(1.5\ \text{L})}{(0.0821\ \text{L}\cdot\text{atm}\cdot\text{mol}^{-1}\cdot\text{K}^{-1})(298\ \text{K})} = 0.589\ \text{mol}$$

The mass of chlorine is

$$\text{mass} = (0.589\ \text{mol})\left(\frac{70.90\ \text{g}}{1\ \text{mol}}\right)$$

$$= 42\ \text{g}$$

5-20. Solving the ideal-gas equation for V, we have

$$V = \frac{nRT}{P}$$

Converting to the proper units we have

$$T = 27 + 273 = 300\ \text{K}$$

$$P = (870\ \text{torr})\left(\frac{1\ \text{atm}}{760\ \text{torr}}\right) = 1.145\ \text{atm}$$

$$n = (100\ \text{g})\left(\frac{1\ \text{mol}}{44.09\ \text{g}\ C_3H_8}\right) = 2.268\ \text{mol}$$

Thus substituting in the values for T, P, n and R, we have

$$V = \frac{(0.0821\ \text{L}\cdot\text{atm}\cdot\text{mol}^{-1}\cdot\text{K}^{-1})(2.268\ \text{mol})(300\ \text{K})}{1.145\ \text{atm}}$$

$$= 48.8\ \text{L}$$

5-22. The ideal-gas equation is

$$n = \frac{PV}{RT}$$

Substituting in the values for P, T, V and R, we have

$$n = \frac{(1.00\ \text{atm})(100\ \text{L})}{(0.0821\ \text{L}\cdot\text{atm}\cdot\text{mol}^{-1}\cdot\text{K}^{-1})(373\ \text{K})} = 3.27\ \text{mol}$$

We compute the number of molecules by multiplying n by Avogadro's number, thus

$$\text{no of molecules} = (3.27\ \text{mol})(6.022 \times 10^{23}\ \text{molecules}\cdot\text{mol}^{-1})$$

$$= 1.97 \times 10^{24}\ \text{molecules}$$

5-24. The form of the ideal-gas equation we need is

$$n = \frac{PV}{RT}$$

Converting to the proper units, we have

$$T = 20 + 273 = 293 \text{ K}$$

$$P = (1.0 \times 10^{-3} \text{ torr})\left(\frac{1 \text{ atm}}{760 \text{ torr}}\right) = 1.32 \times 10^{-6} \text{ atm}$$

$$V = (1.00 \text{ mL})\left(\frac{1 \text{ L}}{1000 \text{ mL}}\right) = 1.00 \times 10^{-3} \text{ L}$$

Substituting in the values for P, V, R and T, we have

$$n = \frac{(1.32 \times 10^{-6} \text{ atm})(1.00 \times 10^{-3} \text{ L})}{(0.0821 \text{ L·atm·mol}^{-1}\text{·K}^{-1})(293 \text{ K})} = 5.49 \times 10^{-11} \text{ mol}$$

The number of molecules is

$$\text{no of molecules} = (5.49 \times 10^{-11} \text{ mol})(6.022 \times 10^{23} \text{ molecules·mol}^{-1})$$

$$= 3.3 \times 10^{13} \text{ molecules}$$

5-26. The number of moles of gas remains fixed but P, V and T change. Thus from the ideal gas equation applied to the initial and final conditions we obtain

$$nR = \frac{P_iV_i}{T_i} = \frac{P_fV_f}{T_f}$$

Solving for V_f

$$V_f = V_i\left(\frac{P_i}{P_f}\right)\left(\frac{T_f}{T_i}\right)$$

$$V_f = (31.5 \text{ L})\left(\frac{1.3 \text{ atm}}{3.00 \times 10^{-3} \text{ atm}}\right)\left[\frac{250}{293}\right] = 1.2 \times 10^4 \text{ L}$$

5-28. We use the ideal gas equation to compute the number of moles of O_2 before and after reaction and then take the difference to find the number of moles reacted. This procedure works because the reaction product is a solid

$$4Al(s) + 3O_2(g) \longrightarrow 2Al_2O_3(s)$$

The number of moles of oxygen before reaction is

$$n_i = \frac{P_iV_i}{RT_i} = \frac{(1.00\text{ atm})(1.00\text{ L})}{(0.0821\text{ L·atm·K}^{-1}\text{·mol}^{-1})(298\text{ K})} = 0.0409\text{ mol}$$

The number of moles of oxygen remaining is

$$n_f = \frac{P_fV_f}{RT_f} = \frac{(0.91\text{ atm})(1.00\text{ L})}{(0.0821\text{ L·atm·K}^{-1}\text{·mol}^{-1})(298\text{ K})} = 0.0372\text{ mol}$$

The number of moles of oxygen reacted is

$$n_i - n_f = (0.0409 - 0.0372)\text{mol} = 0.0037\text{ mol}$$

The number of grams of oxygen reacted is

$$\text{mass of } O_2 = (0.0037\text{ mol})\left(\frac{32.00\text{g}}{1\text{ mol}}\right) = 0.12\text{ g}$$

5-30. The balanced reaction for the combustion of octane is

$$2C_8H_{18}(g) + 25O_2(g) \longrightarrow 16CO_2(g) + 18H_2O(g)$$

The mass of 1 gallon of gasoline is

$$\text{mass} = (1.00\text{ gal})\left(\frac{4\text{ qts}}{1\text{ gal}}\right)\left(\frac{0.946\text{ L}}{1\text{ qt}}\right)\left(\frac{10^3\text{ mL}}{1\text{ L}}\right)\left(\frac{0.70\text{ g}}{1\text{ mL}}\right) = 2.65 \times 10^3\text{ g}$$

The number of moles of oxygen required to react with 2.65×10^3 g of C_8H_{18} is

$$\text{moles of } O_2 = (2.65 \times 10^3\text{ g})\left(\frac{1\text{ mol } C_8H_{18}}{114.22\text{g } C_8H_{18}}\right)\left(\frac{25\text{ mol } O_2}{2\text{ mol } C_8H_{18}}\right)$$

$$= 2.90 \times 10^2\text{ mol}$$

The volume of O_2 required at 20^0C and 1 atm is

$$V = \frac{nRT}{P} = \frac{(2.90 \times 10^2\text{ mol})(0.0821\text{ L·atm·mol}^{-1}\text{·K}^{-1})(273\text{ K})}{1.00\text{ atm}}$$

$$= 6.50 \times 10^3\text{ L}$$

The volume of air required is

$$V = \frac{6.50 \times 10^3 \text{ L}}{0.20} = 3.25 \times 10^4 \text{ L}$$

Conversion of liters to gallons yields

$$V = (3.25 \times 10^4 \text{ L})\left(\frac{1 \text{ qt}}{0.946 \text{ L}}\right)\left(\frac{1 \text{ gal}}{4 \text{ qt}}\right) = 8.6 \times 10^3 \text{ gal}$$

$$= 8600 \text{ gal}$$

5-32. The reaction is

$$MnO_2(s) + 4HCl(aq) \longrightarrow MnCl_2(aq) + 2H_2O(l) + Cl_2(g)$$

The number of moles of chlorine desired is

$$n = \frac{PV}{RT} = \frac{(750 \text{ torr})\left(\frac{1 \text{ atm}}{760 \text{ torr}}\right)(0.500 \text{ L})}{(0.0821 \text{ L}\cdot\text{atm}\cdot\text{mol}^{-1}\cdot\text{K}^{-1})(298 \text{ K})}$$

$$= 0.02017 \text{ mol}$$

The mass of MnO_2 required is

$$\text{mass of } MnO_2 = (0.0202 \text{ mol } Cl_2)\left(\frac{1 \text{ mol } MnO_2}{1 \text{ mol } Cl_2}\right)\left(\frac{86.94 \text{ g } MnO_2}{1 \text{ mol } MnO_2}\right)$$

$$= 1.75 \text{ g}$$

5-34. The density is given by

$$\rho = \frac{MP}{RT}$$

The molar mass of CF_2Cl_2 is 120.9 $\text{g}\cdot\text{mol}^{-1}$. At 0°C and 1.00 atm, the density of CF_2Cl_2 is

$$\rho = \frac{(120.9 \text{ g mol}^{-1})(1.00 \text{ atm})}{(0.0821 \text{ L}\cdot\text{atm}\cdot\text{mol}^{-1}\cdot\text{K}^{-1})(273 \text{ K})}$$

$$= 5.39 \text{ g}\cdot\text{L}^{-1}$$

5-36. We can compute the density of ether from the given mass and volume

$$\rho = \frac{2.97\ \text{g}}{0.500\ \text{L}} = 5.94\ \text{g}\cdot\text{L}^{-1}$$

The molar mass is related to the density by the equation

$$\rho = \frac{MP}{RT}$$

Solving for M, we have

$$M = \frac{\rho RT}{P}$$

$$= \frac{(5.94\ \text{g}\cdot\text{L}^{-1})(0.0821\ \text{L}\cdot\text{atm}\cdot\text{K}^{-1}\cdot\text{mol}^{-1})(368\ \text{K})}{(2.13\ \text{atm})}$$

$$= 84.3\ \text{g}\cdot\text{mol}^{-1}$$

5-38. We first determine the empirical formula of the compound. Taking a 100 g sample, we have

$$92.24\ \text{g C} \Leftrightarrow 7.76\ \text{g H}$$

Dividing the mass of each by its atomic mass, we have

$$7.68\ \text{mol C} \Leftrightarrow 7.70\ \text{mol H}$$

$$1.00\ \text{mol C} \Leftrightarrow 1.00\ \text{mol H}$$

Thus the empirical formula of benzene is CH. The molar mass of benzene is

$$M = \frac{\rho RT}{P}$$

$$= \frac{\left(\frac{2.334\ \text{g}}{0.500\ \text{L}}\right)(0.0821\ \text{L}\cdot\text{atm}\cdot\text{K}^{-1}\cdot\text{mol}^{-1})(373\ \text{K})}{1.83\ \text{atm}}$$

$$= 78.1\ \text{g}\cdot\text{mol}^{-1}$$

The formula mass of CH is 13.02. The molecular mass of benzene is six times the formula mass of CH $\left(\frac{78.1}{13.02} = 6.00\right)$. Therefore, the molecular formula of benzene is C_6H_6.

5-40. Proceeding as in Problem 5-38 we have

$$54.52 \text{ g C} \;\hat{=}\; 9.17 \text{ g H} \;\hat{=}\; 36.31 \text{ g O}$$

$$4.54 \text{ mol C} \;\hat{=}\; 9.10 \text{ mol H} \;\hat{=}\; 2.27 \text{ mol O}$$

$$2.00 \text{ mol C} \;\hat{=}\; 4.01 \text{ mol H} \;\hat{=}\; 1.00 \text{ mol O}$$

Thus the empirical formula is C_2H_4O. The molar mass is

$$M = \rho\frac{RT}{P} = \frac{\left(\frac{1.203 \text{ g}}{0.250 \text{ L}}\right)(0.0821 \text{ L}\cdot\text{atm}\cdot\text{K}^{-1}\cdot\text{mol}^{-1})(368 \text{ K})}{(1.65 \text{ atm})} = 88.1 \text{ g}\cdot\text{mol}^{-1}$$

The formula mass of C_2H_4O is 44.05, thus the molecular mass is twice the formula mass. Therefore the molecular formula of ethyl acetate is $C_4H_8O_2$.

5-42. Air is 20% oxygen. Thus if air is compressed to a total pressure of 4.0 atm, the partial pressure of oxygen would be

$$P_{O_2} = (0.20)(4.0 \text{ atm}) = 0.80 \text{ atm}$$

If we use tank gas with 5.0% oxygen, then the partial pressure of oxygen at a total pressure of 4.0 atm would be

$$P_{O_2} = (0.050)(4.0 \text{ atm}) = 0.20 \text{ atm}$$

5-44. The decomposition of 2 moles of TNT yields 12 + 5 + 3 = 20 mol of product gases. A 1000 g sample of TNT yields

$$(1000 \text{ g TNT})\left(\frac{1 \text{ mol TNT}}{227.14\text{g TNT}}\right)\left(\frac{20 \text{ mol gas}}{2 \text{ mol TNT}}\right) = 44.0 \text{ mol gas}$$

From the ideal gas equation we compute the volume

$$V = \frac{(44.0 \text{ mol})(0.0821 \text{ L}\cdot\text{atm}\cdot\text{mol}^{-1}\cdot\text{K}^{-1})(273 \text{ K})}{(1.0 \text{ atm})}$$

$$= 9.9 \times 10^2 \text{ L} = 990 \text{ L}$$

The pressure developed by 44.0 mol in 50 L at 773 K is

$$P = \frac{(44.0\text{ mol})(0.0821\text{ L·atm·mol}^{-1}\text{·K}^{-1})(773\text{ K})}{(50\text{ L})}$$

$$= 56\text{ atm}$$

5-46. From Dalton's law of partial pressures we have

$$P_{total} = P_{H_2} + P_{H_2O}$$

thus

$$P_{H_2} = P_{total} - P_{H_2O} = 752\text{ torr}-22.4\text{ torr} = 730\text{ torr}$$

The volume at 0°C and 760 torr is calculated from the ideal gas equation applied to the initial and final states with n constant

$$V_f = V_i\left(\frac{P_i}{P_f}\right)\left(\frac{T_f}{T_i}\right)$$

Thus

$$V_f = (0.200\text{ L})\left(\frac{730\text{ torr}}{760\text{ torr}}\right)\left(\frac{273\text{ K}}{297\text{ K}}\right) = 0.177\text{ L}$$

At 0°C and 760 torr one mole of an ideal gas occupies 22.4 L; therefore, the number of moles of H_2 at 0°C and 760 torr is

$$\text{moles of } H_2 = \frac{0.1766\text{ L}}{22.4\text{ L·mol}^{-1}} = 0.00788\text{ mol}$$

We could also compute the number of moles of H_2 from the ideal gas equation.

5-48. Recall that in working kinetic theory problems we use the value of $R = 8.31\ J\cdot mol^{-1}\cdot K^{-1}$. The average speed in $m\cdot s^{-1}$ of a gas molecule is calculated from the equation

$$v_{av} = \left(\frac{3RT}{M_{kg}}\right)^{\frac{1}{2}}$$

The molar mass in kilograms per mole is

$$M_{kg} = \frac{20.18\ g\cdot mol^{-1}}{1000\ g\cdot kg^{-1}} = 0.02018\ kg\cdot mol^{-1}$$

Thus at 298 K, we have for the average speed

$$v_{av} = \left[\frac{(3)(8.31\ J\cdot mol^{-1}\cdot K^{-1})(298\ K)}{0.02018\ kg\cdot mol^{-1}}\right]^{\frac{1}{2}}$$

$$= 607\ m\cdot s^{-1}$$

5-50. Table 5-3 gives average speeds in meter per second. Therefore, we have to convert 760 mph to meters per second. A kilometer is equal to 0.621 miles, thus 760 miles is equal to

$$(760\ \text{miles})\left(\frac{1\ km}{0.621\ \text{mile}}\right)\left(\frac{1000\ m}{1\ km}\right) = 1.22 \times 10^6\ m$$

Converting hours to seconds

$$(1\ hr)\left(\frac{60\ min}{1\ hr}\right)\left(\frac{60\ sec}{1\ min}\right) = 3.60 \times 10^3\ s$$

Thus the speed of sound in air at 20°C at sea level is

$$\text{speed of sound} = \frac{1.22 \times 10^6\ m}{3.60 \times 10^3\ s} = 340\ m\cdot s^{-1}$$

From Table 5-3 we have

$$v_{av,N_2} = 510\ m\cdot s^{-1} \qquad \left(\frac{340}{510}\right) \times 100 = 67\%\ \text{of}\ v_{av,N_2}$$

$$v_{av,O_2} = 480\ m\cdot s^{-1} \qquad \left(\frac{340}{480}\right) \times 100 = 71\%\ \text{of}\ v_{av,O_2}$$

Thus sound travels at about $\frac{2}{3}$ the average speed of air molecules.

5-52. The ratio of the average speeds of H_2 and I_2 in a gas mixture is computed using the average speed equation applied to both H_2 and I_2

$$\frac{v_{av,H_2}}{v_{av,I_2}} = \frac{\left[\frac{3RT}{M_{H_2}}\right]^{1/2}}{\left[\frac{3RT}{M_{I_2}}\right]^{1/2}} = \left[\frac{M_{I_2}}{M_{H_2}}\right]^{1/2}$$

$$= \left[\frac{253.8}{2.016}\right]^{1/2} = 11.22$$

Note that H_2 with a much lower mass than I_2 has a much higher average speed than I_2.

5-54. The mean free path is given by

$$l = (3.1 \times 10^7 \text{ pm}^3\cdot\text{atm}\cdot\text{K}^{-1}) \frac{T}{\sigma^2 P}$$

The value of σ for helium is 210 pm (Table 5-4). At 1.00 torr,

$$P = (1.00 \text{ torr})\left(\frac{1 \text{ atm}}{760 \text{ torr}}\right) = 1.32 \times 10^{-3} \text{ atm}$$

and

$$l = \frac{(3.1 \times 10^7 \text{ pm}^3\cdot\text{atm}\cdot\text{K}^{-1})(293 \text{ K})}{(210 \text{ pm})^2(1.32 \times 10^{-3}\text{atm})}$$

$$= 1.6 \times 10^8 \text{ pm} = 1.6 \times 10^{-4} \text{ m}$$

At 760 torr, P = 1.00 atm and

$$l = \frac{(3.1 \times 10^7 \text{ pm}^3\cdot\text{atm}\cdot\text{K}^{-1})(293 \text{ K})}{(210 \text{ pm})^2(1.00 \text{ atm})}$$

$$= 2.1 \times 10^5 \text{ pm} = 2.1 \times 10^{-7} \text{ m}$$

The value of σ for krypton is 410 pm. At 1.00 torr

$$l = \frac{(3.1 \times 10^7\ \text{pm}^3\cdot\text{atm}\cdot\text{K}^{-1})(293\ \text{K})}{(410\ \text{pm})^2(1.32 \times 10^{-3}\ \text{atm})}$$

$$= 4.1 \times 10^7\ \text{pm} = 4.1 \times 10^{-5}\ \text{m}$$

At 760 torr

$$l = \frac{(3.1 \times 10^7\text{pm}^3\cdot\text{atm}\cdot\text{K}^{-1})(293\ \text{K})}{(410\ \text{pm})^2(1.00\ \text{atm})}$$

$$= 5.4 \times 10^4\ \text{pm} = 5.4 \times 10^{-8}\ \text{m}$$

5-56. The mean free path is given by

$$l = (3.1 \times 10^7\ \text{pm}^3\cdot\text{atm}\cdot\text{K}^{-1})\ \frac{T}{\sigma^2 P}$$

Solving for P, we have

$$P = \frac{(3.1 \times 10^7\text{pm}^3\cdot\text{atm}\cdot\text{K}^{-1})T}{\sigma^2 l}$$

Converting the distance to picometers, we have

$$(1.00\ \mu\text{m})\left(\frac{1\ \text{m}}{10^6\ \mu\text{m}}\right)\left(\frac{10^{12}\text{pm}}{\text{m}}\right) = 1.00 \times 10^6\ \text{pm}$$

$$(1.00\ \text{mm})\left(\frac{1\ \text{m}}{10^3\ \text{mm}}\right)\left(\frac{10^{12}\text{pm}}{1\ \text{m}}\right) = 1.00 \times 10^9\ \text{pm}$$

$$(1.00\ \text{m})\left(\frac{10^{12}\ \text{pm}}{1\ \text{m}}\right) = 1.00 \times 10^{12}\ \text{pm}$$

The mean free path is 1.00 μm when

$$P = \frac{(3.1 \times 10^7\ \text{pm}^3\cdot\text{atm}\cdot\text{K}^{-1})(293\ \text{K})}{(280\ \text{pm})^2(1.00 \times 10^6\ \text{pm})} = 0.12\ \text{atm}$$

The mean free path is 1.00 mm when

$$P = \frac{(3.1 \times 10^7 \text{ pm}^3\cdot\text{atm}\cdot\text{K}^{-1})(293 \text{ K})}{(280 \text{ pm})^2(1.00 \times 10^9 \text{ pm})} = 1.2 \times 10^{-4} \text{ atm}$$

The mean free path is 1.00 m when

$$P = \frac{(3.1 \times 10^7 \text{ pm}^3\cdot\text{atm}\cdot\text{K}^{-1})(293 \text{ K})}{(280 \text{ pm})^2(1.00 \times 10^{12} \text{ pm})} = 1.2 \times 10^{-7} \text{ atm}$$

5-58. The number of collisions per second per molecule (collision frequency) is given by

$$z = \frac{v_{av}}{l}$$

The average speed for N_2 at 20^0C is

$$v_{av} = \left(\frac{3RT}{M_{kg}}\right)^{1/2} = \left[\frac{(3)(8.31 \text{ kg}\cdot\text{m}^2\cdot\text{s}^{-2})(293 \text{ K})}{(0.0280 \text{ kg}\cdot\text{mol}^{-1})}\right]^{1/2} = 511 \text{ m}\cdot\text{s}^{-1}$$

The mean free path of N_2 at 20^0C and 1.0×10^{-3} torr is

(σ_{N_2} = 370 pm from Table 5-4)

$$l = \frac{(3.1 \times 10^7 \text{pm}^3\cdot\text{atm}\cdot\text{K}^{-1})(293 \text{ K})}{(370 \text{ pm})^2(1.0 \times 10^{-3} \text{ torr})\left(\frac{1 \text{ atm}}{760 \text{ torr}}\right)} = 5.0 \times 10^{10} \text{pm}$$

$$= 5.0 \times 10^{-2} \text{ m}$$

Thus

$$z = \frac{511 \text{ m}\cdot\text{s}^{-1}}{5.0 \times 10^{-2} \text{m}} = 1.0 \times 10^4 \text{ collisions}\cdot\text{s}^{-1}$$

5-60. The ratio of the rates of effusion (leaking) of the two gases is given by Graham's law

$$\frac{\text{rate A}}{\text{rate B}} = \left(\frac{M_B}{M_A}\right)^{1/2}$$

If we take A = hydrogen and B = carbon dioxide, then we have

$$\text{rate}_{H_2} = (\text{rate}_{CO_2})\left(\frac{M_{CO_2}}{M_{H_2}}\right)^{1/2}$$

The rate of effusion of carbon dioxide is

$$\text{rate}_{CO_2} = 1.50\ \text{mL}\cdot\text{day}^{-1}$$

Thus the rate of effusion of hydrogen is

$$\text{rate}_{H_2} = (1.50\ \text{mL}\cdot\text{day}^{-1})\left(\frac{44.01}{2.016}\right)^{\frac{1}{2}} = 7.01\ \text{mL}\cdot\text{day}^{-1}$$

In one day, 7.01 mL of H_2 will leak out.

5-62. Graham's law gives

$$\frac{\text{rate A}}{\text{rate B}} = \left(\frac{M_B}{M_A}\right)^{\frac{1}{2}}$$

or

$$\frac{M_B}{M_A} = \left(\frac{\text{rate A}}{\text{rate B}}\right)^2$$

We can find the percentage of CO in the gas mixture from the molecular mass calculated from Graham's law. If we let B = CO-CO_2 mixture and A = N_2, then we have

$$M_{CO\text{-}CO_2} = (M_{N_2})\left(\frac{\text{rate}_{N_2}}{\text{rate}_{CO\text{-}CO_2}}\right)^2$$

The rate of effusion of N_2 is

$$\text{rate}_{N_2} = \frac{1.00\ \text{mL}}{175\ \text{s}} = 5.714\times 10^{-3}\ \text{mL}\cdot\text{s}^{-1}$$

The rate of effusion of the CO-CO_2 mixture is

$$\text{rate}_{CO\text{-}CO_2} = \frac{1.00\ \text{mL}}{200\ \text{s}} = 5.000\times 10^{-3}\ \text{mL}\cdot\text{s}^{-1}$$

Thus

$$M_{CO\text{-}CO_2} = (28.02)\left[\frac{5.714\times 10^{-3}\ \text{mL}\cdot\text{s}^{-1}}{5.00\ \times 10^{-3}\ \text{mL}\cdot\text{s}^{-1}}\right]^2 = 36.59$$

If we let X = the fraction of the mixture that is CO, then 1.00 - X = the fraction that is CO_2. The molecular mass of the mixture is

$$M_{CO-CO_2} = XM_{CO} + (1-X)M_{CO_2}$$

$$36.59 = X(28.02) + (1-X)(44.01)$$

$$= 28.02X + 44.01 - 44.01X$$

Collecting like terms, we have

$$15.99\ X = 7.42$$

or

$$X = \frac{7.42}{15.99} = 0.464$$

The mixture is 46% CO.

5-64. We shall start with the van der Waals equation in the form

$$P = \frac{nRT}{V-nb} - \frac{n^2a}{V^2}$$

At low densities, $\frac{n}{V}$ and thus $\frac{n^2}{V^2}$ are very small and can be neglected. We now have

$$P = \frac{nRT}{V-nb}$$

If we divide the numerator and denominator by V, then we have

$$P = \frac{\frac{nRT}{V}}{1 - \frac{nb}{V}}$$

Again $\frac{n}{V}$ is very small and can be neglected compared to 1, thus we have

$$P = \frac{nRT}{V}$$

or

$$PV = nRT$$

which is the ideal-gas equation.

5-66. The pressure of the gas given by the ideal-gas equation is

$$P = \frac{nRT}{V}$$

The SI unit of volume is the cubic meter

$$V = (2.10 \text{ mL})\left(\frac{1 \text{ L}}{1000 \text{ mL}}\right)\left(\frac{10^{-3} \text{m}^3}{1 \text{ L}}\right) = 2.10 \times 10^{-6} \text{m}^3$$

$$n = (6.15 \text{ mg})\left(\frac{1 \text{ g}}{1000 \text{ mg}}\right)\left(\frac{1 \text{ mol}}{44.01 \text{ g}}\right) = 1.40 \times 10^{-4} \text{ mol}$$

Thus

$$P = \frac{(1.40 \times 10^{-4} \text{ mol})(8.31 \text{ N}\cdot\text{m}\cdot\text{K}^{-1}\cdot\text{mol}^{-1})(348 \text{ K})}{(2.10 \times 10^{-6} \text{ m}^3)}$$

$$= 1.92 \times 10^5 \text{ N}\cdot\text{m}^{-2} = 1.92 \times 10^5 \text{ Pa}$$

5-68. The number of moles of $N_2O(g)$ is given by

$$n = \frac{PV}{RT} = \frac{(4.50 \times 10^4 \text{ Pa})(2.10 \times 10^{-3}\text{m}^3)}{(8.31 \text{ N}\cdot\text{m}\cdot\text{K}^{-1}\cdot\text{mol}^{-1})(288 \text{ K})}$$

$$= 3.95 \times 10^{-2} \text{ mol}$$

Notice that the volume must be units of m^3

$$\text{mass of } N_2O(g) = (3.95 \times 10^{-2} \text{ mol})\left(\frac{44.02 \text{ g}}{1 \text{ mol}}\right) = 1.74 \text{ g}$$

5-70. The SI unit of pressure is the pascal and the SI unit of volume is the cubic meter, thus (see inside back cover)

$$R = (0.08206\ L\cdot atm\cdot K^{-1}\cdot mol^{-1})\left(\frac{1.0133\times 10^{5}Pa}{1\ atm}\right)\left(\frac{1.00\times 10^{-3}m^{3}}{1\ L}\right)$$

but

$$1\ Pa = 1\ N\cdot m^{-2} \qquad \text{and} \qquad 1\ J = 1\ N\cdot m$$

Therefore

$$1\ Pa\cdot m^{3} = (1N\cdot m^{-2})(m^{2}) = 1\ N\cdot m = 1\ J$$

and

$$R = 8.314\ J\cdot K^{-1}\cdot mol^{-1}$$

CHAPTER 6

SOLUTIONS TO THE EVEN-NUMBERED PROBLEMS

6-2. The reaction is

$$Ba(s) + Cl_2(g) \longrightarrow BaCl_2(s)$$

The number of moles of $BaCl_2$ produced is

$$\text{moles of } BaCl_2 = (2.46 \text{ g Ba})\left(\frac{1 \text{ mol Ba}}{137.3 \text{ g Ba}}\right)\left(\frac{1 \text{ mol } BaCl_2}{1 \text{ mol Ba}}\right) = 0.0179 \text{ mol}$$

The amount of heat that is evolved when one mole of $BaCl_2$ is formed is

$$\Delta H_{rxn} = \frac{15.4 \text{ kJ}}{0.0179 \text{ mol}} = 860 \text{ kJ}\cdot\text{mol}^{-1}$$

6-4. The reaction is

$$2Mg(s) + O_2(g) \longrightarrow 2MgO(s)$$

The number of moles of MgO produced is

$$\text{moles of MgO} = (0.165 \text{ g Mg})\left(\frac{1 \text{ mol Mg}}{24.31 \text{ g Mg}}\right)\left(\frac{2 \text{ mol MgO}}{2 \text{ mol Mg}}\right) = 0.00679 \text{ mol}$$

The heat evolved per mole of MgO is

$$\Delta H_{rxn} = \frac{4.08 \text{ kJ}}{0.00679 \text{ mol}} = 601 \text{ kJ}\cdot\text{mol}^{-1}$$

6-6. a) $\Delta H^0_{rxn} = 2\Delta\overline{H}^0_f[H_2O(l)] + \Delta\overline{H}^0_f[O_2(g)] - 2\Delta\overline{H}^0_f[H_2O_2(l)]$

Using the data from Table 6-1, we have

$$\Delta H^0_{rxn} = (2 \text{ mol})(-285.8 \text{ kJ}\cdot\text{mol}^{-1}) + (1 \text{ mol})(0 \text{ kJ}\cdot\text{mol}^{-1}) - (2 \text{ mol})(-187.8 \text{ kJ}\cdot\text{mol}^{-1})$$

$$= -196.0 \text{ kJ} \qquad \text{exothermic}$$

b) $\Delta H^0_{rxn} = \Delta\overline{H}^0_f[MgCO_3(s)] - \Delta\overline{H}^0_f[MgO(s)] - \Delta\overline{H}^0_f[CO_2(g)]$

$= (1 \text{ mol})(-1096 \text{ kJ}\cdot\text{mol}^{-1}) - (1 \text{ mol})(-601.7 \text{ kJ}\cdot\text{mol}^{-1})$

$\quad -(1 \text{ mol})(-393.5 \text{ kJ}\cdot\text{mol}^{-1})$

$= -101 \text{kJ}$ exothermic

c) $\Delta H^0_{rxn} = 4\Delta\overline{H}^0_f[NO(g)] + 6\Delta\overline{H}^0_f[H_2O(g)] - 4\Delta\overline{H}^0_f[NH_3(g)] - 5\Delta\overline{H}^0_f[O_2(g)]$

$= (4 \text{ mol})(90.37 \text{ kJ}\cdot\text{mol}^{-1}) + (6 \text{ mol})(-241.8 \text{ kJ}\cdot\text{mol}^{-1})$

$\quad -(4 \text{ mol})(-46.19 \text{ kJ}\cdot\text{mol}^{-1}) - (5 \text{ mol})(0 \text{ kJ}\cdot\text{mol}^{-1})$

$= -904.6 \text{ kJ}$ exothermic

6-8. a) $\Delta H^0_{rxn} = \Delta\overline{H}^0_f[CO_2(g)] + 2\Delta\overline{H}^0_f[H_2O(g)] - \Delta\overline{H}^0_f[CH_3OH(l)] - \frac{3}{2}\Delta\overline{H}^0_f[O_2(g)]$

$= (1 \text{ mol})(-393.5 \text{ kJ}\cdot\text{mol}^{-1}) + (2 \text{ mol})(-241.8 \text{ kJ}\cdot\text{mol}^{-1})$

$\quad -(1 \text{ mol})(-238.7 \text{ kJ}\cdot\text{mol}^{-1}) - 0$

$= -638.4 \text{ kJ}$

The heat of combustion of CH_3OH per gram is

$$\left(\frac{-638.4 \text{ kJ}}{1\text{mol}}\right)\left(\frac{1 \text{ mol}}{32.04 \text{ g}}\right) = -19.93 \text{ kJ}\cdot\text{g}^{-1}$$

b) $\Delta H^0_{rxn} = \Delta\overline{H}^0_f[N_2(g)] + 2\Delta\overline{H}^0_f[H_2O(g)] - \Delta\overline{H}^0_f[N_2H_4(l)] - \Delta\overline{H}^0_f[O_2(g)]$

$= 0 + (2 \text{ mol})(-241.8 \text{ kJ}\cdot\text{mol}^{-1}) - (1 \text{ mol})(50.6 \text{ kJ}\cdot\text{mol}^{-1}) - 0$

$= -534.2 \text{ kJ}$

The heat of combustion of N_2H_4 per gram is

$$\left(\frac{-534.2 \text{ kJ}}{1 \text{ mol}}\right)\left(\frac{1 \text{ mol}}{32.05 \text{ g}}\right) = -16.67 \text{ kJ}\cdot\text{g}^{-1}$$

The combustion of one gram of CH_3OH produces 1.2 times as much heat per gram as does N_2H_4.

6-10. $\Delta H^0_{rxn} = 12\Delta\overline{H}^0_f[CO_2(g)] + 11\Delta\overline{H}^0_f[H_2O(l)] - \Delta\overline{H}^0_f[C_{12}H_{22}O_{11}(s)] - 12\Delta\overline{H}^0_f[O_2(g)]$

In this case we are given ΔH^0_{rxn} and must determine $\Delta\overline{H}^0_f[C_{12}H_{22}O_{11}(s)]$. Using the data in Table 6-1

$$-5646.7\ kJ = (12\ mol)(-393.5\ kJ\cdot mol^{-1}) + (11\ mol)(-285.8\ kJ\cdot mol^{-1}) - (1\ mol)\ \Delta\overline{H}^0_f[C_{12}H_{22}O_{11}(s)]$$

Solving for $\Delta\overline{H}^0_f[C_{12}H_{22}O_{11}(s)]$

$$(1\ mol)\Delta\overline{H}^0_f[C_{12}H_{22}O_{11}(s)] = +5646.7\ kJ - 4722\ kJ - 3144\ kJ$$

$$\Delta\overline{H}^0_f[C_{12}H_{22}O_{11}(s)] = -2219\ kJ\cdot mol^{-1}$$

6-12. $\Delta H^0_{rxn} = \Delta\overline{H}^0_f[Cu_2O(s)] - \Delta\overline{H}^0_f[CuO(s)] - \Delta\overline{H}^0_f[Cu(s)]$

$= (1\ mol)(-169.0\ kJ\cdot mol^{-1}) - (1\ mol)(-157.3\ kJ\cdot mol^{-1}) - 0$

$= -11.7\ kJ$

6-14. a) $\Delta H^0_{rxn} = 2\Delta\overline{H}^0_f[HF(g)] - \Delta\overline{H}^0_f[H_2(g)] - \Delta\overline{H}^0_f[F_2(g)]$

$\Delta H^0_{rxn} = (2\ mol)(-271.1\ kJ\cdot mol^{-1}) - 0 - 0$

$= -542.2\ kJ$ exothermic

b) $\Delta H^0_{rxn} = 2\Delta\overline{H}^0_f[CO_2(g)] - 2\Delta\overline{H}^0_f[CO(g)] - \Delta\overline{H}^0_f[O_2(g)]$

$= (2\ mol)(-393.5\ kJ\cdot mol^{-1}) - (2\ mol)(-110.5\ kJ\cdot mol^{-1}) - 0$

$= -566.0\ kJ$ exothermic

c) $\Delta H^0_{rxn} = 2\Delta\overline{H}^0_f[NH_3(g)] - 3\Delta\overline{H}^0_f[H_2(g)] - \Delta\overline{H}^0_f[N_2(g)]$

$= (2\ mol)(-46.19\ kJ\cdot mol^{-1}) - 0 - 0$

$= -92.38\ kJ$ exothermic

d) $\Delta H^0_{rxn} = 2\Delta\overline{H}^0_f[NO_2(g)] - 2\Delta\overline{H}^0_f[NO(g)] - \Delta\overline{H}^0_f[O_2(g)]$

$= (2\ mol)(33.85\ kJ\cdot mol^{-1}) - (2\ mol)(90.37\ kJ\cdot mol^{-1}) - 0$

$= -113.04\ kJ$ exothermic

6-16. The reaction is

$$H_2O(l) \longrightarrow H_2O(g)$$

The heat required to vaporize 1.00 mol of water is equal to ΔH^0_{rxn}. Thus

$$\text{heat} = \Delta\overline{H}^0_{rxn} = \Delta\overline{H}^0_f[H_2O(g)] - \Delta\overline{H}^0_f[H_2O(l)]$$

$$= (1\text{ mol})(-241.8\text{ kJ}\cdot\text{mol}^{-1}) - (1\text{ mol})(-285.8\text{ kJ}\cdot\text{mol}^{-1})$$

$$= 44.0\text{ kJ}$$

6-18. For the reaction

$$Ca(OH)_2(s) \longrightarrow CaO(s) + H_2O(l)$$

$$\Delta H_{rxn} = -(-56.27\text{ kJ}) = 56.27\text{ kJ}$$

The heat required to convert one gram of $Ca(OH)_2$ is

$$\left(\frac{56.27\text{ kJ}}{1\text{ mol}}\right)\left(\frac{1\text{ mol }Ca(OH)_2}{74.10\text{ g }Ca(OH)_2}\right) = 0.7594\text{ kJ}\cdot\text{g}^{-1}$$

6-20. Reverse the first equation and add it to the second equation

$CuCl_2(s) \longrightarrow Cu(s) + Cl_2(g)$	$\Delta H^0_{rxn} = +206\text{ kJ}$
$2Cu(s) + Cl_2(g) \longrightarrow 2CuCl(s)$	$\Delta H^0_{rxn} = -136\text{ kJ}$
$Cu(s) + CuCl_2(s) \longrightarrow 2CuCl(s)$	$\Delta H^0_{rxn} = +70\text{ kJ}$

6-22. To obtain the third equation, multiply the first equation by 2, reverse the second equation, and then add the two together.

$$2H_2(g) + 2F_2(g) \longrightarrow 4HF(g) \qquad \Delta H^0_{rxn} = (2)(-542.2\ kJ) = -1084.4\ kJ$$

$$2H_2O(l) \longrightarrow 2H_2(g) + O_2(g) \qquad \Delta H^0_{rxn} = 571.6\ kJ$$

$$2H_2(g) + 2F_2(g) + 2H_2O(l) \longrightarrow 4HF(g) + 2H_2(g) + O_2(g)$$

$$2F_2(g) + 2H_2O(l) \longrightarrow 4HF(g) + O_2(g)$$

$$\Delta H_{rxn} = -1084.4\ kJ + 571.6\ kJ = -512.8\ kJ$$

6-24. The equations that correspond to the combustion reactions are

(1) $(CH_3)_2C_6H_4 + 10\frac{1}{2}\ O_2(g) \longrightarrow 8CO_2(g) + 5H_2O(l)$

m-xylene

$$\Delta H^0_{rxn} = -4553.9\ kJ$$

(2) $(CH_3)_2C_6H_4 + 10\frac{1}{2}\ O_2(g) \longrightarrow 8CO_2(g) + 5H_2O(l)$

p-xylene

$$\Delta H^0_{rxn} = -4556.8\ kJ$$

To obtain the desired equation reverse equation (2) and then add to equation (1)

$(CH_3)_2C_6H_4 + 10\frac{1}{2}\ O_2(g) \longrightarrow 8CO_2(g) + 5H_2O(l)$

m-xylene

$$\Delta H^0_{rxn} = -4553.9\ kJ$$

$8CO_2(g) + 5H_2O(l) \longrightarrow (CH_3)_2C_6H_4 + 10\frac{1}{2}\ O_2(g)$

p-xylene

$$\Delta H^0_{rxn} = 4556.8\ kJ$$

$(CH_3)_2C_6H_4 \longrightarrow (CH_3)_2C_6H_4$

m-xylene → p-xylene

$$\Delta H^0_{rxn} = -4553.9\ kJ + 4556.8\ kJ$$

$$= 2.9\ kJ$$

6-26. To obtain the desired equation, reverse the first equation and add to the second equation

$$XeF_2(s) \longrightarrow Xe(g) + F_2(g) \qquad \Delta H^0_{rxn} = 164\ kJ$$

$$Xe(g) + 2F_2(g) \longrightarrow XeF_4(s) \qquad \Delta H^0_{rxn} = -262\ kJ$$

$$XeF_2(s) + F_2(g) \longrightarrow XeF_4(s)$$

$$\Delta H^0_{rxn} = 164\ kJ - 262\ kJ = -98\ kJ$$

6-28. The reaction involves the rupture of two O-F bonds, thus

$$\Delta H^0_{rxn} \approx 2\overline{H}(O\text{-}F)$$

We are given the value of ΔH^0_{rxn}; therefore, we can calculate the value of $\overline{H}(O\text{-}F)$.

$$368\ kJ \approx (2\ mol)\overline{H}(O\text{-}F)$$

$$\overline{H}(O\text{-}F) \approx \frac{368\ kJ}{2\ mol} = 184\ kJ\cdot mol^{-1}$$

6-30. There are three C-H bonds and one C-Cl bond in CH_3Cl and two C-H bonds, one C-F bond, and one C-Cl bond in CH_2FCl; thus

$$\Delta H_{rxn} \approx 3\overline{H}(C\text{-}H)+\overline{H}(C\text{-}Cl)+\overline{H}(F\text{-}F)-2\overline{H}(C\text{-}H)-\overline{H}(C\text{-}F)-\overline{H}(C\text{-}Cl)-\overline{H}(H\text{-}F)$$

$$\approx (3\ mol)(414\ kJ\cdot mol^{-1})+(1\ mol)(331\ kJ\cdot mol^{-1})+(1\ mol)(155\ kJ\cdot mol^{-1})$$

$$-(2\ mol)(414\ kJ\cdot mol^{-1})-(1\ mol)(439\ kJ\cdot mol^{-1})-(1\ mol)(331\ kJ\cdot mol^{-1})$$

$$-(1\ mol)(565\ kJ\cdot mol^{-1})$$

$$\approx -435\ kJ$$

6-32. $\Delta H^0_{rxn} \approx \overline{H}(N_2) + 3\overline{H}(H\text{-}H) - 6\overline{H}(N\text{-}H)$

$$\approx (1\ mol)\overline{H}(N_2) + (3\ mol)(435\ kJ\cdot mol^{-1})-(6\ mol)(390\ kJ\cdot mol^{-1})$$

$$= (1\ mol)\overline{H}(N_2) - 1035\ kJ$$

Also

$$\Delta H^0_{rxn} = 2\Delta\overline{H}_f[NH_3(g)] - \Delta\overline{H}^0_f[N_2(g)] - 3\Delta\overline{H}^0_f[H_2(g)]$$

$$= (2\text{ mol})(-46.19\text{ kJ}\cdot\text{mol}^{-1})$$

$$= -92.38\text{ kJ}$$

Thus we can write

$$-92.38\text{ kJ} \approx (1\text{ mol})\overline{H}(N_2) - 1035\text{ kJ}$$

or

$$(1\text{ mol})\overline{H}(N_2) \approx 1035\text{ kJ} - 92.38\text{ kJ}$$

$$\approx 943\text{ kJ}$$

$$\overline{H}(N_2) \approx 943\text{ kJ}\cdot\text{mol}^{-1}$$

6-34. The relevant reaction is

$$CCl_4(g) \longrightarrow C(g) + 4Cl(g)$$

$$\Delta H^0_{rxn} \approx 4\overline{H}(C\text{-}Cl)$$

and

$$\Delta H^0_{rxn} = \Delta\overline{H}^0_f[C(g)] + 4\Delta\overline{H}^0_f[Cl(g)] - \Delta\overline{H}^0_f[CCl_4(g)]$$

$$= (1\text{ mol})(709\text{ kJ}\cdot\text{mol}^{-1}) + (4\text{ mol})(128\text{ kJ}\cdot\text{mol}^{-1}) - (1\text{ mol})(-103\text{ kJ}\cdot\text{mol}^{-1})$$

$$= 1324\text{ kJ}$$

Therefore

$$\overline{H}(C\text{-}Cl) \approx \frac{1324\text{ kJ}}{4\text{ mol}} = 331\text{ kJ}\cdot\text{mol}^{-1}$$

6-36. Use Equation (6-10)

$$C_P = \frac{\Delta H}{\Delta T}$$

In this case, $\Delta T = (28.74 - 25.00)^0C = 3.74^0C = 3.74$ K. Thus

$$C_P = \frac{285\text{ J}}{3.74\text{ K}} = 76.2\text{ J}\cdot\text{K}^{-1}$$

for the 33.6 g sample of C_6H_{14}. The specific heat is the heat capacity per gram or

$$c_{sp} = \frac{76.2\ J\cdot K^{-1}}{33.6\ g} = 2.27\ J\cdot K^{-1}\cdot g^{-1}$$

The molar heat capacity is

$$\overline{C}_P = (2.27\ J\cdot K^{-1}\cdot g^{-1})\left(\frac{86.17\ g}{1\ mol}\right) = 195\ J\cdot K^{-1}\cdot mol^{-1}$$

6-38. The total heat capacity is given by Equation (6-10)

$$C_P = \frac{\Delta H}{\Delta T} = \frac{1.00 \times 10^6\ J}{10\ K} = 1.00 \times 10^5\ J\cdot K^{-1}$$

The heat capacity is also given by

$$C_P = (\text{mol of Na})\overline{C}_P$$

Therefore

$$\text{moles of Na} = \frac{C_P}{\overline{C}_P} = \frac{1.00 \times 10^5\ J\cdot K^{-1}}{30.8\ J\cdot K^{-1}\cdot mol^{-1}} = 3.247\times 10^3\ mol$$

The minimum mass of sodium that is needed is

$$\text{mass of Na} = (3.25 \times 10^3\ mol)\left(\frac{22.99\ g}{1\ mol}\right)$$

$$= 7.46 \times 10^4\ g$$

6-40. The sample of copper, being at a higher temperature than the water, will lose heat to the water and so decrease its temperature. As heat is absorbed by the water, it will heat up. This process will continue until the copper and the water are at the same temperature. The key fact here is that the heat lost by the copper must be equal to the heat gained by the water. In an equation, we have

$$-\Delta H_{Cu} = \Delta H_{H_2O}$$

Using Equation (6-10), the equation becomes

$$-C_{P,Cu}\Delta T_{Cu} = C_{P,H_2O}\Delta T_{H_2O}$$

For the sample of copper

$$-\Delta T_{Cu} = 100 - t_f$$

and for the water

$$\Delta T_{H_2O} = t_f - 0$$

where t_f is the final temperature. The value of $C_{P,Cu}$ is

$$C_{P,Cu} = (0.385\ J{\cdot}K^{-1}{\cdot}g^{-1})(50.0\ g) = 19.25\ J{\cdot}K^{-1}$$

and the value of C_{P,H_2O} is

$$C_{P,H_2O} = (4.18\ J{\cdot}K^{-1}{\cdot}g^{-1})(250\ mL)(1.00\ g{\cdot}mL^{-1})$$

$$= 1045\ J{\cdot}K^{-1}$$

Putting all this together, we have

$$-\Delta H_{Cu} = (19.25\ J{\cdot}K^{-1})(100 - t_f)$$

$$= \Delta H_{H_2O} = (1045\ J{\cdot}K^{-1})t_f$$

$$1925 - 19.25t_f = 1045\ t_f$$

$$1925 = 1064t_f$$

$$t_f = \frac{1925}{1064} = 1.81^{\circ}C$$

6-42. As in Problem 6-40, the basic equation is

$$C_{P,Al}\Delta T_{Al} = -C_{P,H_2O}\Delta T_{H_2O}$$

For the aluminum

$$\Delta T_{Al} = (52.1 - 25.0)\,^{0}C = 27.1\ K$$

For the water

$$C_{P,H_2O} = (4.18\ J\cdot K^{-1}\cdot g^{-1})(99.9\ g) = 418\ J\cdot K^{-1}$$

$$-\Delta T_{H_2O} = (55.0 - 52.1)\,^{0}C = 2.9\ K$$

Putting all this into the heat balance equation gives

$$C_{P,Al}(27.1\ K) = (418\ J\cdot K^{-1})(2.9\ K)$$

or

$$C_{P,Al} = \frac{1210\ J}{27.1\ K} = 44.6\ J\cdot K^{-1}$$

The specific heat is

$$c_{sp} = \frac{44.6\ J\cdot K^{-1}}{50.0\ g} = 0.89\ J\cdot K^{-1}\cdot g^{-1}$$

6-44. The heat evolved by the reaction is given by Equation (6-13)

$$\Delta H_{rxn} = -C_{P,\ calorimeter}\ \Delta T$$

$$= -(4\ 55\ J\cdot K^{-1})(1.43\ K) = -651\ J$$

This amount of heat is evolved when 0.0500 L of 0.500 M solutions react. The number of moles that react is given by

$$n = MV = (0.500\ mol\cdot L^{-1})(0.0500\ L) = 0.0250\ mol$$

The heat of reaction for one mole of reactants is

$$\Delta H^{0}_{rxn} = \frac{-651\ J}{0.0250\ mol} = -26.0 \times 10^{3}\ J\cdot mol^{-1} = -26.0\ kJ\cdot mol^{-1}$$

6-46. The value of ΔH_{rxn} is

$$\Delta H_{rxn} = -(4.37\ kJ \cdot K^{-1})(2.70\ K) = -11.8\ kJ$$

The 10.0 g sample of CaO is equivalent to

$$\text{moles of CaO} = (10.0\ g)\left(\frac{1\ mol}{56.08\ g}\right) = 0.178\ mol$$

and the 1.00 L of H_2O is equivalent to

$$\text{moles of } H_2O = (1000\ mL)\left(\frac{1.0\ g}{1\ mL}\right)\left(\frac{1\ mol}{18.02\ g}\right) = 55.5\ mol$$

Thus the H_2O is in excess and the CaO is the limiting reactant. From the reaction 0.178 mol of CaO yields 0.178 mol of $Ca(OH)_2$ and so the heat evolved in the formation of one mole of $Ca(OH)_2$ is

$$\Delta H_{rxn} = \frac{-11.8\ kJ}{0.178\ mol} = -66.3\ kJ \cdot mol^{-1}$$

6-48. The value of ΔH_{rxn} is

$$\Delta H_{rxn} = -(5.16\ kJ \cdot K^{-1})(0.511\ K) = -2.64\ kJ$$

The 5.00 g of nitric acid corresponds to

$$(5.00\ g)\left(\frac{1\ mol}{63.02\ g}\right) = 0.0793\ mol\ HNO_3$$

The heat evolved per mole of HNO_3 that dissolves or the molar heat of solution is

$$\Delta H^0_{rxn} = \frac{-2.64\ kJ}{0.0793\ mol} = -33.3\ kJ \cdot mol^{-1}$$

6-50. $$\Delta H_{rxn} = -(29.7\ kJ \cdot K^{-1})(2.635\ K) = -78.26\ kJ$$

$$\Delta H_{rxn} = \frac{-78.26\ kJ}{5.00\ g} = -15.65\ kJ \cdot g^{-1}$$

$$\Delta H_{rxn} = (-15.65\ kJ \cdot g^{-1})\left(\frac{180.16\ g}{1\ mol}\right) = -2820\ kJ \cdot mol^{-1}$$

An energy of 15.7 kJ is released when 1 g of fructose is burned.

6-52. $\Delta H_{rxn} = -(21.7 \text{ kJ}\cdot\text{K}^{-1})(1.800 \text{ K}) = -39.06 \text{ kJ}$

$$\Delta H_{rxn} = \frac{-39.06 \text{ kJ}}{2.62 \text{ g}} = -14.91 \text{ kJ}\cdot\text{g}^{-1}$$

$$\Delta H^0_{rxn} = (-14.91 \text{ kJ}\cdot\text{g}^{-1})\left(\frac{90.08 \text{ g}}{1 \text{ mol}}\right) = -1343 \text{ kJ}\cdot\text{mol}^{-1}$$

The reaction for the combustion of lactic acid is

$$C_3H_6O_3(s) + 3O_2(g) \longrightarrow 3CO_2(g) + 3H_2O(l)$$

$$\Delta H^0_{rxn} = 3\Delta\overline{H}^0_f[CO_2(g)] + 3\Delta\overline{H}^0_f[H_2O(l)] - \Delta\overline{H}^0_f[C_3H_6O_3(s)]$$
$$-3\Delta\overline{H}^0_f[O_2(g)]$$

Using the above value for ΔH^0_{rxn} and the data in Table 6-1, we have

$$-1343 \text{ kJ} = (3 \text{ mol})(-393.5 \text{ kJ}\cdot\text{mol}^{-1}) + (3 \text{ mol})(-285.8 \text{ kJ}\cdot\text{mol}^{-1})$$
$$-(1 \text{ mol})\Delta\overline{H}^0_f[C_3H_6O_3(s)]$$
$$= -2037.9 \text{ kJ} - (1 \text{ mol})\Delta\overline{H}^0_f[C_3H_6O_3(s)]$$

Solving for $\Delta\overline{H}^0_f[C_3H_6O_3(s)]$ we get

$$\Delta\overline{H}^0_f[C_3H_6O_3(s)] = -695 \text{ kJ}\cdot\text{mol}^{-1}$$

6-54. We assume that the energy consumed in riding is due to the combustion of body fat. Given that one kilogram of body fat yields 33,000 kJ, we have

$$1 \text{ lb} = 0.454 \text{ kg} \quad (33{,}000 \text{ kJ}\cdot\text{kg}^{-1})(0.454 \text{ kg}) \approx$$
$$15{,}000 \text{ kJ} = (2800 \text{ kJ}\cdot\text{hr}^{-1})(\text{number of hours})$$

$$\text{number of hours of riding} = \frac{15000 \text{ kJ}}{2800 \text{ kJ}\cdot\text{hr}^{-1}} = 5.4 \text{ hrs}$$

The distance traveled in 5.4 hrs is

$$\text{distance} = (13 \text{ mi}\cdot\text{hr}^{-1})(5.4 \text{ hr}) = 70 \text{ mi}$$

6-56. The heat required to raise the temperature of 82 kg of water by $3^{0}C$ is

$$\text{heat} = (4.18\ J\cdot K^{-1}\cdot g^{-1})(82 \times 10^{3}\ g)(3\ K)$$

$$= 1.03 \times 10^{6}\ J \quad = \quad 1030\ kJ$$

This heat is provided by the burning of glucose in the body

$$(2820\ kJ\cdot mol^{-1})(\text{moles of glucose}) = 1030\ kJ$$

$$\text{moles of glucose} = \frac{1030\ kJ}{2820\ kJ\cdot mol^{-1}} = 0.365\ mol$$

The mass of glucose that must be burned is

$$\text{mass} = (0.365\ mol)\left(\frac{180.16\ g}{1\ mol}\right) = 66\ g$$

6-58. We use the equation (Problem 6-57)

$$\bar{C}_P = N \times 25\ J\cdot K^{-1}\cdot mol^{-1}$$

Let the formula of the compound be Rb_xO_y. Then the formula mass of the compound is

$$\text{formula mass} = 85.47x + 16.00y$$

and N is the equation of Dulong and Petit's rule is

$$N = x + y$$

Thus we have

$$\bar{C}_P = \left(\frac{(85.47x + 16.00y)\,g}{1\ mol}\right) 0.64\ J\cdot K^{-1}\cdot g^{-1}$$

$$= (x+y)25\ J\cdot K^{-1}\cdot g^{-1}$$

By trial and error, we find that x=1 and y=2, so that the formula of the oxide of rubidium is RbO_2.

6-60. We follow the same procedure as in Problem 6-58. Let the formula of matlockite be $Pb_xF_yCl_z$. The formula mass is

$$\text{formula mass} = 207.2x + 19.00y + 35.45z$$

and Dulong and Petit's rule gives

$$\overline{C}_P = \left\{\frac{(207.2x + 19.00y + 35.45z)\text{g}}{1 \text{ mol}}\right\} 0.290 \text{ J}\cdot\text{K}^{-1}\cdot\text{mol}^{-1}$$

$$\approx (x+y+z)25 \text{ J}\cdot\text{K}^{-1}\cdot\text{mol}^{-1}$$

By trial and error, we find that x=1, y=1 and z=1, and so the formula for matlockite is PbFCl.

CHAPTER 7

SOLUTIONS TO THE EVEN-NUMBERED PROBLEMS

7-2. Of the four species listed, Be^{2+} has the largest ionization energy because it is doubly positively charged. Of the remaining three species, Li^{+} has the largest ionization energy because it is positively charged. Because ionization energies increase going across a row in the periodic table, the order is F, Ne, Li^{+}, Be^{2+}.

7-4. Using Table 7-1, we have for a beryllium atom

n	I_n/MJ·mol^{-1}	log(I_n/MJ·mol^{-1})
1	0.90	-0.046
2	1.76	0.25
3	14.86	1.17
4	21.01	1.32

If we plot log[I_n/MJ·mol^{-1}] versus n, the number of electrons removed, t

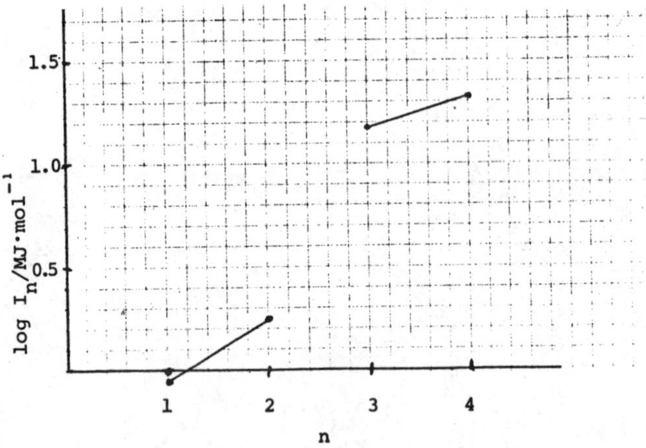

The first two electrons are relatively easy to remove, suggesting that the four electrons in a beryllium atom are arranged in two shells. The inner shell, which contains two electrons, is relatively stable.

7-6. The noble gases are in Group 8 and have eight outer electrons, except for helium which has two. The Lewis electron-dot formulas of the noble gases are

$$\cdot\mathrm{He}\cdot \qquad :\ddot{\underset{\cdot\cdot}{\mathrm{Ne}}}: \qquad :\ddot{\underset{\cdot\cdot}{\mathrm{Ar}}}: \qquad :\ddot{\underset{\cdot\cdot}{\mathrm{Kr}}}: \qquad :\ddot{\underset{\cdot\cdot}{\mathrm{Xe}}}: \qquad :\ddot{\underset{\cdot\cdot}{\mathrm{Rn}}}:$$

7-8. Aluminum (Group 3) has three outer electrons, so we write Al^{3+} indicating that Al has lost its three outer electrons.

Aluminum (Group 3) has three outer electrons; Al^{3+} has eight outer electrons and so we write $:\ddot{\underset{\cdot\cdot}{\mathrm{Al}}}:^{3+}$ or Al^{3+} indicating that Al has lost its three outer electrons.

Chlorine (Group 7) has seven outer electrons, Cl^- has eight outer electrons and so we write $:\ddot{\underset{\cdot\cdot}{\mathrm{Cl}}}:^-$.

7-10. Solve Equation (7-1) for ν.

$$\nu = \frac{c}{\lambda} = \frac{3.00 \times 10^{8}\ \mathrm{m\cdot s^{-1}}}{589.2 \times 10^{-9}\ \mathrm{m}}$$

$$= 5.09 \times 10^{14}\ \mathrm{s^{-1}}$$

7-12. Substitute Equation (7-1) for ν into Equation (7-2) to obtain for the energy of an 80 pm photon

$$E = \frac{hc}{\lambda} = \frac{(6.626 \times 10^{-34}\ \mathrm{J\cdot s})(3.00\times10^{8}\ \mathrm{m\cdot s^{-1}})}{(80 \times 10^{-12}\ \mathrm{m})} = 2.48 \times 10^{-15}\ \mathrm{J}$$

To obtain the ionization energy of one neon atom, we divide the 2.08 $\mathrm{MJ\cdot mol^{-1}}$ by Avogadro's number

$$I = \frac{2.08 \times 10^{6}\ \mathrm{J\cdot mol^{-1}}}{6.022 \times 10^{23}\ \mathrm{atom\cdot mol^{-1}}} = 3.45 \times 10^{-18}\ \mathrm{J\cdot atom^{-1}}$$

We see that one photon has more than enough energy to ionize one neon atom.

7-14. Substitute Equation (7-1) for ν into Equation (7-2) to write for the energy of one X-ray photon

$$E = \frac{hc}{\lambda} = \frac{(6.626 \times 10^{-34} J\cdot s)(3.00 \times 10^{8} m\cdot s^{-1})}{(1.00 \times 10^{-9} m)} = 1.99 \times 10^{-16} J$$

To obtain the energy of one bond, we divide 300 $kJ\cdot mol^{-1}$ by Avogadro's number

$$\text{energy per bond} = \frac{300 \times 10^{3} J\cdot mol^{-1}}{6.022 \times 10^{23} \text{bonds}\cdot mol^{-1}} = 4.98 \times 10^{-19} J\cdot \text{bond}^{-1}$$

The number of bonds that can be broken is

$$\text{number of bonds} = \frac{1.99 \times 10^{-16} J}{4.98 \times 10^{-19} J\cdot \text{bond}^{-1}} = 400 \text{ bonds}$$

7-16. Solving Equation (7-2) for ν, we have

a) $$\nu = \frac{E}{h} = \frac{1 \times 10^{-19} J}{6.626 \times 10^{-34} J\cdot s^{-1}} = 1.5 \times 10^{14} s^{-1}$$

The wavelength is

b) $$\lambda = \frac{c}{\nu} = \frac{3.00 \times 10^{8} m\cdot s^{-1}}{1.5 \times 10^{14} s^{-1}} = 2.0 \times 10^{-6} \text{ m} = 2.0 \text{ μm}$$

7-18. The energy per photon of green light is given by Equations (7-1) and (7-2)

$$E = h\nu = \frac{hc}{\lambda} = \frac{(6.626 \times 10^{-34} J\cdot s)(3.00 \times 10^{8} m\cdot s^{-1})}{510 \times 10^{-9} m}$$

$$= 3.90 \times 10^{-19} J\cdot \text{photon}^{-1}$$

The number of photons in 2.35×10^{-18} J of light is

$$\frac{2.35 \times 10^{-18} J}{3.90 \times 10^{-19} J\cdot \text{photon}^{-1}} = 6 \text{ photons}$$

7-20. The energy of one photon is given by

$$E = \frac{hc}{\lambda} = \frac{(6.626 \times 10^{-34} J\cdot s)(3.00 \times 10^{8} m\cdot s^{-1})}{10.6 \times 10^{-6} m} = 1.88 \times 10^{-20} J$$

The number of photons in a pulse of 1 J is

$$\frac{1\ J}{1.88 \times 10^{-20} J\cdot photon^{-1}} = 5 \times 10^{19}\ photons$$

7-22. Using the formula $\lambda_{max}T = 2.89 \times 10^{-3} m\cdot K$, we have

a) $$T = \frac{2.89 \times 10^{-3} m\cdot K}{\lambda_{max}} = \frac{2.89 \times 10^{-3} m\cdot K}{800 \times 10^{-9} m} = 3600\ K = 3300^{\circ}C$$

b) $$T = \frac{2.89 \times 10^{-3} m\cdot K}{680 \times 10^{-9} m} = 4300\ K = 4000^{\circ}C$$

c) $$T = \frac{2.89 \times 10^{-3} m\cdot K}{600 \times 10^{-9} m} = 4800\ K = 4500^{\circ}C$$

d) $$T = \frac{2.89 \times 10^{-3} m\cdot K}{510 \times 10^{-9} m} = 5700\ K = 5400^{\circ}C$$

7-24. The maximum intensity of a red giant occurs in the red to infrared region of the spectrum. We assume that $\lambda_{max} = 680$ nm (see Problem 7-22). Using the equation that is given in Problem 7-21,

$$T = \frac{2.89 \times 10^{-3} m\cdot K}{\lambda_{max}} = \frac{2.89 \times 10^{-3} m\cdot K}{680 \times 10^{-9} m} = 4300\ K \simeq 4000^{\circ}C$$

7-26. The mass of the baseball in kilograms is

$$m = (5.0\ oz)\left(\frac{1\ lb}{16\ oz}\right)\left(\frac{1\ kg}{2.20 lb}\right) = 0.14\ kg$$

The speed of the baseball in meters per second is

$$v = \left(\frac{90\ mile}{1\ hr}\right)\left(\frac{1\ km}{0.62\ mile}\right)\left(\frac{10^{3} m}{1\ km}\right)\left(\frac{1\ hr}{60\ min}\right)\left(\frac{1\ min}{60\ s}\right) = 40\ m\cdot s^{-1}$$

We use Equation (7-3)

$$\lambda = \frac{h}{mV} = \frac{6.626 \times 10^{-34} \text{J}\cdot\text{s}}{(0.14 \text{ kg})(40 \text{ m}\cdot\text{s}^{-1})} = 1.2 \times 10^{-34} \text{m}$$

7-28. We have used the fact that a joule is equal to a $\text{kg}\cdot\text{m}^2\cdot\text{s}^{-2}$.
We use Equation (5-14) to calculate the average speed of the neutrons

$$v_{av} = \left(\frac{3RT}{Mkg}\right)^{1/2} = \left[\frac{(3)(8.31 \text{ J}\cdot\text{mol}^{-1}\cdot\text{K}^{-1})(1000 \text{ K})}{(1.67\text{x}10^{-27}\text{kg}\cdot\text{neutron}^{-1})(6.022\text{x}10^{23}\text{neutrons}\cdot\text{mol}^{-1})}\right]^{1/2}$$

$$= 4.98 \times 10^{3} \text{ m}\cdot\text{s}^{-1}$$

Therefore

$$\lambda = \frac{h}{mv} = \frac{6.626 \times 10^{-34} \text{J}\cdot\text{s}}{(1.67 \times 10^{-27} \text{kg})(4.98 \times 10^{3} \text{m}\cdot\text{s}^{-1})} = 7.97 \times 10^{-11} \text{m}$$

$$= 79.7 \text{ pm}$$

7-30. Again we use Equation (7-3). In this problem we solve for the speed v

$$v = \frac{h}{\lambda m} = \frac{6.626 \times 10^{-34} \text{J}\cdot\text{s}}{(9.11 \times 10^{-31} \text{kg})(100 \times 10^{-12} \text{m})} = 7.27 \times 10^{6} \text{m}\cdot\text{s}^{-1}$$

We have used the fact that a joule is equal to a $\text{kg}\cdot\text{m}^2\cdot\text{s}^{-2}$.

7-32. The same lines appear in both absorption spectra and emission spectra because the lines are due to transitions from one level to another and the frequency is directly proportional to the difference in energies of the two levels.

7-34. The Lyman series is due to transitions to the n = 1 state. The frequency of the lines are given by Equation (7-8)

$$\nu_{n\to 1} = (3.29 \times 10^{15}\text{s}^{-1}) \left[\frac{1}{1^2} - \frac{1}{n^2}\right]$$

The frequency of the line is

$$\nu = \frac{c}{\lambda} = \frac{3.00 \times 10^8\text{m}\cdot\text{s}^{-1}}{1.03 \times 10^{-7}\text{m}} = 2.91 \times 10^{15}\text{s}^{-1}$$

Substituting this value for $\nu_{n\to 1}$ into Equation (7-8), we have

$$2.91 \times 10^{15}\text{s}^{-1} = (3.29 \times 10^{15}\text{s}^{-1})\left[1 - \frac{1}{n^2}\right]$$

Solving for n, we have

$$1 - \frac{1}{n^2} = \frac{2.91 \times 10^{15}\text{s}^{-1}}{3.29 \times 10^{15}\text{s}^{-1}} = 0.8845$$

$$\frac{1}{n^2} = 1 - 0.8845 = 0.1155$$

$$n^2 = 8.66$$

$$n = 2.94 \approx 3$$

The original energy level of the electron is the n = 3 state.

7-36. The energy required to ionize a hydrogen atom that is in the first excited state is given by

$$\text{IE} = E_\infty - E_2 = 0 - \left(\frac{-2.18 \times 10^{-18}\text{J}}{2^2}\right)$$

$$= 5.45 \times 10^{-19}\text{J}$$

7-38. For transitions from higher states to a lower state, we can derive formulas such as

$$\nu_{n\to 1} = (1.32 \times 10 \quad s^{-1})\left(\frac{1}{1^2} - \frac{1}{n^2}\right) \qquad n = 2,3,...$$

$$\nu_{n\to 2} = (1.32 \times 10 \quad s^{-1})\left(\frac{1}{2^2} - \frac{1}{n^2}\right) \qquad n = 3,4,...$$

which give analogs of the Lyman and Balmer series for atomic hydrogen.

7-40. When the contours are closely spaced, the property that is plotted, such as the probability of finding an electron within a region or the altitude of the land, is changing rapidly. When the contours are widely separated, the property is changing slowly.

7-42. In the 4f subshell, $n = 4$ and $l = 3$. Thus $m_l = -3, -2, -1, 0, 1, 2, 3$. There are 7 orbitals in the 4f subshell.

7-44. a) $n = 3$, $l = 3$ (not possible)

b) $n = 6$, $l = 1$ (possible)

c) $n = 6$, $l = 4$ (possible)

d) $n = 5$, $f = 3$ (possible)

e) $n = 4$, $l = 2$ (possible)

7-46. For a 4d orbital, $n = 4$ and $l = 2$

n	l	m_l	m_s
4	2	-2	$+\frac{1}{2}$
			$-\frac{1}{2}$
		-1	$+\frac{1}{2}$
			$-\frac{1}{2}$
		0	$+\frac{1}{2}$
			$-\frac{1}{2}$
		1	$+\frac{1}{2}$
			$-\frac{1}{2}$
		2	$+\frac{1}{2}$
			$-\frac{1}{2}$

There are 10 possible sets.

7-48. a) 3d b) 3d c) 2p d) 4f

7-50. The subshell 4p exists because $n = 4$ and $l = 1$ is possible. The subshell 2d does not exist because $n = 2$ and $l = 2$ is not possible. The subshell 3s exists because $n = 3$ and $l = 0$ is possible. The subshell 1f does not exist because $n = 1$ and $l = 3$ is not possible. The subshell 2p exists because $n = 2$ and $l = 1$ is possible.

7-52. a) not allowed; m_l cannot equal -1 if $l = 0$.

b) allowed

c) allowed

d) not allowed; l cannot equal 2 if $n = 2$.

e) not allowed; m_s cannot equal 0.

7-54. We must use Hess's law to do this problem. From Table 7-1, we find that

a)
$$B(g) \longrightarrow B^{+}(g) + e^{-}(g) \qquad \Delta H_{rxn} = I_1 = 0.80 \text{ MJ}$$
$$B^{+}(g) \longrightarrow B^{2+}(g) + e^{-}(g) \qquad \Delta H_{rxn} = I_2 = 2.42 \text{ MJ}$$
$$B^{2+}(g) \longrightarrow B^{3+}(g) + e^{-}(g) \qquad \Delta H_{rxn} = I_3 = 3.66 \text{ MJ}$$

$$B(g) \longrightarrow B^{3+}(g) + 3e^{-}(g) \qquad \Delta H_{rxn} = 6.88 \text{ MJ}$$

b)
$$2Li(g) \longrightarrow 2Li^{+}(g) + 2e^{-}(g) \qquad \Delta H_{rxn} = 2I_1 = 1.04 \text{ MJ}$$
$$Mg^{2+}(g) + e^{-}(g) \longrightarrow Mg^{+}(g) \qquad \Delta H_{rxn} = -I_2 = -1.45 \text{ MJ}$$
$$Mg^{+}(g) + e^{-}(g) \longrightarrow Mg(g) \qquad \Delta H_{rxn} = -I_1 = -0.74 \text{ MJ}$$

$$2Li(g) + Mg^{2+}(g) \longrightarrow 2Li^{+}(g) + Mg(g) \qquad \Delta H_{rxn} = -1.15 \text{ MJ}$$

c)
$$O^{2+}(g) + e^{-}(g) \longrightarrow O^{+}(g) \qquad \Delta H_{rxn} = -I_2 = -3.39 \text{ MJ}$$
$$O^{+}(g) + e^{-}(g) \longrightarrow O(g) \qquad \Delta H_{rxn} = -I_1 = -1.31 \text{ MJ}$$

$$O^{2+}(g) + 2e^{-}(g) \longrightarrow O(g) \qquad \Delta H_{rxn} = -4.70 \text{ MJ}$$

CHAPTER 8

SOLUTIONS TO THE EVEN-NUMBERED PROBLEMS

8-2. To work problems of this type, you have to know the order of the oribtal energies as shown in Figure 8-3.

a) 5d
b) 4p
c) 4f
d) 5s
e) 6p

8-4. For a 4f orbital, $n = 4$ and $l = 3$, thus m_l can be -3, -2, -1, 0, 1, 2, 3. For each value of m_l, m_s can be $+\frac{1}{2}$ or $-\frac{1}{2}$. Therefore, the 14 possible sets of quantum numbers are

n	l	m_l	m_s
4	3	-3	$+\frac{1}{2}$
			$-\frac{1}{2}$
		-2	$+\frac{1}{2}$
			$-\frac{1}{2}$
		-1	$+\frac{1}{2}$
			$-\frac{1}{2}$
		0	$+\frac{1}{2}$
			$-\frac{1}{2}$
		1	$+\frac{1}{2}$
			$-\frac{1}{2}$
		2	$+\frac{1}{2}$
			$-\frac{1}{2}$
		3	$+\frac{1}{2}$
			$-\frac{1}{2}$

8-6. Note that for each value of m_l, m_s can have the values $+\frac{1}{2}$ or $-\frac{1}{2}$. Thus for each shell

K shell $n = 1,\ l = 0,\ m_l = 0;\ m_s = +\frac{1}{2}, -\frac{1}{2}$ — 2 electrons

L shell $n = 2,\ l = 0,\ m_l = 0;\ m_s = +\frac{1}{2}, -\frac{1}{2}$ — 2 x 1

$l = 1,\ m_l = -1,0,1;\ m_s = +\frac{1}{2}, -\frac{1}{2}$ — 2 x 3

8 electrons

M shell $n = 3$ $l = 0,\ m_l = 0;\ m_s = +\frac{1}{2}, -\frac{1}{2}$ — 2 x 1

$l = 1,\ m_l = -1,0,1;\ m_s = +\frac{1}{2},-\frac{1}{2}$ — 2 x 3

$l = 2,\ m_l = -2,-1,0,1,2;\ m_s = +\frac{1}{2},-\frac{1}{2}$ — 2 x 5

18 electrons

N shell $n = 4$ $l = 0, m_l = 0; m_s = +\frac{1}{2},-\frac{1}{2}$ 2 x 1

$l = 1, m_l = -1,0,1; m_s = +\frac{1}{2},-\frac{1}{2}$ 2 x 3

$l = 2, m_l = -2,-1,0,1,2; m_s = +\frac{1}{2},-\frac{1}{2}$ 2 x 5

$l = 3, m_l = -3,-2,-1,0,1,2,3; m_s = +\frac{1}{2},-\frac{1}{2}$ 2 x 7

32 electrons

Note that the maximum number of electrons with the same value of n in an atom is equal to $2n^2$.

8-8. In an f-transition series, the seven f orbitals are being filled. A set of seven f orbitals can hold up to 14 electrons.

8-10. a) Ruled out; the 2p orbitals cannot hold 7 electrons.

b) Allowed, an excited state.

c) Ruled out; the 3d orbitals cannot hold 14 electrons.

d) Allowed.

8-12. a) 2s can hold a maximum of 2 electrons.

b) 3s can hold a maximum of 2 electrons; also 2p must be filled before 3s.

c) 2p can hold maximum of 6 electrons.

d) 3p must be filled before higher energy orbitals and 3d can hold maximum of 10 electrons.

8-14. a) $1s^2 2s^2 2p^6 3s^2 3p^1$ 13 electrons, aluminum

b) $1s^2 2s^2 2p^6 3s^2 3p^6 4s^2 3d^3$ 23 electrons, vanadium

c) $1s^2 2s^2 2p^5$ 9 electrons, fluorine

d) $1s^2 2s^2 2p^6 3s^2 3p^6 4s^2 3d^{10} 4p^1$ 31 electrons, gallium

e) $1s^2 2s^2 2p^4$ 8 electrons, oxygen

8-16. a) 26 electrons $1s^2 2s^2 2p^6 3s^2 3p^6 4s^2 3d^6$

b) 74 electrons $1s^2 2s^2 2p^6 3s^2 3p^6 4s^2 3d^{10} 4p^6 5s^2 4d^{10} 5p^6 6s^2 4f^{14} 5d^4$

c) 33 electrons $1s^2 2s^2 2p^6 3s^2 3p^6 4s^2 3d^{10} 4p^3$

d) 10 electrons $1s^2 2s^2 2p^6$

e) 19 electrons $1s^2 2s^2 2p^6 3s^2 3p^6 4s^1$

8-18. a) 34 electrons $1s^22s^22p^63s^23p^64s^23d^{10}4p^4$

b) 25 electrons $1s^22s^22p^63s^23p^64s^23d^5$

c) 50 electrons $1s^22s^22p^63s^23p^64s^23d^{10}4p^65s^24d^{10}5p^2$

d) 79 electrons $1s^22s^22p^63s^23p^64s^23d^{10}4p^65s^24d^{10}5p^66s^14f^{14}5d^{10}$

e) 63 electrons $1s^22s^22p^63s^23p^64s^23d^{10}4p^65s^24d^{10}5p^66s^24f^7$

8-20. Refer to Figure 8-5.

a) The outer electron configuration $3s^23p^1$ corresponds to the group 3 element in the third row or aluminum.

b) The outer electron configuration $5s^14d^5$ corresponds to the fourth member of the 4d transition metal series or molybdenum.

c) The outer electron configuration $4s^23d^6$ corresponds to the sixth member of the 3d transition metal series or iron.

d) The outer electron configuration $6s^25d^{10}$ corresponds to the tenth member of the 5d transition metal series or mercury.

8-22. a) N (half filled p), Group 5; Mn (half filled d).

b) He, Ne, Group 8.

c) As (Group 5), Mn.

8-24. a) The first inner transition metal series to begin to fill f orbitals is the lanthanide series, thus the element is lanthanum, La. Note, however, that La is an exception and is actually a $5d^1$ metal.

b) The elements in the fourth row of the periodic table fill the 3d orbitals. Thus the element is the third member of the transition metal series, which is vanadium, V.

c) The next to last element in the 3d transition metals series has filled d orbitals; the element is copper, Cu. (Outer configuration $4s^13d^{10}$).

d) Elements in the third row of the periodic table have a filled 2p orbital. The element that has four electrons in the 3p orbital is sulfur, S.

e) The element must be in the fourth row of the periodic table. The first element to have two electrons in the 4s orbital is calcium, Ca.

8-26. The electron configurations of the elements would be

$$1s^2 2s^2 2p^6 3s^2 3p^6 4s^2 3d^{10} 4p^6 5s^2 4d^{10} 5p^6 6s^2 4f^{14} 5d^{10} 6p^6 7s^2 5f^{14} 6d^{10} 7p^6 8s^2 5g^{1-18}$$

The atomic numbers of the elements are Z = 121 through Z = 138. There would be 18 elements in this series.

8-28. Recall that we show only the outer shell electrons in the Lewis electron-dot formula. For example [ ]· indicates that the atom has one outer electron and so must belong to Group 1.

a) 1 b) 6 c) 5 d) 3 e) 4

8-30. a) excited state; there are paired electrons in two p orbitals and one empty p orbital.

b) excited state; the two electrons in the 3p orbital are paired.

c) ground state.

d) excited state; there are paired electrons in two 3p orbitals and one empty 3p orbital.

e) ground state.

8-32. We first write the ground-state electron configuration, then using Hund's Rule we determine the number of unpaired electrons in the partially filled subshell.

	ground-state configuration	number of unpaired electrons
a)	[Ne]$3s^23p^6$	0
b)	[He]$2s^22p^3$	3
c)	[Ne]	0
d)	[Kr]$5s^24d^{10}5p^5$	1
e)	[He]$2s^22p^4$	2

8-34. Recall that the Lewis electron-dot formula shows only the outer shell electrons. For neutral main group elements, the number of outer-shell electrons is equal to the group number

·Be· ·Mg· ·Ca· ·Sr· ·Ba· ·Ra·

8-36. In the formation of positive ions, the corresponding noble-gas core is that for the noble gas in the preceding row of the periodic table.

a) 2 electrons, Ca^{2+}, argon

b) 1 electron, Li^+, helium

c) 1 electron, Na^+, neon

d) 2 electrons, Mg^{2+}, neon

e) 3 electrons, Al^{3+}, neon

8-38. First we determine the number of electrons in the ion and then we write the ground-state electron configuration

a) 34 + 2 = 36 electrons: $1s^22s^22p^63s^23p^64s^23d^{10}4p^6$ or [Kr]

b) 4 - 1 = 3 electrons: $1s^22s^1$

c) 31 - 1 = 30 electrons: $1s^22s^22p^63s^23p^64s^23d^{10}$

d) 82 - 2 = 80 electrons: [Xe]$6s^24f^{14}5d^{10}$

e) 15 + 3 = 18 electrons: $1s^22s^22p^63s^23p^6$ or [Ar]

8-40. We first determine the total number of electrons in the ion, then we figure out the ground-state electron configuration, and finally we count the number of unpaired electrons.

	species	ground-state configuration	number of unpaired electrons
a)	K^+	[Ar]	0
b)	Sc^{3+}	[Ar]	0
c)	F^+	[He]$2s^22p^4$	2
d)	Cl^-	[Ar]	0
e)	O^-	[He]$2s^22p^5$	1

8-42. Isoelectronic means having the same number of electrons. We thus determine the number of electrons for each species using the atomic number together with the ionic charge.

2 electrons: H^-, Li^+, C^{4+}

10 electrons: Mg^{2+}, Ne, N^{3-}

18 electrons: K^+, Ca^{2+}, P^{3-}

36 electrons: Sr^{2+}, Br^-, Kr

8-44. In this problem we work out the electronic configurations of the reactants and the products.

a) $I(g) + e^- \longrightarrow I^-(g)$

$[Kr]5s^24d^{10}5p^5 + e^- \longrightarrow [Kr]5s^24d^{10}5p^6$ (or [Xe])

b) $K(g) + F(g) \longrightarrow K^+(g) + F^-(g)$

$[Ar]4s^1 + [He]2s^22p^5 \longrightarrow [Ar] + [He]2s^22p^6$ (or [Ne])

8-46. First we figure out the electron configuration of the ground state, then we promote one electron from the highest energy subshell that is occupied to the lowest energy subshell that is unoccupied.

a) The ground state of Be^{2+} is $1s^2$. The lowest energy subshell that is unoccupied is the 2s, thus the first excited state is obtained by promoting an electron from the 1s shell to the 2s subshell or $1s^12s^1$.

b) The ground state of Mg^{2+} (10 electrons) is $1s^22s^22p^6$. The first excited state is $1s^22s^22p^53s^1$.

c) The ground state of Al^{3+} (10 electrons) is $1s^22s^22p^6$. The first excited state is $1s^22s^22p^53s^1$.

d) The ground state of O^{2-} (10 electrons) is $1s^22s^22p^6$. The first excited state is $1s^22s^22p^53s^1$.

8-48. a) Ground state; all of the electrons are in the lowest energy subshells.

b) Excited state; the ground state is $1s^22s^22p^63s^1$.

c) Excited state; the ground state is $1s^22s^22p^63s^2$.

d) Excited state; the ground state is $1s^22s^22p^63s^23p^64s^1$.

8-50. a) He < Ne < Ar < Kr

b) Li < Na < K < Rb

c) Ne < F < N < B < Be

8-52. For a given nuclear charge, the fewer the number of electrons, the smaller is the species. The electrons are pulled closer to the nucleus.

8-54. Recall that ionization energies increase as we move from left to right across the periodic table and they decrease as we move down a group.

a) B < O < F < Ne

b) Sb < Te < I < Xe

c) Cs < Rb < K < Ca

d) Na < Al < S < Ar

8-56. Ba < Sr < Ca < Se < S < O

CHAPTER 9

SOLUTIONS TO THE EVEN-NUMBERED PROBLEMS

9-2. In order to emphasize the attainment of noble-gas configurations for ions, we shall write the electron configurations in terms of noble-gas configurations.

a) $Ca([Ar]4s^2)+2F([He]2s^22p^5) \longrightarrow Ca^{2+}([Ar])+2F^-([Ne])$

The product is $CaF_2(s)$.

b) $2Na([Ne]3s^1)+O([He]2s^22p^4) \longrightarrow 2Na^+([Ne])+O^{2-}([Ne])$

The product is $Na_2O(s)$.

c) $Mg([Ne]3s^2)+S([Ne]3s^23p^4) \longrightarrow Mg^{2+}([Ne])+S^{2-}([Ar])$

The product is $MgS(s)$.

9-4. a) $Ca([Ar]4s^2)+2Br([Ar]4s^23d^{10}4p^5) \longrightarrow Ca^{2+}([Ar])+2Br^-([Kr])$

The product is $CaBr_2$.

b) $3Mg([Ne]3s^2)+2N([He]2s^22p^3) \longrightarrow 3Mg^{2+}([Ne])+2N^{3-}([Ne])$

The product is Mg_3N_2.

c) $2Cs([Xe]6s^1)+Se([Ar]4s^23d^{10}4p^4) \longrightarrow 2Cs^+([Xe])+Se^{2-}([Kr])$

The product is Cs_2Se.

9-6. a) If we write the Lewis electron-dot formulas for K and I, then the reaction can be written

$$K\cdot \; + \; :\!\ddot{\underset{..}{I}}\!\cdot \; \longrightarrow \; \underbrace{K^+ \; + \; :\!\ddot{\underset{..}{I}}\!:^-}_{KI}$$

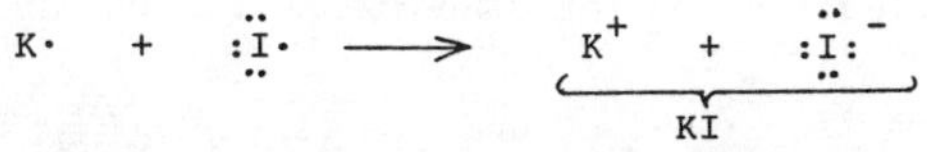

The potassium atom gives up one outer electron to achieve a noble-gas electron configuration. The iodine atom accepts one electron to achieve a noble-gas electron configuration.

b) If we write the Lewis electron-dot formulas for Ba and O, then the reaction can be written

$$\cdot Ba\cdot \; + \; \cdot\ddot{\underset{..}{O}}\!\cdot \; \longrightarrow \; \underbrace{Ba^{2+} \; + \; :\!\ddot{\underset{..}{O}}\!:^{2-}}_{BaO}$$

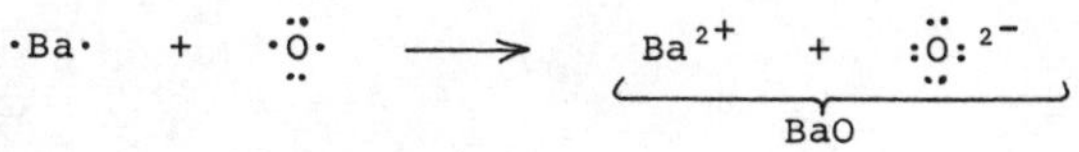

Barium gives up two outer electrons to achieve a noble-gas electron configuration. Oxygen accepts two electrons to achieve a noble-gas electron configuration.

c) If we write the Lewis electron-dot formulas for Be and Se, then the reaction can be written

$$\cdot\text{Be}\cdot \quad + \quad \cdot\ddot{\underset{\cdot\cdot}{\text{Se}}}\cdot \quad \longrightarrow \quad \underbrace{\text{Be}^{2+} \quad + \quad :\ddot{\underset{\cdot\cdot}{\text{Se}}}:^{2-}}_{\text{BeSe}}$$

Beryllium gives up two outer electrons to achieve a noble-gas electron configuration. Selenium accepts two electrons to achieve a noble-gas electron configuration.

9-8. a) The Lewis electron-dot formulas for Na and O are

$$\text{Na}\cdot \qquad \text{and} \qquad \cdot\ddot{\underset{\cdot\cdot}{\text{O}}}\cdot$$

The sodium atom can achieve a noble-gas electron configuration by losing one electron. The oxygen atom can achieve a noble-gas electron configuration by gaining two electrons and so it requires two sodium atoms for each oxygen atom. The product is Na_2O.

b) The Lewis electron-dot formulas for Sc and Cl are

$$\cdot\underset{\cdot}{\text{Sc}}\cdot \qquad \text{and} \qquad :\ddot{\underset{\cdot\cdot}{\text{Cl}}}\cdot$$

The scandium atom can achieve a noble-gas electron configuration by losing three electrons. The chlorine atom can achieve a noble-gas electron configuration by gaining one electron and so it requires three chlorine atoms for each scandium atom. The product is $ScCl_3$.

c) The Lewis electron-dot formulas for Al and S are

$$\cdot\underset{\cdot}{\mathrm{Al}}\cdot \quad \text{and} \quad \cdot\overset{\cdot\cdot}{\underset{\cdot\cdot}{\mathrm{S}}}\cdot$$

Each aluminum atom can achieve a noble-gas electron configuration by losing three electrons. Each sulfur atom can achieve a noble-gas electron configuration by gaining two electrons and so it requires two aluminum atoms for three sulfur atoms. The product is Al_2S_3.

9-10. Recall that for neutral atoms the (n+1)s orbitals are of lower energy than the nd orbitals, for example, 4s < 3d. However, for ions, the reverse is true; that is, 4s > 3d. Thus we first determine the number of electrons in the ion and then we write the electron configuration filling the orbitals in the order 1s, 2s, 2p, 3s, 3p, 3d, 4s, and so forth.

a) Ru^{2+}([Kr]$4d^6$)

b) W^{3+}([Xe]$4f^{14}5d^3$)

c) Pd^{2+}([Kr]$4d^8$)

d) Ag^{2+}([Kr]$4d^9$)

e) Ir^{3+}([Xe]$4f^{14}5d^6$)

9-12. a) Fe^{2+}([Ar]$3d^6$), 6 d electrons

b) Zn^{2+}([Ar]$3d^{10}$), 10 d electrons

c) V^{2+}([Ar]$3d^3$), 3 d electrons

d) Ni^{2+}([Ar]$3d^8$), 8 d electrons

e) Ti^{2+}([Ar]$3d^2$), 2 d electrons

For d transition series +2 ions, the number of d electrons is equal to the position of the metal in the series found by counting off from the start of the series.

9-14. a) For n=3 or 4, a pseudo-noble-gas outer electron configuration is one of the type $ns^2np^6nd^{10}$. For n=3, the corresponding noble gas is Ar (the third noble gas), thus we have Cu^+([Ar]$3d^{10}$)

b) Ga^{3+}([Ar]$3d^{10}$)

c) In this case n=5 and the corresponding pseudo-noble-gas outer electron configuration is of the type $ns^2np^6(n-1)f^{14}nd^{10}$ or $5s^25p^64f^{14}5d^{10}$

Hg^{2+}([Xe]$4f^{14}5d^{10}$)

d) Au^+([Xe]$4f^{14}5d^{10}$)

9-16. In this problem we simply determine the number of electrons in the positive ion and in the negative ion. If the numbers are the same, then the ions are isoelectronic. The number of electrons is shown in parentheses.

a) Na^+(10), Cl^-(18), not isoelectronic

b) RbBr, isoelectronic ions, each has 36 electrons

c) Sr^{2+}(36), Cl^-(18), not isoelectronic

d) $SrBr_2$, isoelectronic ions, each has 36 electrons

e) MgF_2, isoelectronic ions, each has 10 electrons

f) K^+(18), I^-(54), not isoelectronic

9-18. Figure 9-1 gives the ionic charges of various ions that correspond to noble-gas electron configurations. We use these charges, together with the requirement that the ionic compound must be electrically neutral, to work out the formula.

a) AlI_3

b) NaF

c) CaO

d) $BaBr_2$

e) K_2S

9-20. In this problem, the charge on the cation or metal ion is indicated by the Roman numeral after its name.

a) $PbCl_2$ (Figure 9-3)

b) Bi_2S_3 (Figure 9-3)

c) FeO

d) MoF_3

e) Co_2O_3

9-22. Note that these ions are all isoelectronic. Thus the relative sizes are

$$Al^{3+} < Mg^{2+} < Na^{+} < F^{-} < O^{2-}$$

9-24. We simply add the appropriate ionic radii given in Table 9-1.

a) 181 pm + 31 pm = 212 pm   Be-Cl distance

212 pm + 212 pm = 424 pm   Cl-Cl distance

b) 78 pm + 140 pm = 218 pm

c) 95 pm + 154 pm = 249 pm

9-26. The radius of an iodine atom is 140 pm (Figure 8-11). The volume of an iodine atom is

$$V = \frac{4}{3}\pi r^3 = \left(\frac{4}{3}\right)(3.14)(140\text{ pm})^3 = 1.15 \times 10^7\text{ pm}^3$$

The radius of an iodide ion is 216 pm (Table 9-1). The volume of an iodide ion is

$$V = \frac{4}{3}\pi r^3 = \left(\frac{4}{3}\right)(3.14)(216\text{ pm})^3 = 4.22 \times 10^7\text{ pm}^3$$

$$\text{increase in volume} = 4.22 \times 10^7\text{pm}^3 - 1.15 \times 10^7\text{ pm}^3$$
$$= 3.07 \times 10^7\text{ pm}^3$$

$$\%\text{ increase in volume} = \frac{3.07 \times 10^7\text{pm}^3}{1.15 \times 10^7\text{pm}^3} \times 100$$
$$= 267\%$$

9-28. a) K^+ and I^- ions

b) Cd^{2+} and Br^- ions

c) Ca^{2+} and OH^- ions

d) Ni^{2+} and Cl^- ions

e) Li^+ and Br^- ions

9-30. a) C_3H_7OH, nonelectrolyte - an organic compound

b) KNO_3, strong electrolyte - a salt

c) HClO, weak electrolyte - not an acid that is a strong electrolyte

d) Na_2SO_3, strong electrolyte - a salt

9-32. See Figure 8-12 for ionization energies. The lower the ionization energy of an atom, the easier it is to remove an electron. Also it requires a greater amount of energy to remove an electron from an ion than from an atom. Thus

$$K > Na > Li > Ca \simeq Al$$

9-34. Recall that metals lose electrons to form noble-gas or pseudo-noble-gas electron configurations, which are relatively stable electron configurations. Metals cannot form such electron configurations by gaining an electron and thus all metals have negative electron affinities.

9-36. The value of ΔH_{rxn} for a reaction in which separated gaseous ions are formed is equal to the sum of the ionization energy and the electron affinity of the nonmetal.

a) $K(g) + 419\ kJ \longrightarrow K^+(g) + e^-$

$Cl(g) + e^- \longrightarrow Cl^-(g) + 348\ kJ$

If we add these two equations, then we have

$$K(g) + Cl(g) + 71\ kJ \longrightarrow K^+(g) + Cl^-(g) \qquad \Delta H_{rxn} = 71\ kJ$$

b) $Na(g) + 496\ kJ \longrightarrow Na^+(g) + e^-$

$I(g) + e^- \longrightarrow I^-(g) + 295\ kJ$

If we add these two equations, then we have

$Na(g) + I(g) + 201\ kJ \longrightarrow Na^+(g) + I^-(g)$ $\Delta H_{rxn} = 201\ kJ$

c) $Li(g) + 520\ kJ \longrightarrow Li^+(g) + e^-$

$Br(g) + e^- \longrightarrow Br^-(g) + 324\ kJ$

If we add these two equations, then we have

$Li(g) + Br(g) + 196\ kJ \longrightarrow Li^+(g) + Br^-(g)$ $\Delta H_{rxn} = 196\ kJ$

9-38. a) $Au(g) + 890\ kJ \longrightarrow Au^+(g) + e^-$

$I(g) + e^- \longrightarrow I^-(g) + 295\ kJ$

$Au(g) + I(g) + 595\ kJ \longrightarrow Au^+(g) + I^-(g)$ $\Delta H_{rxn} = 595\ kJ$

b) $F^-(g) + 333\ kJ \longrightarrow F(g) + e^-$

$Cl(g) + e^- \longrightarrow Cl^-(g) + 348\ kJ$

$F^-(g) + Cl(g) \longrightarrow F(g) + Cl^-(g) + 15\ kJ$ $\Delta H_{rxn} = -15\ kJ$

c) $In(g) + 558\ kJ \longrightarrow In^+(g) + e^-$

$Br(g) + e^- \longrightarrow Br^-(g) + 324\ kJ$

$In(g) + Br(g) + 234\ kJ \longrightarrow In^+(g) + Br^-(g)$ $\Delta H_{rxn} = 234\ kJ$

9-40. radius of Na^+ = 95 pm

radius of F^- = 136 pm

The distance between the centers of the ions is

95 pm + 136 pm = 231 pm

From Coulomb's law we have

$$E = \frac{2.31 \times 10^{-16}\ Z_1 Z_1}{d} \qquad \begin{pmatrix} \text{E in joules} \\ \text{d in picometers} \end{pmatrix}$$

where

$$Z_1 = +1 \text{ for } Na^+ \quad \text{and} \quad Z_2 = -1 \text{ for } F^-$$

Thus

$$E = \frac{2.31 \times 10^{-16}(+1)(-1)}{231}$$

$$E = -1.00 \times 10^{-18} J$$

9-42. a) For the ionization of Ag (Figure 8-12)

$$Ag(g) + 731 \text{ kJ} \longrightarrow Ag^+(g) + e^-$$

b) For the addition of an electron to Cl (Table 9-2)

$$Cl(g) + e^- \longrightarrow Cl^-(g) + 348 \text{ kJ}$$

Adding steps a and b, we have

$$Ag(g) + Cl(g) + 383 \text{ kJ} \longrightarrow Ag^+(g) + Cl^-(g)$$

c) We now bring together Ag^+ and Cl^- to their ion-pair separation. In Table 9-1 we find the radius of Ag^+ = 126 pm and the radius of Cl^- = 181 pm, thus the separation as ion-pair = 307 pm. The energy released by the formation of one ion pair is given by Coulomb's law

$$E = 2.31 \times 10^{-16} \frac{Z_1 Z_2}{d} \quad \text{(E in J, d in pm)}$$

$$E = \frac{2.31 \times 10^{-16}\ (+1)(-1)}{307}$$

$$E = -7.52 \times 10^{-19}$$

To compute the energy released on formation of one mole of ion pairs, we multiply by Avogadro's number

$$E = \left(-7.52 \times 10^{-19} \frac{J}{\text{ion-pair}}\right)\left(6.02 \times 10^{23} \frac{\text{ion-pair}}{\text{mol}}\right)$$

$$E = -453 \text{ kJ}\cdot\text{mol}^{-1}$$

Thus

$$Ag^+(g) + Cl^-(g) \longrightarrow Ag^+Cl^-(g) + 453 \text{ kJ}$$

$$(d = 307 \text{ pm})$$

We combine with

$$Ag(g) + Cl(g) + 383 \text{ kJ} \longrightarrow Ag^+(g) + Cl^-(g)$$

to obtain the final result

$$Ag(g) + Cl(g) \longrightarrow Ag^+Cl^-(g) + 70 \text{ kJ}$$

9-44. a) For the ionization of Cu (Figure 8-12)

$$Cu(g) + 745 \text{ kJ} \longrightarrow Cu^+(g) + e^-$$

b) For the addition of an electron to I (Table 9-2)

$$I(g) + e^- \longrightarrow I^-(g) + 295 \text{ kJ}$$

Adding steps a and b, we have

$$Cu(g) + I(g) + 450 \text{ kJ} \longrightarrow Cu^+(g) + I^-(g)$$

c) We now bring together Cu^+ and I^- to their ion-pair separation

radius of Cu^+	=	96 pm
radius of I^-	=	216 pm
separation	=	312 pm

We use Coulomb's law to calculate the energy released by the formation of one ion-pair

$$E = 2.31 \times 10^{-16} \frac{Z_1 Z_2}{d} \qquad (E \text{ in J, d in pm})$$

$$E = \frac{2.31 \times 10^{-16} (-1)(+1)}{312}$$

$$E = -7.40 \times 10^{-19} \text{ J}$$

For one mole

$$E = \left(-7.40 \times 10^{-19} \frac{\text{J}}{\text{ion-pair}}\right)\left(6.02 \times 10^{23} \frac{\text{ion-pair}}{\text{mol}}\right)$$

$$= -446 \text{ kJ}\cdot\text{mol}^{-1}$$

Thus

$$Cu^+(g) + I^-(g) \longrightarrow CuI(g) + 446 \text{ kJ}$$
$$(d = 312 \text{ pm})$$

Combining the above result with

$$Cu(g) + I(g) + 450 \text{ kJ} \longrightarrow Cu^+(g) + I^-(g)$$

we have

$$Cu(g) + I(g) + 4 \text{ kJ} \longrightarrow CuI(g)$$
$$(d = 312 \text{ pm})$$

9-46.

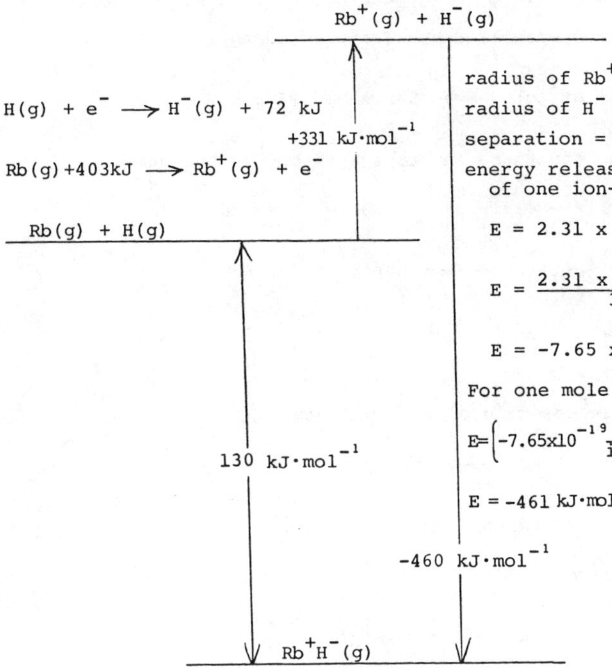

radius of Rb^+ = 148 pm

radius of H^- = 154 pm

separation = 302 pm

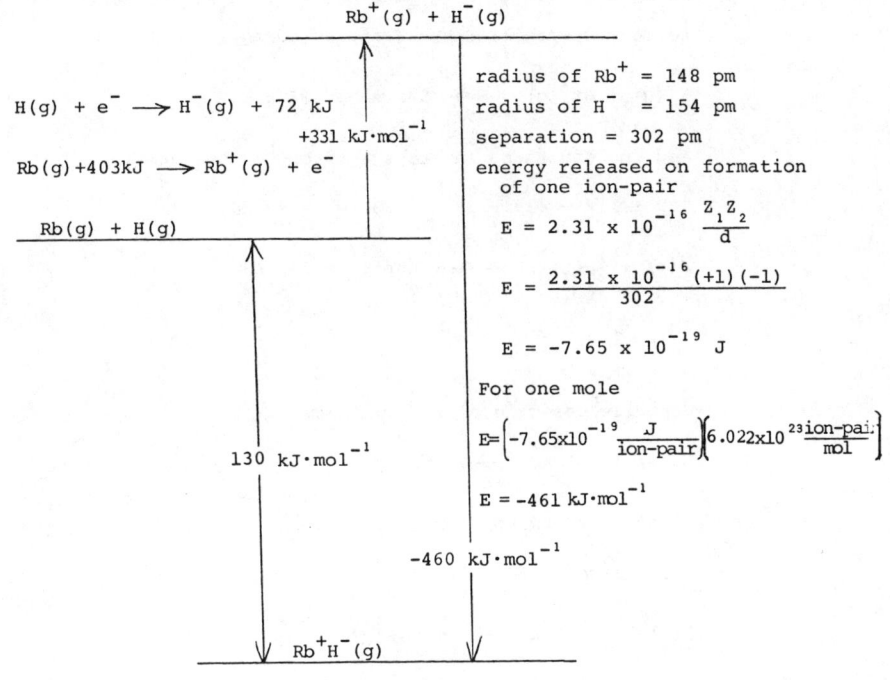

energy released on formation of one ion-pair

$$E = 2.31 \times 10^{-16} \frac{Z_1 Z_2}{d}$$

$$E = \frac{2.31 \times 10^{-16} (+1)(-1)}{302}$$

$$E = -7.65 \times 10^{-19} \text{ J}$$

For one mole

$$E = \left(-7.65\times10^{-19} \frac{\text{J}}{\text{ion-pair}}\right)\left(6.022\times10^{23} \frac{\text{ion-pair}}{\text{mol}}\right)$$

$$E = -461 \text{ kJ}\cdot\text{mol}^{-1}$$

9-48. a) Vaporize one mole of potassium metal

$K(s) + 89 \text{ kJ} \longrightarrow K(g)$

b) Vaporize and dissociate one half mole of $Br_2(l)$

$(223 \text{ kJ}\cdot\text{mol}^{-1})(\frac{1}{2} \text{ mol}) = 112 \text{ kJ}$

$\frac{1}{2}Br_2(l) + 112 \text{ kJ} \longrightarrow Br(g)$

c) Ionize one mole of K(g) (Figure 8-12)

$K(g) + 419 \text{ kJ} \longrightarrow K^+(g) + e^-$

d) Attach one mole of electrons to one mole of Br(g) (Table 9-2)

$Br(g) + e^- \longrightarrow Br^-(g) + 324 \text{ kJ}$

e) Bring one mole of $K^+(g)$ and one mole of $Br^-(g)$ together to form crystalline KBr (lattice energy)

$$K^+(g) + Br^-(g) \longrightarrow KBr(s) + 688\text{ kJ}$$

Add the equations in steps a through e to obtain

$(89 + 112 + 419 - 324 - 688 = -392\text{ kJ})$

$$K(s) + \tfrac{1}{2}Br_2(l) \longrightarrow KBr(s) + 392\text{ kJ}$$

$$\Delta H_{rxn} = -392\text{ kJ}$$

9-50. a) Vaporize one mole of lithium metal

$$Li(s) + 161\text{ kJ} \longrightarrow Li(g)$$

b) Dissociate one half mole of H_2

$$(436\text{ kJ}\cdot\text{mol}^{-1})(\tfrac{1}{2}\text{ mol}) = 218\text{ kJ}$$

$$\tfrac{1}{2}H_2(g) + 218\text{ kJ} \longrightarrow H(g)$$

c) Ionize one mole of Li(g) (Figure 8-12)

$$Li(g) + 520\text{ kJ} \longrightarrow Li^+(g) + e^-$$

d) Attach one mole of electrons to one mole of H(g) (Table 9-2)

$$H(g) + e^- \longrightarrow H^-(g) + 72\text{ kJ}$$

e) Bring one mole of $Li^+(g)$ and one mole of $H^-(g)$ together to form crystalline LiH (lattice energy)

$$Li^+(g) + H^-(g) \longrightarrow LiH(s) + 917\text{ kJ}$$

Add the reactions in steps a through e to obtain

$(161 + 218 + 520 - 72 - 917 = -90)$

$$Li(s) + \tfrac{1}{2}H_2(g) \longrightarrow LiH(s) + 90\text{ kJ}$$

$$\Delta H_{rxn} = -90\text{ kJ}$$

9-52. a) Vaporization of K(s)

$K(s) + 89\ kJ \longrightarrow K(g)$

b) Vaporization and dissociation of one half mole $Br_2(l)$

$\frac{1}{2} Br_2(l) + 112\ kJ \longrightarrow Br(g)$

c) We are given that

$K(g) + Br(g) \longrightarrow KBr(g) + 329\ kJ$

d) We are given that

$KBr(s) + 392\ kJ \longrightarrow K(s) + \frac{1}{2} Br_2(l)$

Adding steps a through d, we obtain

(89 + 112 - 329 + 392) = 264)

$KBr(s) + 264\ kJ \longrightarrow KBr(g)$

Thus the enthalpy of vaporization for KBr(s) is

$\Delta H^0_{vap} = 264\ kJ \cdot mol^{-1}$

CHAPTER 10

SOLUTIONS TO THE EVEN-NUMBERED PROBLEMS

10-2. a) The electron-dot formulas of the chlorine and sulfur atoms are

$$:\ddot{C}l\cdot \qquad \cdot\ddot{S}\cdot \qquad \cdot\ddot{C}l:$$

If we join these atoms together by sharing a pair of electrons, then both the chlorine atoms and the sulfur atom will have a noble-gas electron configuration and the Lewis formula is

$$:\ddot{C}l-\ddot{S}-\ddot{C}l:$$

Note that there are 20 valence electrons in the three atoms and 10 electron pairs in the molecule.

b) Because the germanium atom is the unique atom in $GeCl_4$, we assume that the germanium atom will be a central atom. The electron-dot formulas of the atoms are

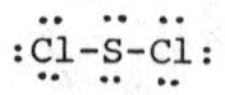

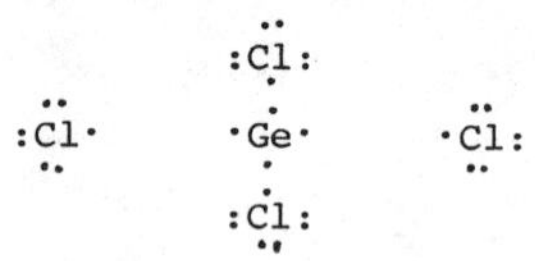

$$\begin{array}{c} :\ddot{C}l: \\ :\ddot{C}l\cdot \quad \cdot Ge\cdot \quad \cdot \ddot{C}l: \\ :\ddot{C}l: \end{array}$$

If we join these atoms together by sharing a pair of electrons, then each atom will have a noble-gas electron configuration and the Lewis formula is

$$\begin{array}{c} :\ddot{C}l: \\ | \\ :\ddot{C}l-Ge-\ddot{C}l: \\ | \\ :\ddot{C}l: \end{array}$$

Note that there are 32 valence electrons in the five atoms and 16 electron pairs in the molecule.

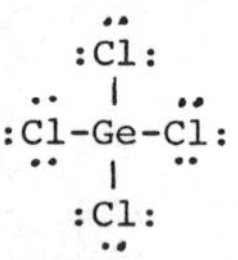

c) Because the arsenic atom is the unique atom in AsBr$_3$, we assume that the arsenic atom is a central atom. The electron-dot formulas of the atoms are

$$:\!\ddot{\underset{..}{Br}}\!\cdot \qquad \cdot\ddot{As}\cdot \qquad \cdot\ddot{\underset{..}{Br}}\!:$$

$$:\!\underset{..}{Br}\!:$$

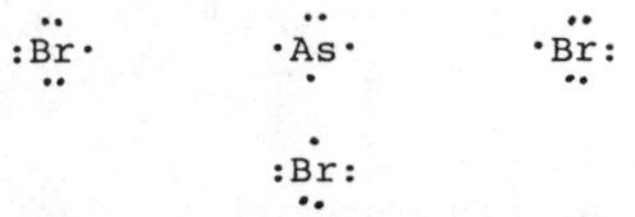

If we join these atoms together by sharing a pair of electrons, then each atom will have a noble-gas electron configuration and the Lewis formula is

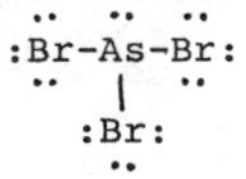

$$:\!\ddot{\underset{..}{Br}}\!-\!\ddot{As}\!-\!\ddot{\underset{..}{Br}}\!:$$
$$|$$
$$:\!\underset{..}{Br}\!:$$

Note that there are 26 valence electrons in the four atoms and 13 electron pairs in the molecule.

10-4. We assume that the two nitrogen atoms are central atoms and write

$$\qquad\qquad :\!\ddot{\underset{..}{F}}\!: \quad :\!\ddot{\underset{..}{F}}\!:$$
$$:\!\ddot{\underset{..}{F}}\!\cdot \quad \cdot\ddot{N}\cdot \quad \cdot\ddot{N}\cdot \quad \cdot\ddot{\underset{..}{F}}\!:$$

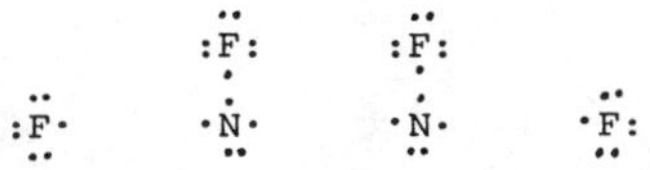

We join the atoms by sharing electron pairs and obtain

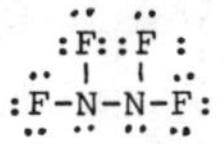

$$:\!\ddot{\underset{..}{F}}\!:\!:\!\ddot{\underset{..}{F}}\!:$$
$$:\!\ddot{\underset{..}{F}}\!-\!N\!-\!N\!-\!\ddot{\underset{..}{F}}\!:$$

Note that there are 38 valence electrons in the constituent atoms and 19 electron pairs in the molecule.

10-6. a) We write

$$H\cdot \qquad \cdot\ddot{\underset{..}{Br}}\!:$$

By joining the atoms by sharing an electron pair, we get

$$H\!-\!\ddot{\underset{..}{Br}}\!:$$

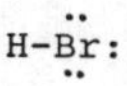

There are 8 valence electrons in the constituent atoms and 4 electron pairs in the molecule.

b) Because the silicon atom is unique in SiH_4, we assume that it is a central atom and write

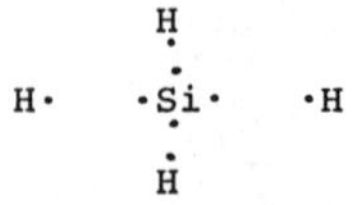

$$\begin{array}{c} H \\ H\cdot \quad \cdot Si \cdot \quad \cdot H \\ H \end{array}$$

By joining the atoms by sharing an electron pair, we get

$$\begin{array}{c} H \\ | \\ H-Si-H \\ | \\ H \end{array}$$

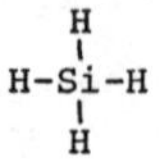

There are 8 valence electrons in the constituent atoms and 4 electron pairs in the molecule.

c) The hydrogen atoms must be terminal and so we write

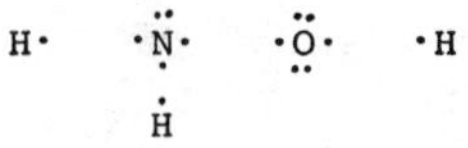

$$H\cdot \quad \cdot \ddot{N}\cdot \quad \cdot \ddot{O}\cdot \quad \cdot H$$
$$\overset{\cdot}{H}$$

By joining the atoms by sharing an electron pair, we get

$$\begin{array}{c} H-\ddot{N}-\ddot{O}-H \\ | \\ H \end{array}$$

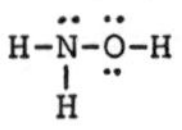

There are 14 valence electrons in the constituent atoms and 7 electron pairs in the molecule.

10-8. We assume that the titanium atom is a central atom and so we write

$$\begin{array}{c} :\ddot{Cl}: \\ :\ddot{Cl}\cdot \quad \cdot Ti\cdot \quad \cdot \ddot{Cl}: \\ :\ddot{Cl}: \end{array}$$

The resulting Lewis formula is

$$\begin{array}{c} :\ddot{Cl}: \\ | \\ :\ddot{Cl}-Ti-\ddot{Cl}: \\ | \\ :\ddot{Cl}: \end{array}$$

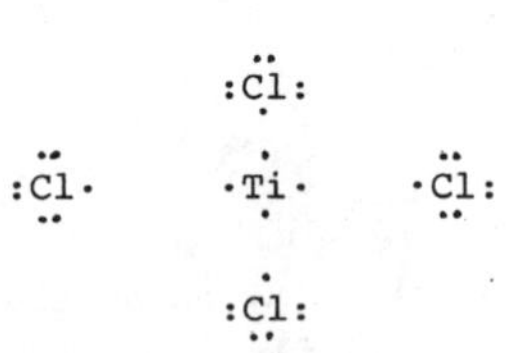

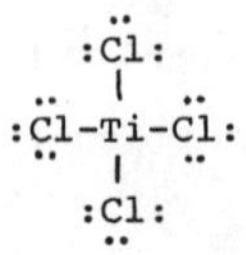

There are 32 valence electrons in the constituent atoms and 16 electron pairs in the molecule. Note that titanium has 4 valence electrons.

10-10. The Lewis formula for NH_3 is

$$H-\overset{..}{N}-H$$
$$|$$
$$H$$

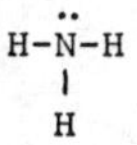

When a proton is removed, the electron remains with the nitrogen atom; thus the Lewis formula for the amide ion is

$$H-\overset{\ominus}{\underset{..}{N}}-H$$

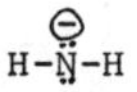

$NaNH_2$ sodium amide

$Ba(NH_2)_2$ barium amide

10-12. Each sulfur atom is covalently bonded to two other sulfur atoms to form an eight-membered ring. The Lewis formula for solid sulfur is

[Eight-membered ring of S atoms with lone pairs]

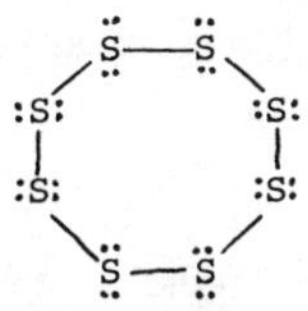

10-14. a) The hydrogen atoms must be terminal atoms and so

$$\begin{array}{c} H \\ | \\ H-C-\overset{..}{\underset{..}{S}}-H \\ | \\ H \end{array}$$

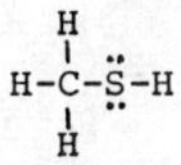

b) Line up the carbon and oxygen atoms as they appear in the chemical formula and place the hydrogen atoms on the carbon atoms in terminal positions to get

$$\begin{array}{c} H \quad\quad H \\ | \quad\quad | \\ H-C-\overset{..}{\underset{..}{O}}-C-H \\ | \quad\quad | \\ H \quad\quad H \end{array}$$

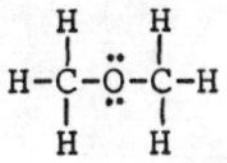

c) The hydrogen atoms must be terminal atoms and so

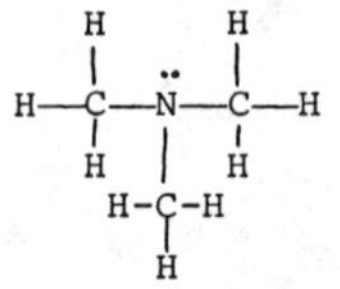

10-16. a) The hydrogen atoms must be terminal atoms and one of the carbon atoms is a central atom

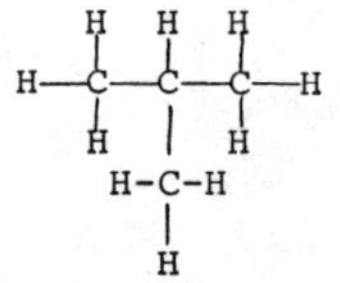

b) For $(CH_3)_4C$ we have

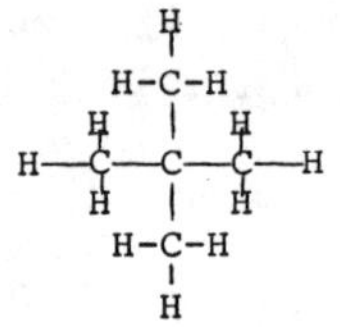

c) For $C_2H_5CH(CH_3)_2$, we have

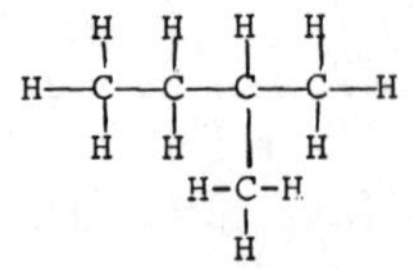

10-18. The carbon atom is a central atom in each molecule.

$$\ddot{\underset{\cdot\cdot}{:}Cl:} \\ :\ddot{Cl}-\underset{|}{C}-\ddot{Cl}: \\ :\ddot{Cl}:$$

$$H \\ | \\ :\ddot{Cl}-\underset{|}{C}-\ddot{Cl}: \\ :\ddot{Cl}:$$

$$H \\ | \\ :\ddot{Cl}-\underset{|}{C}-\ddot{Cl}: \\ H$$

$$H \\ | \\ H-\underset{|}{C}-\ddot{Cl}: \\ H$$

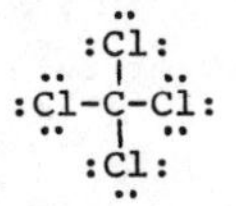

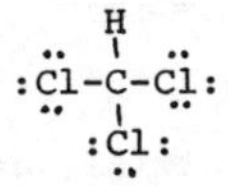

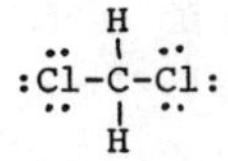

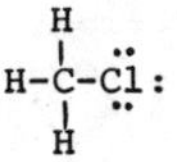

10-20. When a proton is removed from the oxygen atom in CH$_3$OH, we get

$$H \\ | \\ H-\underset{|}{C}-\ddot{\underset{\cdot\cdot}{O}}:^{\ominus} \\ H$$

    KOCH$_3$ potassium methoxide

    Al(OCH$_3$)$_3$ aluminum methoxide

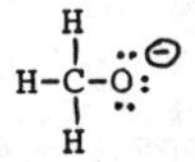

10-22. a) The carbon atom is central and there are 24 valence electrons. Thus we have

$$:\ddot{Cl}: \\ | \\ :\ddot{Cl}-C=\ddot{O}:$$

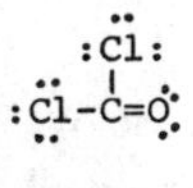

b) There are 10 valence electrons in HCN. The best way to satisfy the octet rule is to write

$$H-C\equiv N:$$

c) There are 34 valence electrons in HOOCCOOH. The only way to satisfy the octet rule for the carbon and oxygen atoms is to write

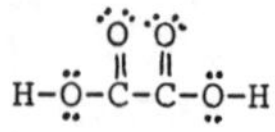

10-24. The hydrogen atoms are terminal atoms and so we write

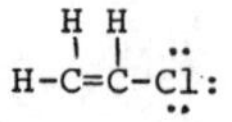

10-26. There are 24 valence electrons or 12 electron pairs in CH_3COOH. The only way to satisfy the octet rule for the carbon and oxygen atoms using 12 electron pairs is to write

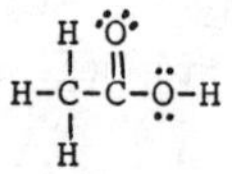

The -COOH group, which has the Lewis formula

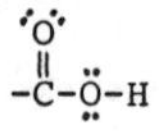

is characteristic of organic acids (see Problem 10-25).

10-28. The only way to satisfy the octet rule for the carbon and oxygen atoms using 9 electron pairs is to write

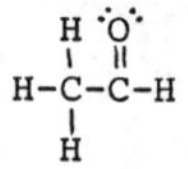

10-30. Line up the four carbon atoms in a row and add the hydrogen atoms. There are 11 electron pairs to distribute and each carbon atom must satisfy the octet rule. Therefore, we obtain

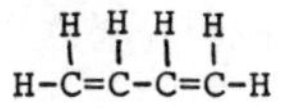

and

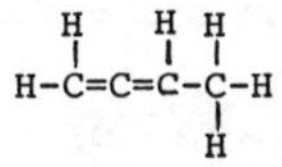

10-32. The hydrogen atom is a terminal atom. There are 16 valence electrons or 8 electron pairs in HN_3. The nitrogen atoms must satisfy the octet rule and so

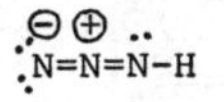

The azide ion, N_3^-, is obtained when the proton is removed from HN_3 and so we write

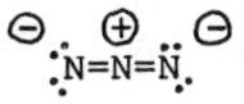

10-34. The two resonance forms are

The superposition of these two resonance forms gives

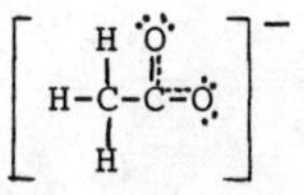

The two carbon-oxygen bonds in CH_3COO^- are equivalent; they have the same bond length and the same energy.

10-36. The two resonance forms are

$$:\!\overset{..}{O}\!=\!\overset{\oplus}{\underset{..}{O}}\!-\!\overset{\ominus}{\underset{..}{\overset{..}{O}}}\!: \quad \longleftrightarrow \quad \overset{\ominus}{:\!\underset{..}{\overset{..}{O}}}\!-\!\overset{\oplus}{\underset{..}{O}}\!=\!\overset{..}{\underset{..}{O}}\!:$$

The superposition of these two resonance forms is

$$:\!\overset{..}{\underset{.}{O}}\; =\!=\; \overset{..}{O}\; =\!=\; \overset{..}{\underset{..}{O}}\!:$$

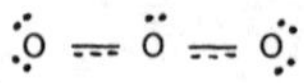

The two bonds in O_3 are equivalent; they have the same bond length and the same energy.

10-38. If nitrogen is the central atom, then we have

$$:\!\overset{..}{\underset{..}{Cl}}\!-\!\overset{..}{N}\!=\!\overset{..}{\underset{..}{O}}\!:$$

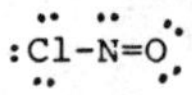

If oxygen is the central atom, then we have

$$:\!\overset{..}{\underset{..}{Cl}}\!-\!\overset{\oplus}{\underset{..}{O}}\!=\!\overset{\ominus}{\underset{..}{N}}\!:$$

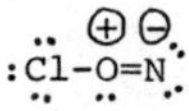

The formal charges in ClON are larger than in ClNO and so we shall reject the ClON. The structure ClNO is predicted to be the correct arrangement of the atoms.

10-40.
formula	Lewis formula
N_2O	$:N\!\equiv\!\overset{\oplus}{N}\!-\!\overset{\ominus}{\underset{..}{\overset{..}{O}}}\!: \longleftrightarrow \overset{\ominus}{:\!\underset{..}{N}}\!=\!\overset{\oplus}{N}\!=\!\overset{..}{\underset{..}{O}}\!:$
NO	$\cdot\overset{..}{N}\!=\!\overset{..}{\underset{..}{O}}\!:$
N_2O_3	$:\!\overset{..}{\underset{..}{O}}\!=\!N\!-\!\overset{\oplus}{\underset{\underset{\underset{\ominus}{:\!\overset{..}{\underset{..}{O}}\!:}}{|}}{N}}\!=\!\overset{..}{\underset{..}{O}}\!: \longleftrightarrow \overset{\ominus}{:\!\underset{..}{\overset{..}{O}}}\!-\!N\!=\!\overset{\oplus}{\underset{\underset{\underset{}{\overset{..}{\underset{..}{O}}\!.}}{||}}{N}}\!-\!\overset{..}{\underset{..}{O}}\!:$
NO_2	$:\!\overset{..}{\underset{.}{O}}\!=\!\overset{\oplus}{N}\!-\!\overset{\ominus}{\underset{..}{\overset{..}{O}}}\!: \longleftrightarrow \overset{\ominus}{:\!\underset{..}{\overset{..}{O}}}\!-\!\overset{\oplus}{N}\!=\!\overset{..}{\underset{.}{O}}\!:$

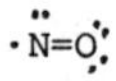

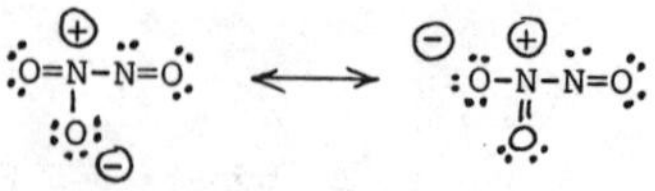

N₂O₄ $\quad\quad\quad$ $^{\ominus}{:}\overset{..}{\underset{..}{O}}{-}\overset{\oplus}{N}{-}\overset{\oplus}{N}{-}\overset{..}{\underset{..}{O}}{:}^{\ominus}$ $\quad$ + other resonance forms
$\quad\quad\quad\quad\quad\quad\quad\quad\quad$ $\overset{..}{\underset{..}{O}}$ $\overset{..}{\underset{..}{O}}$

N₂O₅ $\quad\quad\quad$ $^{\ominus}{:}\overset{..}{\underset{..}{O}}{-}\overset{\oplus}{N}{-}\overset{..}{\underset{..}{O}}{-}\overset{\oplus}{N}{-}\overset{..}{\underset{..}{O}}{:}^{\ominus}$ $\quad$ + other resonance forms
$\quad\quad\quad\quad\quad\quad\quad\quad\quad\quad$ ∥ $\quad\quad$ ∥
$\quad\quad\quad\quad\quad\quad\quad\quad\quad$ $\overset{..}{\underset{..}{O}}$ $\quad\;\;$ $\overset{..}{\underset{..}{O}}$

10-42. a) BrO_3 contains 59 electrons or 25 valence electrons. The Lewis formula is

$\quad\quad\quad\quad\quad\quad$ $:\overset{..}{\underset{..}{O}}{=}\overset{\cdot\;\oplus}{Br}{=}\overset{..}{\underset{..}{O}}:$ $\quad$ + other resonance forms
$\quad\quad\quad\quad\quad\quad\quad\quad\quad\quad$ $|$
$\quad\quad\quad\quad\quad\quad\quad\quad\quad$ $:\overset{..}{\underset{..}{O}}:^{\ominus}$

BrO_3 contains an odd electron.

b) SO_3 contains 32 electrons or 24 valence electrons. The Lewis formula is

$\quad\quad\quad\quad\quad\quad$ $:\overset{..}{\underset{..}{O}}{=}\overset{\oplus}{S}{=}\overset{..}{\underset{..}{O}}:$ $\quad$ + other resonance forms
$\quad\quad\quad\quad\quad\quad\quad\quad\quad$ $|$
$\quad\quad\quad\quad\quad\quad\quad\quad$ $:\overset{..}{\underset{..}{O}}:^{\ominus}$

c) HNO contains 16 electrons or 12 valence electrons. The Lewis formula is

$\quad\quad\quad\quad\quad\quad\quad\quad$ $H-\overset{..}{N}{=}\overset{..}{\underset{..}{O}}:$

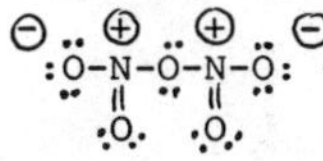

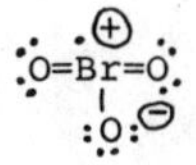

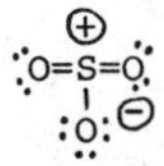

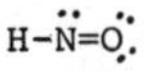

d) HO_2 contains 17 electrons or 13 valence electrons. The Lewis formula is

$\quad\quad\quad\quad\quad\quad\quad\quad$ $H-\overset{..}{\underset{..}{O}}-\overset{..}{\underset{..}{O}}\cdot$

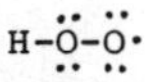

HO_2 contains an odd electron.

e) SO_4^- contains 49 electrons or 31 valence electrons. The Lewis formula is

$\quad\quad\quad\quad\quad\quad\quad\quad$ $:\overset{..}{\underset{..}{O}}:$
$\quad\quad\quad\quad\quad\quad\quad\quad\quad$ $|$
$\quad\quad\quad\quad\quad\quad$ $:\overset{..}{\underset{..}{O}}{=}S{=}\overset{..}{\underset{..}{O}}:$ $\quad$ + other resonance forms
$\quad\quad\quad\quad\quad\quad\quad\quad\quad$ $|$
$\quad\quad\quad\quad\quad\quad\quad\quad$ $^{\ominus}:\overset{..}{\underset{..}{O}}:$

SO_4^- contains an odd electron.

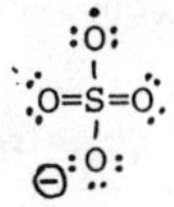

10-44. NO

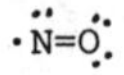 free radical

NO_2

 free radical

HO

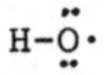 free radical

HNO_3

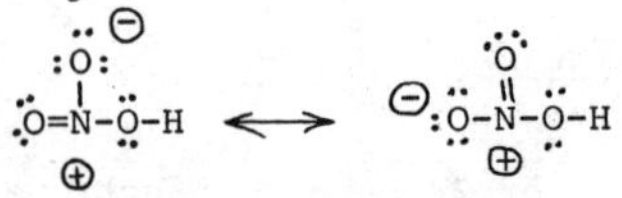

10-46. a) 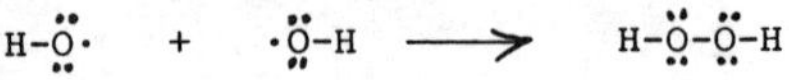

b) 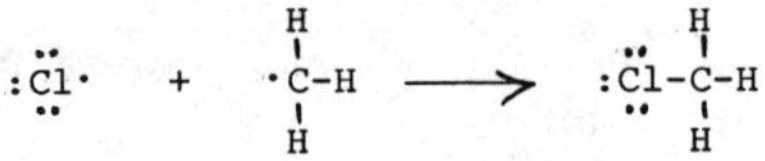

10-48. a) Some resonance forms are

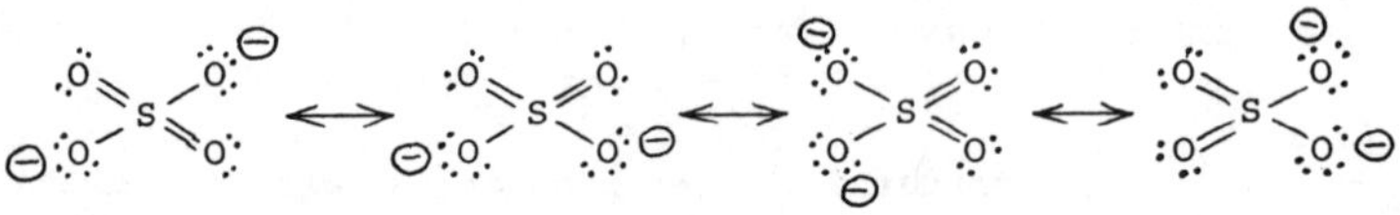

The superposition of these resonance forms gives

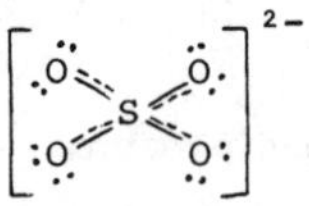

All four bonds are equivalent; they have the same bond lengths and the same bond energy.

b) Some resonance forms are

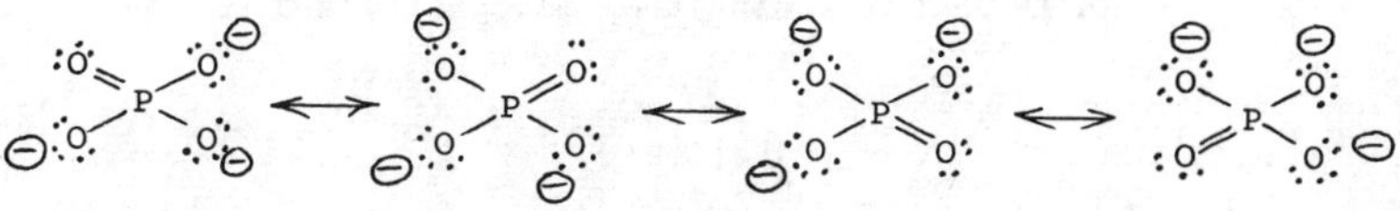

The superposition of these resonance forms is

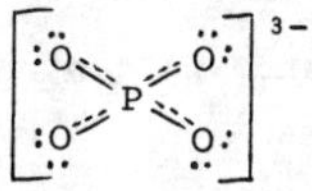

All four bonds are equivalent; they have the same bond lengths and the same bond energy.

c)

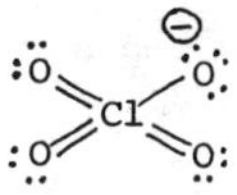

+ 3 resonance forms

The superposition formula is

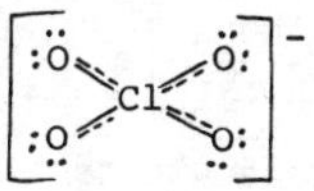

All four bonds are equivalent; they have the same bond length and the same bond energy.

10-50. a) There are 42 valence electrons or 11 electron pairs in ClF_5. The chlorine atom is central and so

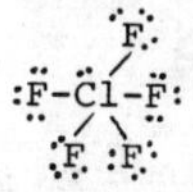

b) There are 28 valence electrons or 14 electron pairs in IF_3. The iodine atom is central and so

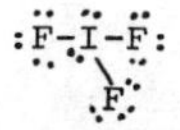

c) There are 56 valence electrons or 28 electron pairs in IF_7. The iodine atom is central and so

d) There are 34 valence electrons or 17 electron pairs in IF_4^+. The iodine atom is central and so

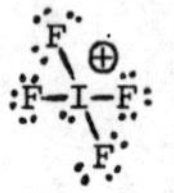

e) There are 48 valence electrons or 24 electron pairs in BrF_6^+. The bromine atom is central and so

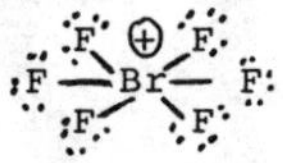

10-52. a) 32 valence electrons, 16 electron pairs, chlorine atom central

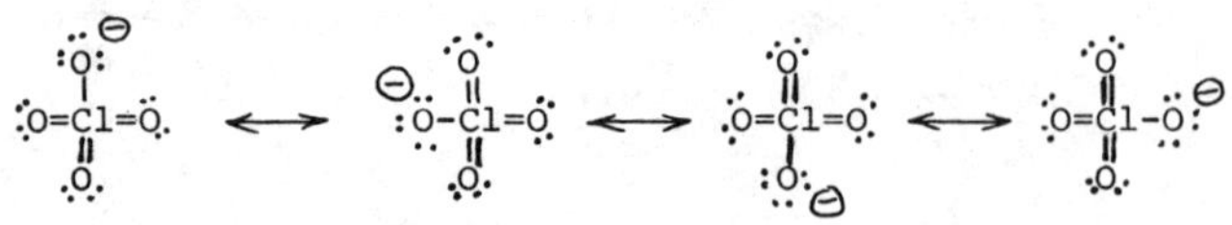

or

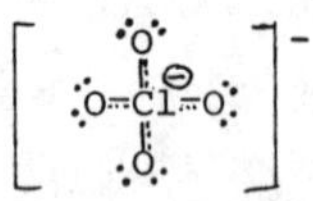

b) 13 valence electrons, 6 electron pairs

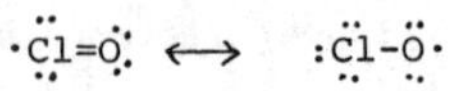

c) 26 valence electrons, 13 electron pairs, chlorine atom central

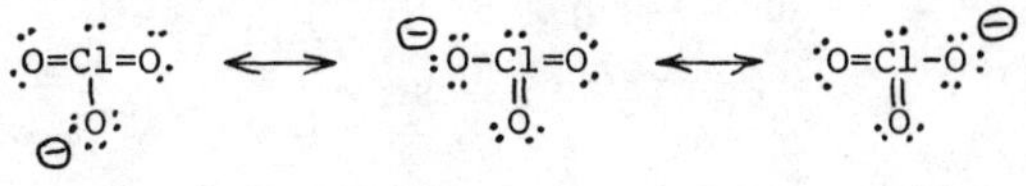

or

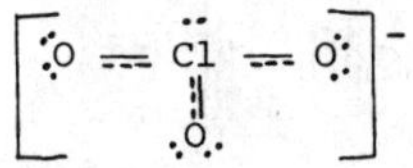

d) 19 valence electrons, 9 electron pairs, one unpaired electron, chlorine atom central

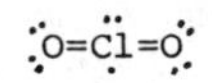

e) 14 valence electrons, 7 electron pairs

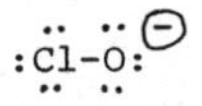

10-54. a) 32 valence electrons, 16 electron pairs, sulfur atom central

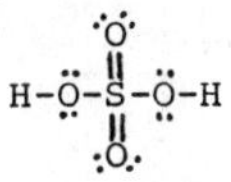

b) 32 valence electrons, 16 electron pairs, one sulfur atom central

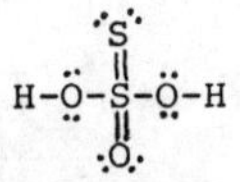

c) 56 valence electrons, 28 electron pairs

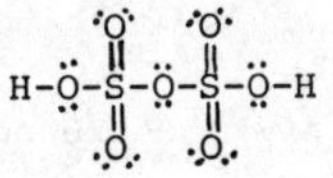

d) 50 valence electrons, 25 electron pairs, the sulfur atoms central

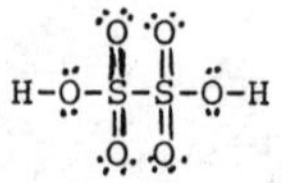

e) 62 valence electrons, 31 electron pairs

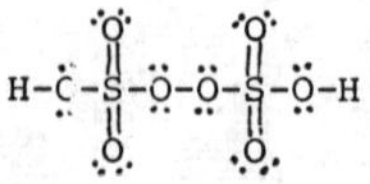

10-56. In each case, there are 32 valence electrons or 16 electron pairs and the phosphorus atom is the central atom.

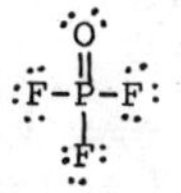

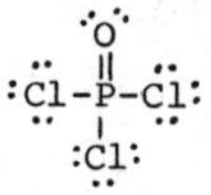

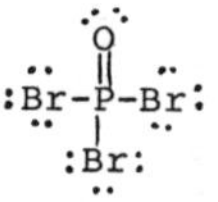

10-58.

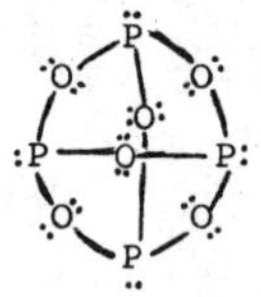

10-60. a) Because electronegativity decreases as we go down a column in the periodic table, the dipole moments increase in the order

$$HI < HBr < HCl < HF$$

b) See(a)

$$\xrightarrow{\text{increasing}}$$

AsH_3 PH_3 NH_3

c) See (a)

$$\xrightarrow{\text{increasing}}$$

F_2O Cl_2O H_2O

d) See (a)

$$\xrightarrow{\text{increasing}}$$

ClF_3 BrF_3 IF_3

e) See (a)

$$\xrightarrow{\text{increasing}}$$

H_2Te H_2Se H_2S H_2O

10-62. a) Fluorine is more electronegative than hydrogen and so we have

$$\overset{\delta+}{H}—\overset{\delta-}{\ddot{\underset{\cdot\cdot}{F}}}:$$

b) The electronegativities of phosphorus and hydrogen are about the same and so we have

$$H-\underset{\displaystyle\overset{|}{H}}{\ddot{P}}-H$$

c) Sulfur is more electronegative than hydrogen and so we have

$$\overset{\delta+}{H}—\overset{\delta-}{\underset{\cdot\cdot}{\ddot{S}}}—\overset{\delta+}{H}$$

CHAPTER 11

SOLUTIONS TO THE EVEN-NUMBERED PROBLEMS

11-2. In this problem we first use VSEPR theory to predict the molecular shape and from the shape we determine if there are any 90^0 bond angles. (See Table 11-1).

a) NH_4^+ is an AX_4 tetrahedral ion and thus has no 90^0 bond angles.

b) $AlCl_3$ is an AX_3 trigonal planar molecule and thus has no 90^0 bond angles.

c) AlF_6^{3-} is an AX_6 octahedral molecule and thus has 90^0 bond angles.

d) $SiCl_4$ is an AX_4 tetrahedral molecule and thus has no 90^0 bond angles.

e) PCl_5 is an AX_5 trigonal bipyramidal molecule and thus has 90^0 bond angles.

11-4. a) SeF_6 is an AX_6 octahedral molecule and has 180^0 bond angles between oppositely placed ligands (adjacent ligands have 90^0 bond angles).

b) CdI_2 is an AX_2 linear molecule and has a 180^0 bond angle.

c) $AsCl_5$ is an AX_5 trigonal bipyramidal molecule and has a 180^0 bond angle between axial ligands.

d) $SiCl_4$ is an AX_4 tetrahedral molecule and has no 180^0 bond angles.

e) ZnI_2 is an AX_2 linear molecule and has a 180^0 bond angle.

11-6. See Figure 11-12 or Table 11-3.

a) 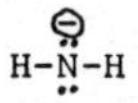 AX_2E_2 bent

b) 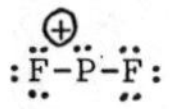 AX_2E bent

c) 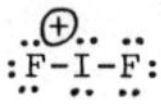 AX_2E_2 bent

d) 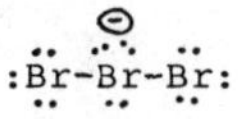 AX_2E_3 linear

e) 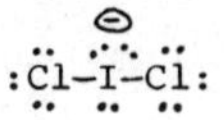 AX_2E_3 linear

11-8. a) 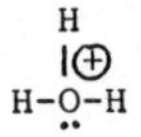 AX_3E trigonal pyramidal

b) 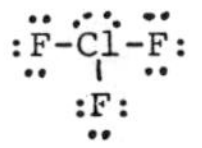 AX_3E_2 T-shaped

c) 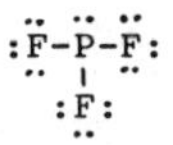 AX_3E trigonal pyramidal

d)  AX_3 trigonal planar

e) 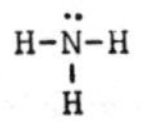 AX_3E trigonal pyramidal

11-10. a)  AX_4E seesaw

b) 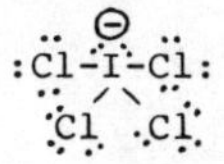 AX_4E_2 square planar

c) 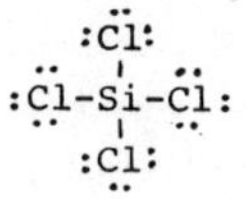 AX_4 tetrahedral

d) 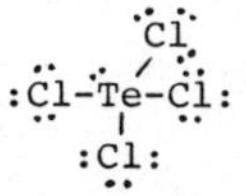 AX_4E seesaw

11-12.

a) 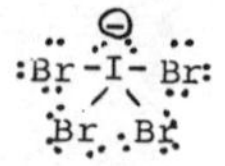 AX_4E_2 square planar

b)  AX_4 tetrahedral

c) 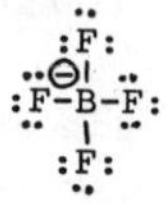 AX_4 tetrahedral

d) 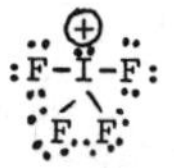 AX_4E seesaw

e) 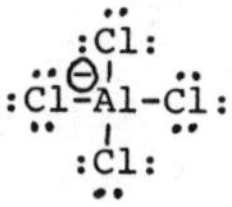 AX_4 tetrahedral

11-14.

a) 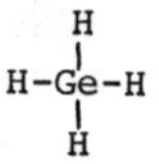 AX_4 tetrahedral 109.5^0 (2)

b) 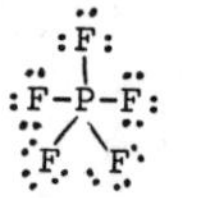 AX_5 trigonal bipyramidal 90^0, 120^0 (1, 3)

c) 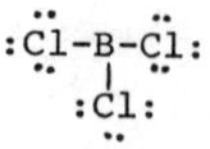 AX_3 trigonal planar 120^0 (3)

d) 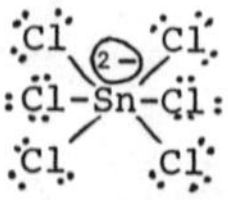 AX_6 octahedral 90^0 (1)

11-16. a) 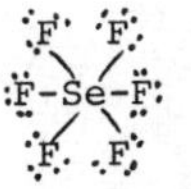 AX_6 octahedral

b) 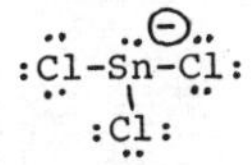 AX_3E trigonal pyramidal

c) 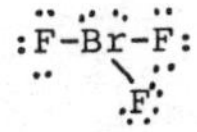 AX_3E_2 T-shaped

d) 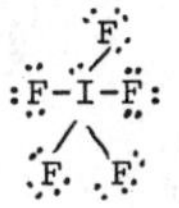 AX_5E square pyramidal (Figure 11-21)

1) a
2) c, d
3) none
4) b
5) none
6) none
7) none
8) a
9) c, d

11-18. a) 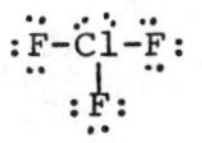 AX_3E_2 T-shaped (Figure 11-17)

b) 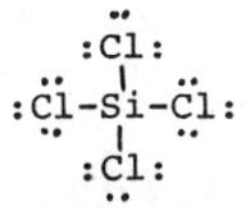 AX_4 tetrahedral

c) 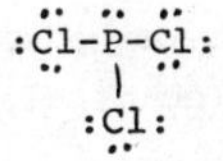 AX_3E trigonal pyramidal (Figure 11-11)

d) 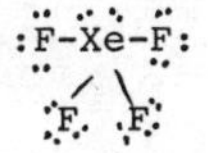 AX_4E_2 square planar (Figure 11-22)

e) 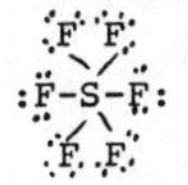 AX_6 octahedral

1) d,e	6) none
2) a	7) none
3) b	8) d,e
4) c	9) a
5) none	

11-20. a) 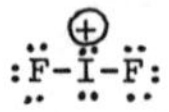 AX_2E_2 bent

b)  AX_2E_3 linear

c) 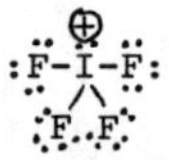 AX_4E seesaw

d) 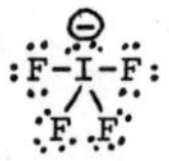 AX_4E_2 square planar

e) 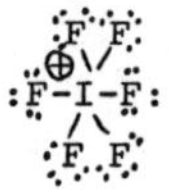 AX_6 octahedral

11-22. Consult the text or problems.

11-24. a) 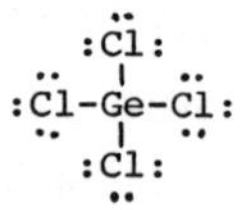 AX_4 tetrahedral (no dipole moment)

b) 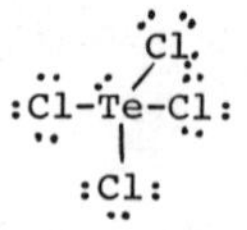 AX_4E seesaw (has a dipole moment)

c) 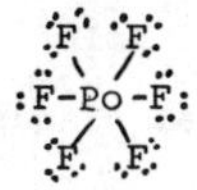 AX_6 octahedral
(no dipole moment)

d) 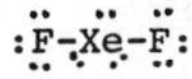 AX_2E_3 linear
(no dipole moment)

e) 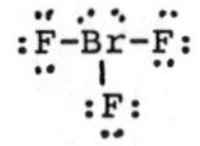 AX_3E_2 T-shaped
(has a dipole moment)

11-26. a) 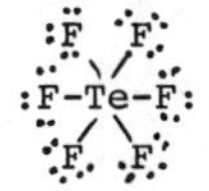 AX_6 octahedral
(no dipole moment)

b) 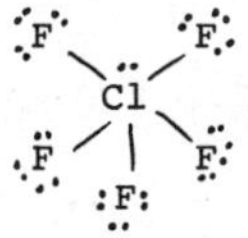 AX_5E square pyramidal
(has a dipole moment)

c) 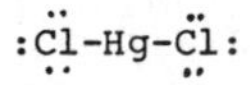 AX_2 linear
(no dipole moment)

d) 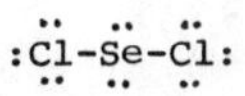 AX_2E_2 bent
(has a dipole moment)

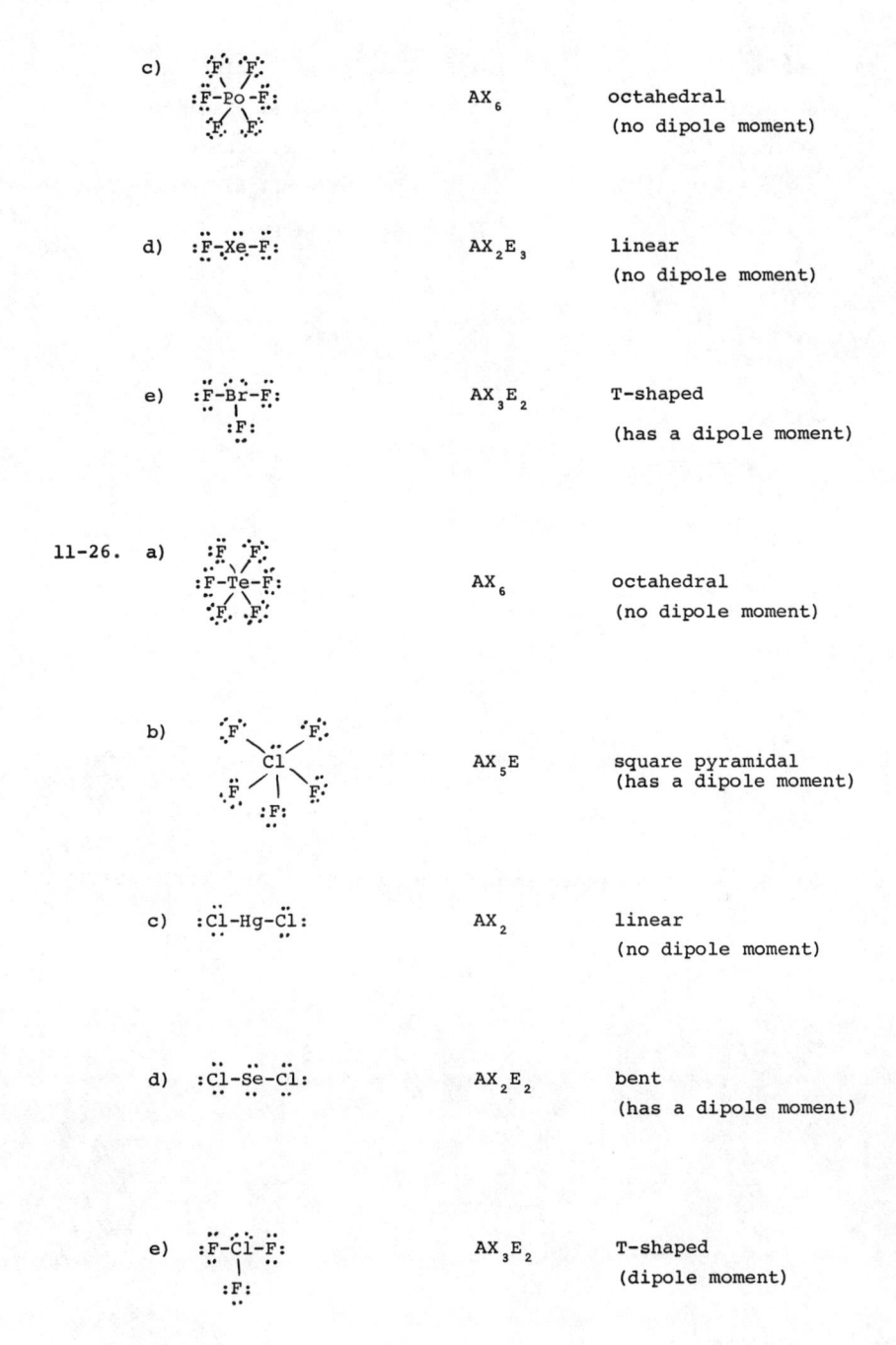

e) 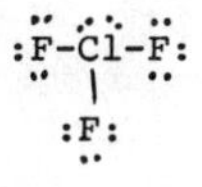 AX_3E_2 T-shaped
(dipole moment)

11-28. a) $TeCl_4$ — AX_4E — seesaw, polar

b) BCl_3 — AX_3 — trigonal planar, nonpolar

c) $CdCl_2$ — AX_2 — linear, nonpolar

d) PCl_5 — AX_5 — trigonal bipyramidal, nonpolar

e) $PbCl_2$ — AX_2E — bent, polar

11-30. a) 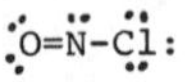 AX_2E bent

b) 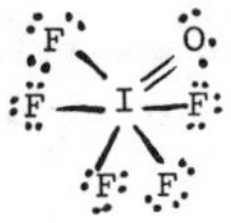 AX_6 octahedral

c)  + resonance forms give 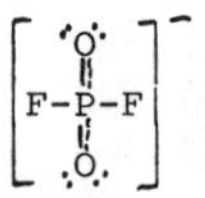 AX_4 tetrahedral

d) 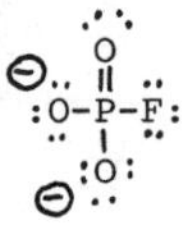 + resonance forms give 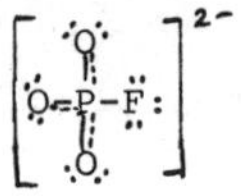 AX_4 tetrahedral

e) 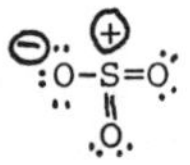 + resonance forms give  AX_3 trigonal planar

11-32. a) 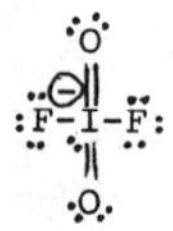 AX_4E seesaw

b) 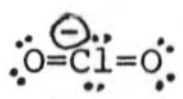 AX_2E_2 bent

c) 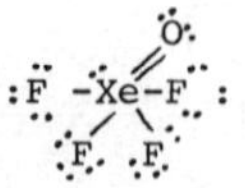 AX_5E square pyramidal

d) :Cl-Se=O: AX$_3$E   trigonal pyramidal
   |
  :Cl:

e) :Ö:
  ‖
:O=Cl-F: AX$_4$ tetrahedral
  ‖
  :O:

11-34. a) :O:
   ‖
  :F-S-F: AX$_5$ trigonal bipyramidal
   / \
   :F :F:

b) :O:
   ‖
 :Cl-S-Cl: AX$_4$ tetrahedral
   ‖
   :O:

c) :O:
   ‖
 :F-As-F: AX$_4$ tetrahedral
   ‖
   :O:

d) :O=Xe=O: AX$_2$E$_2$ bent

e)   :O:
   ‖
 :Cl-As-Cl: AX$_4$ tetrahedral
   |
   :Cl:

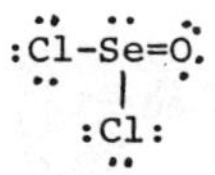

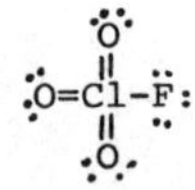

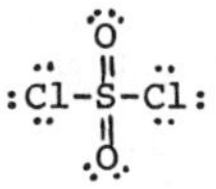

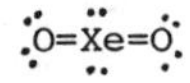

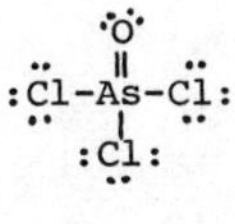

11-36. a) 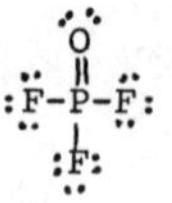 AX_4 tetrahedral

b) 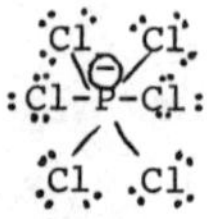 AX_6 octahedral

c) 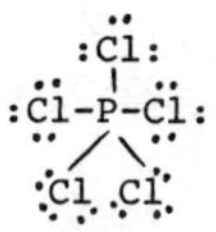 AX_5 trigonal bipyramidal

d) 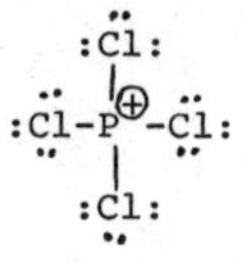 AX_4 tetrahedral

e) 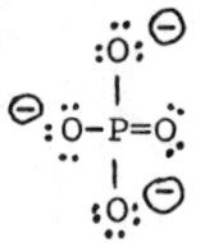 + other resonance forms give 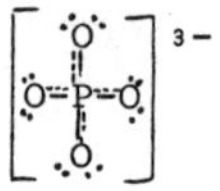 AX_4 tetrahedral

11-38. a) 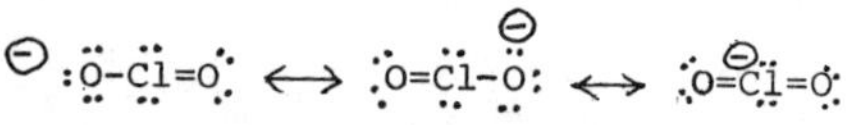 give 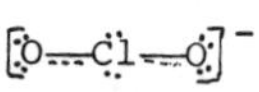 AX_2E_2 bent

b) 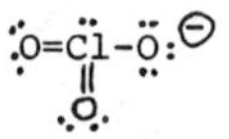 + other resonance forms give 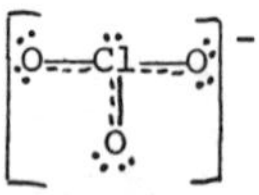 AX_3E trigonal pyramidal

c) 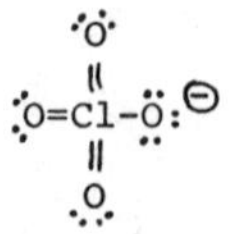 + other resonance forms give 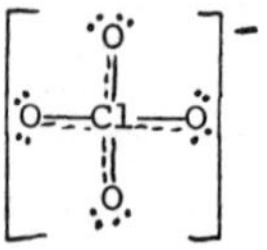 AX_4 tetrahedral

11-40. a) $^{\ominus}\!:\!\ddot{O}\!-\!C\!-\!\ddot{O}\!:^{\ominus}$ + other resonance forms give $\left[\begin{array}{c}:\!\ddot{O}\!=\!=\!C\!=\!=\!\ddot{O}\!:\\ \overset{\shortparallel}{\underset{:\ddot{O}:}{}}\end{array}\right]^{2-}$ AX_3 trigonal planar
$\quad\quad\quad\;\;\;\overset{\shortparallel}{\underset{:\ddot{O}:}{}}$

b) $:\!\ddot{O}\!=\!\overset{\ominus}{\underset{}{I}}\!=\!\ddot{O}\!:$ $AX_2 E_2$ bent

c) $^{\ominus}\!:\!\ddot{S}\!-\!C\!=\!\ddot{S}\!:$ + other resonance forms give $\left[\begin{array}{c}:\!\ddot{S}\!-\!-\!C\!-\!-\!\ddot{S}\!:\\ \overset{\shortmid}{\underset{:\ddot{S}:}{}}\end{array}\right]^{2-}$ AX_3 trigonal planar
$\quad\quad\quad\;\;\;\underset{\ominus}{:\ddot{S}:}$

d) $^{\ominus}\!:\!\ddot{O}\!-\!\overset{:\ddot{O}:}{\underset{:\ddot{O}:}{\overset{\shortparallel}{S}}}\!-\!\ddot{O}\!:^{\ominus}$ + other resonance forms give $\left[\begin{array}{c}:\ddot{O}:\\ :\ddot{O}\!=\!\overset{\shortparallel}{S}\!-\!\ddot{O}\!:\\ \underset{:\ddot{O}:}{}\end{array}\right]^{2-}$ AX_4 tetrahedral

e) $\overset{:\ddot{C}l:}{\underset{\oplus}{^{\ominus}\!:\!\ddot{O}\!-\!N\!=\!\ddot{O}\!:}} \longleftrightarrow \overset{:\ddot{C}l:}{\underset{\oplus}{:\ddot{O}\!=\!N\!-\!\ddot{O}\!:^{\ominus}}}$ give $\left[\overset{:\ddot{C}l:}{:\ddot{O}\!=\!N\!=\!\ddot{O}\!:}\right]$ AX_3 trigonal planar

11-42. a) $:\ddot{O}\!=\!\ddot{N}\!-\!\ddot{O}\!:^{\ominus} \longleftrightarrow {}^{\ominus}\!:\!\ddot{O}\!-\!\ddot{N}\!=\!\ddot{O}\!:$ give $\left[:\ddot{O}\!-\!-\!\ddot{N}\!-\!-\!\ddot{O}\!:\right]^{-}$ $AX_2 E$ bent

b) $^{\ominus}\!:\!\ddot{O}\!-\!\overset{\oplus}{\underset{:\ddot{O}:}{\overset{\shortparallel}{N}}}\!-\!\ddot{O}\!:^{\ominus}$ + resonance forms to give $\left[\begin{array}{c}:\ddot{O}\!=\!N\!=\!O\\ \underset{:\ddot{O}:}{\shortparallel}\end{array}\right]^{-}$ AX_3 trigonal planar

c) $:\!\ddot{O}\!=\!\overset{\oplus}{N}\!=\!\ddot{O}\!:$ AX_2 linear

d) $^{\ominus}\!:\!\overset{:\ddot{O}:^{\ominus}}{\underset{:\ddot{O}:^{\ominus}}{\overset{\shortmid}{\underset{\shortmid}{\overset{\oplus}{N}}}}}\!-\!\ddot{O}\!:^{\ominus}$ AX_4 tetrahedral

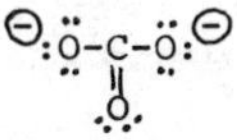

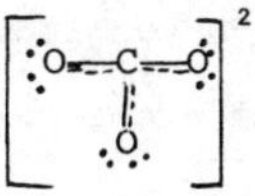

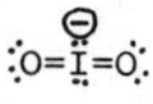

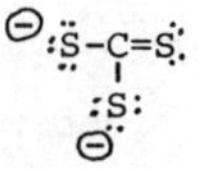

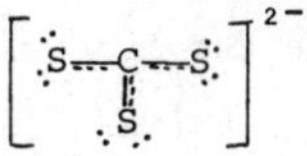

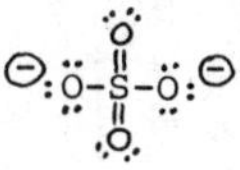

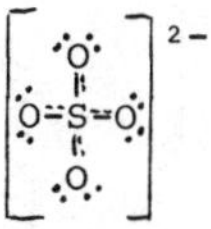

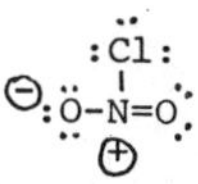

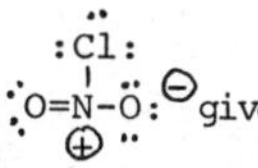

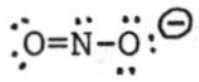

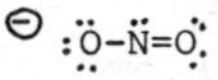

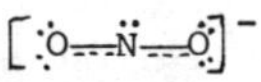

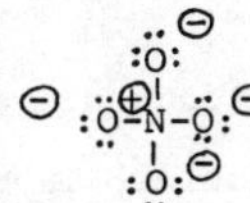

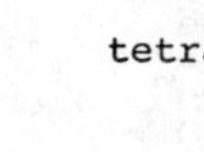

11-44. There are two isomers of $Pt(NH_3)_2Cl_2$. We can place the two chlorine atoms next to one another or opposite to one another

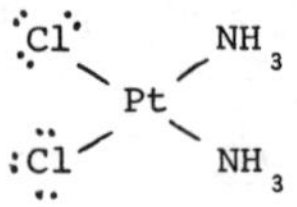
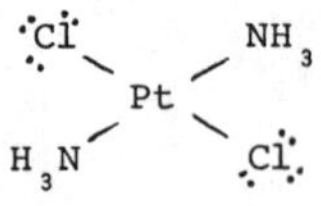

11-46. a) There are two isomers, one with the two chlorine atoms adjacent to one another and one with the chlorine atoms opposite to one another (see 11-45 b)

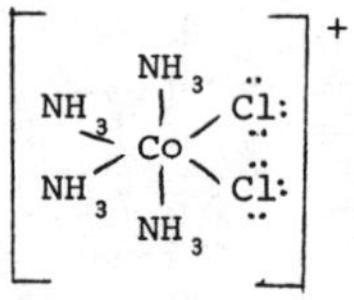

Cl's adjacent

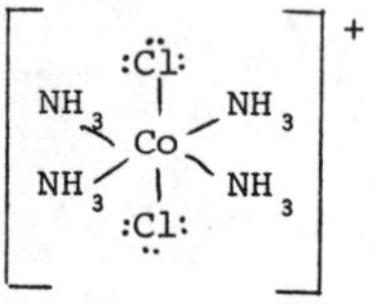

Cl's opposite

b) There are two isomers, one with the three chlorine atoms along three of the four vertices of a square and one with the three chlorine atoms on an octahedral face (see 11-45 c)

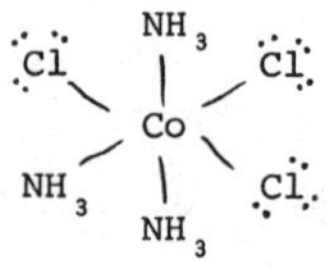

Cl's on the vertices of a square

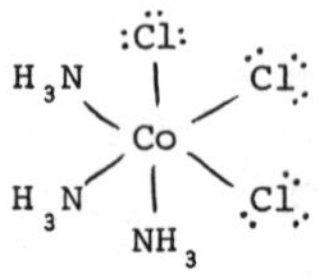

Cl's on a face

11-48. a) Recall that the axial and equatorial positions are not equivalent in a trigonal bipyramidal molecule. Thus we have two isomers

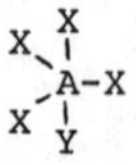

Y axial

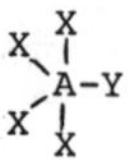

Y equatorial

b) Three isomers

$$\begin{array}{c} X \\ | \\ Y\diagdown A-Y \\ X \diagup | \\ X \end{array} \qquad \begin{array}{c} Y \\ | \\ X\diagdown A-Y \\ X \diagup | \\ X \end{array} \qquad \begin{array}{c} Y \\ | \\ X\diagdown A-X \\ X \diagup | \\ Y \end{array}$$

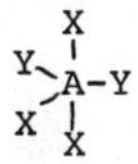

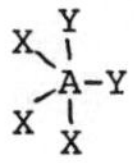

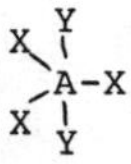

both Y's equatorial   one Y equatorial   both Y's axial
                                   one Y axial

c) Three isomers

$$\begin{array}{c} Y \\ | \\ X\diagdown A-X \\ Y \diagup | \\ Y \end{array} \qquad \begin{array}{c} X \\ | \\ Y\diagdown A-X \\ Y \diagup | \\ Y \end{array} \qquad \begin{array}{c} X \\ | \\ Y\diagdown A-Y \\ Y \diagup | \\ X \end{array}$$

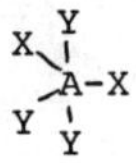

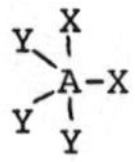

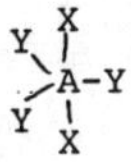

both X's equatorial   one X equatorial   both X's axial
                                   one X axial

SOLUTIONS TO THE EVEN-NUMBERED PROBLEMS

12-2. When the two hydrogen atoms are separated by less than the equilibrium distance (74 pm), they repel each other, which is reflected by positive values of the energy.

12-4. a) The Lewis formula for PH_3 is

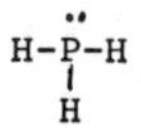

$$H-\overset{..}{\underset{\underset{H}{|}}{P}}-H$$

There are 3 localized bonds and 1 lone pair.

b) The Lewis formula for H_2O_2 is

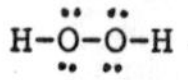

$$H-\overset{..}{\underset{..}{O}}-\overset{..}{\underset{..}{O}}-H$$

There are 3 localized bonds and 4 lone pairs.

c) The Lewis formula for H_3O^+ is

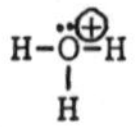

$$H-\overset{..\oplus}{\underset{\underset{H}{|}}{O}}-H$$

There are 3 localized bonds and 1 lone pair.

d) The Lewis formula for Cl_2O is

$$:\overset{..}{\underset{..}{Cl}}-\overset{..}{\underset{..}{O}}-\overset{..}{\underset{..}{Cl}}:$$

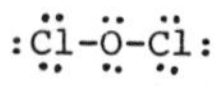

There are 2 localized bonds and 8 lone pairs.

e) The Lewis formula for CH_3OH is

$$H-\overset{\overset{H}{|}}{\underset{\underset{H}{|}}{C}}-\overset{..}{\underset{..}{O}}-H$$

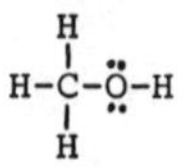

There are 5 localized bonds and 2 lone pairs.

12-6. There are 13 localized bonds in butane. There are 4 x 4 = 16 valence electrons from the four carbon atoms and 10 x 1 = 10 valence electrons from the ten hydrogen atoms, giving a total of 26 valence electrons. The 26 valence electrons occupy the 13 localized bond orbitals.

12-8. The Lewis formula for chloroform is

$$\begin{array}{c} :\ddot{C}l: \\ | \\ :\ddot{C}l-C-H \\ | \\ :\ddot{C}l: \end{array}$$

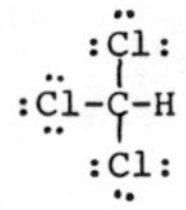

We learned that VSEPR theory predicts that $HCCl_3$ is tetrahedral and we know that the sp^3 hybrid orbitals on a carbon atom will point to the vertices of a tetrahedron. A chlorine atom has a $[Ne](3s)^2(3p_x)^2(3p_y)^2(3p_z)^1$ electron configuration, indicating that one of its 3p orbitals is occupied by only one electron. We can form four equivalent localized bond orbitals by combining each sp^3 orbital on the carbon atom with a 1s orbital on the hydrogen atom and with a 3p orbital on each chlorine atom. There are (4x1)+(7x3)+(1x1) = 26 valence electrons in $HCCl_3$. Two valence electrons of opposite spin occupy each of the four bond orbitals. The remaining 18 valence electrons are lone electron pairs on the three chlorine atoms. Each chlorine atom has three lone electron pairs.

12-10. The Lewis formula for OF_2 is

$$:\ddot{F}-\ddot{O}-\ddot{F}:$$

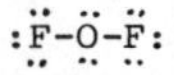

We shall use sp^3 hybrid orbitals on the oxygen atom. A fluorine atom has a $1s^2 2s^2(2p_x)^2(2p_y)^2(2p_z)^1$ electron configuration, indicating that one of its 2p orbitals is occupied by only one electron. We form two equivalent localized bond orbitals by combining each sp³ orbital on the oxygen atom with a 2p orbital on each fluorine atom. Two valence electrons of opposite spin occupy each of the two localized bond orbitals. The two lone electron pairs on the oxygen occupy the remaining two sp³ orbitals on the oxygen atom. The remaining 12 valence electrons are lone electron pairs on the fluorine atoms.

12-12. The Lewis formula for NF_3 is

$$:\!\ddot{F}\!-\!\ddot{N}\!-\!\ddot{F}\!: \\ | \\ :\!\ddot{F}\!:$$

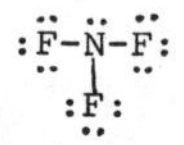

The three bonds are formed by combining a sp³ orbital on the nitrogen atom with a 2p orbital on a fluorine atom. We use a 2p orbital on each fluorine atom because the electron configuration of a fluorine atom is $(1s)^2(2s)^2(2p_x)^2(2p_y)^2(2p_z)^1$, indicating that one of its 2p orbitals is occupied by only one electron. The lone electron pair on the nitrogen atom occupies the remaining sp³ orbital. There are $(1 \times 5)+(3 \times 7) = 26$ valence electrons in NF_3.

12-14. The Lewis formula for hydrogen peroxide is

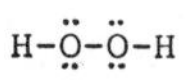

$$H\!-\!\ddot{\underset{..}{O}}\!-\!\ddot{\underset{..}{O}}\!-\!H$$

We shall use sp³ orbitals on each oxygen atom. There are $(2 \times 1)+(2 \times 6) = 14$ valence electrons. There are three localized bonds. The H-O bond is formed by combining an sp³ oxygen orbital with a 1s hydrogen orbital. The O-O bond is formed by combining an sp³ orbital on one oxygen atom with an sp³ orbital on the other oxygen atom. Six of the valence electrons occupy bond orbitals and eight valence electrons constitute the four lone pairs, two on each oxygen atom.

12-16. The Lewis formula for dimethylamine is

$$\begin{array}{c} \text{H} \quad\;\; \text{H} \\ |\;\;\;\;\;\; \ddot{}\;\;\;\; | \\ \text{H-C-N-C-H} \\ |\;\;\;\; |\;\;\;\; | \\ \text{H} \;\; \text{H} \;\; \text{H} \end{array}$$

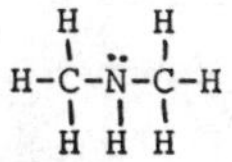

There are 9 σ bonds and 1 lone electron pair. There are (7x1)+(2x4)+(1x5) = 20 valence electrons. Two bonds on the nitrogen atom are formed by combining an sp³ orbital on the nitrogen atom with an sp³ orbital on a carbon atom. One bond on the nitrogen atom is formed by combining an sp³ orbital on the nitrogen atom with a 1s orbital on a hydrogen atom. The lone electron pair occupies the remaining sp³ orbital on the nitrogen atom. Each of the remaining six σ bonds is formed by combining an sp³ orbital on a carbon atom and a 1s orbital on a hydrogen atom. The shape around the nitrogen atom is trigonal pyramidal. The shape around each carbon atom is tetrahedral.

12-18. The Lewis formula for ethyl alcohol is

$$\begin{array}{c} \text{H} \;\; \text{H} \\ |\;\;\;\; |\;\;\;\; \ddot{} \\ \text{H-C-C-O-H} \\ |\;\;\;\; |\;\;\;\; \ddot{} \\ \text{H} \;\; \text{H} \end{array}$$

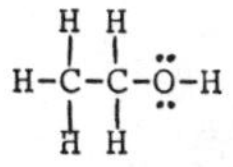

The two σ bonds on the oxygen atom are formed by combining an sp³ oxygen orbital with a hydrogen 1s orbital and an sp³ oxygen orbital with a carbon sp³ orbital. The two lone electron pairs occupy the remaining two sp³ orbitals on the oxygen atom. The σ sigma bonds on the carbon atoms are formed by combining sp³ carbon orbitals with 1s hydrogen orbitals, sp³ carbon orbitals or an sp³ oxygen orbital. The shape of ethyl alcohol is tetrahedral around each carbon atom and bent around the oxygen atom. There are 8 σ bonds, (1x6)+(2x4)+(6x1) = 20 valence electrons, and two lone electron pairs on the oxygen atom.

12-20. There are two covalent bonds and two lone electron pairs on the oxygen atom. Each C-O bond is formed by combining an sp^3 orbital on the oxygen atom with an sp^3 orbital on a carbon atom. Each lone electron pair occupies an sp^3 orbital on the oxygen atom. Each of the eight C-H bond is formed by combining an sp^3 orbital on the carbon atom with a 1s orbital on a hydrogen atom. The C-C bond is formed by combining an sp^3 orbital on one carbon atom with an sp^3 orbital on the other carbon atom.

12-22. The electron configuration of a tellurium atom is
$[Kr](5s)^2(4d)^{10}(5p_x)^2(5p_y)^1(5p_z)^1$, indicating that two of the 5p orbitals are occupied by only one electron. To describe the bonding in H_2Te, use the 1s orbitals on the hydrogen atom and two of the 5p orbitals on the tellurium atom.

12-24. The three Al-Cl bonds in $AlCl_3$ are σ bonds. Each σ bond is formed by combining an sp^2 hybrid orbital on the aluminum atom with a 3p chlorine orbital.

12-26. The Lewis formula for 2-butene is

$$\begin{array}{c} \quad\quad H \quad\quad H \\ \quad\quad | \quad\quad | \\ H-C-C=C-C-H \\ \quad | \quad | \quad | \quad | \\ \quad H \quad H \quad H \quad H \end{array}$$

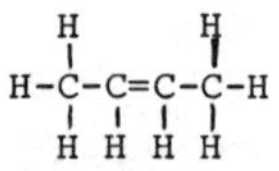

There are 11 σ bonds and 1 π bond in 2-butene. There are (8x1)+(4x4) = 24 valence electrons, which occupy the twelve bond orbitals.

12-28. a) The Lewis formula for $F_2C=CF_2$ is

$$\begin{array}{c} :\ddot{F}-C=C-\ddot{F}: \\ \;\;\ddot{} \quad | \quad | \quad \ddot{} \\ \quad :\ddot{F}::\ddot{F}: \end{array}$$

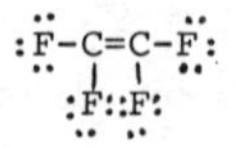

There are 5 σ bonds and 1 π bond.

b) The Lewis formula for $H_2C=CHCH_3$ is

$$\begin{array}{c} \text{H} \\ | \\ \text{H-C=C-C-H} \\ |\quad |\quad | \\ \text{H H H} \end{array}$$

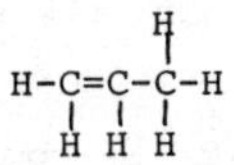

There are 8 σ bonds and 1 π bond.

c) The Lewis formula for $H_2C=C=CCl_2$ is

$$\begin{array}{c} \text{H-C=C=C-}\ddot{\underset{..}{\text{Cl}}}\text{:} \\ |\qquad\quad | \\ \text{H}\quad\;\;:\!\ddot{\underset{..}{\text{Cl}}}\text{:} \end{array}$$

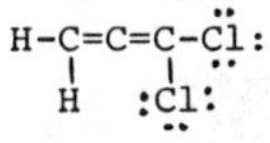

There are 6 σ bonds and 2 π bonds.

d) There are 8 σ bonds and 2 π bonds.

12-30. From VSEPR theory we conclude that vinyl chloride is planar. Each of the three C-H bonds is formed by combining an sp^2 orbital on a carbon atom and a 1s orbital on a hydrogen atom. The C-Cl bond is formed by combining an sp^2 orbital on the carbon atom and a 3p orbital on the chlorine atom. There are three lone electron pairs on the chlorine atom. The C=C double bond is formed by combining sp^2 orbitals on each carbon atom and the remaining p orbitals on each carbon atom.

12-32. The central carbon atom in acetone forms two σ bonds to the two other carbon atoms and a σ bond and a π bond to the oxygen atom. Each carbon-carbon σ bond is formed by combining an sp^2 orbital on the central carbon atom with an sp^3 orbital on the other carbon atom. The carbon-oxygen σ bond is formed by combining the remaining sp^2 orbital of the central carbon atom with an sp^2 orbital on the oxygen atom. The π bond is formed by combining the remaining 2p orbital on the central carbon atom with the remaining 2p orbital on the oxygen atom. The two lone electron pairs on the oxygen atom occupy the other two sp^2 orbitals. The shape of acetone is trigonal planar around the central carbon atom

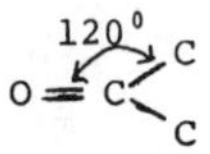

12-34. The σ bond framework of urea is

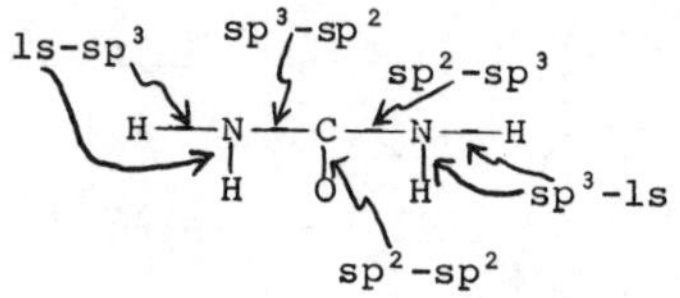

The remaining bond between the carbon atom and the oxygen atom is a π bond that is formed from the 2p orbitals on each atom.

12-36. The Lewis formula for methyl cyanide is

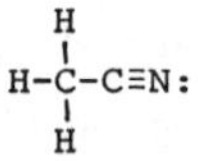

There are five σ bonds and two π bonds. There are (3x1)+(2x4)+(1x5) = 16 valence electrons, which occupy the seven bond orbitals and one lone pair orbital on the nitrogen atom.

12-38. The Lewis formula for the acetylide ion is

$$\ominus\ominus$$
$$:C\equiv C:$$

The σ bond between the carbon atoms is formed by combining an sp orbital on each of the carbon atoms. The two π bonds are formed by combining the 2p orbitals on each of the carbon atoms. Each lone electron pair occupies an sp orbital on a carbon atom.

12-40. The σ-bond framework of dimethylacetylene ($CH_3C\equiv CCH_3$) is

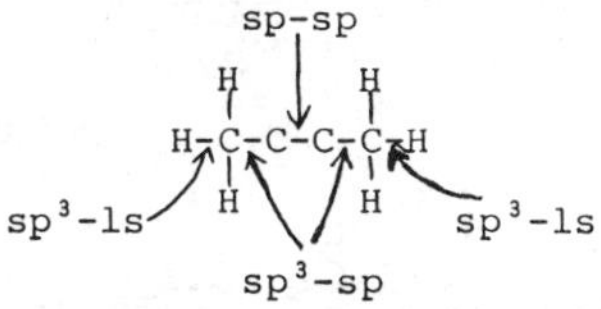

The remaining bonds are two π bonds between the carbon atoms using 2p orbitals. The shape of the molecule is linear around the triple bond

$$C-C\equiv C-C$$

There are nine σ bonds and two π bonds. There are (6x1)+(4x4) = 22 valence electrons which occupy the eleven bond orbitals.

12-42. The Lewis formula for aniline is

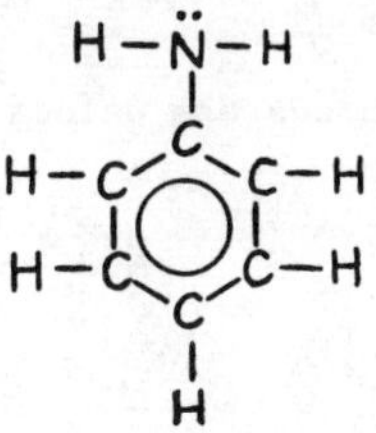

The σ bond framework is

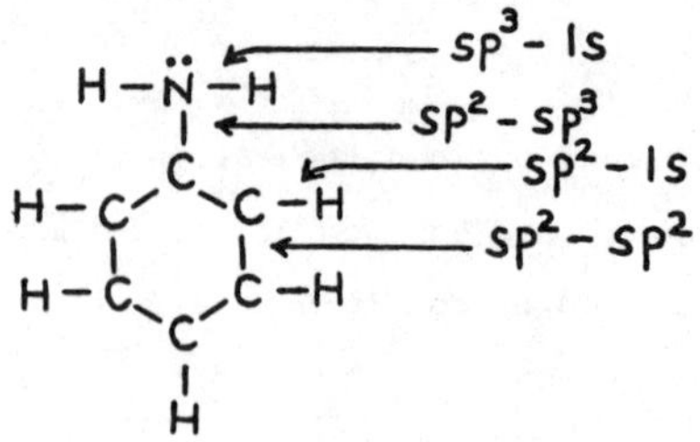

The three π bonds are delocalized over the entire molecule.

12-44. The Lewis formula for anthracene is

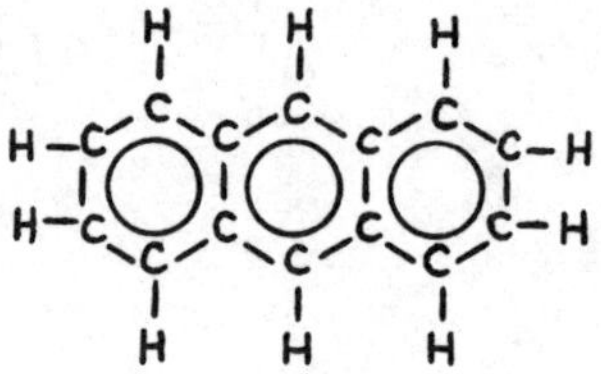

which is a combination of several resonance forms.

The σ bond framework is

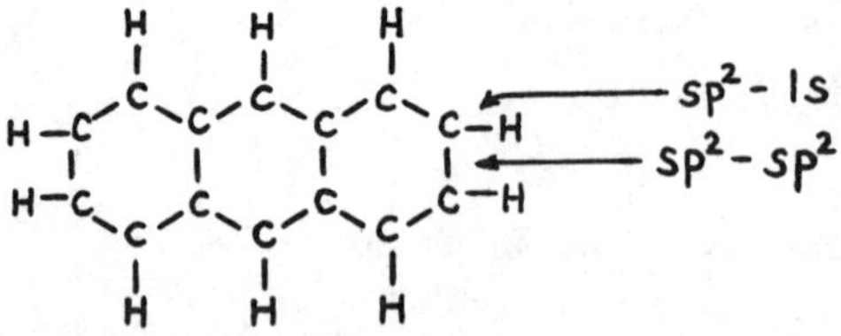

The **seven** π bonds are delocalized over the three rings.

12-46. The Lewis formula for pyridine is

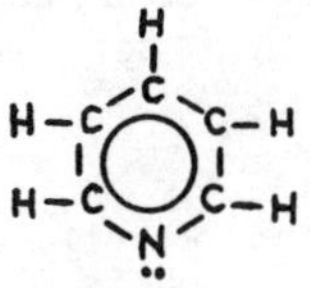

The sigma bond framework of pyridine is

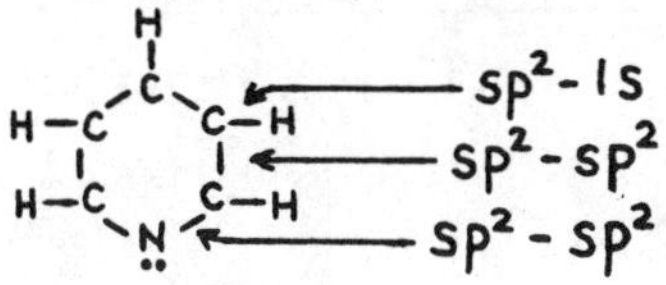

The three π bonds are delocalized over the entire ring. The lone-electron pair on the nitrogen atom occupies the remaining sp^2 orbital.

12-48. There are 10 electrons in diatomic boron. Using Figure 12-27, we see that the ground-state electron configuration of B_2 is $(1\sigma)^2(1\sigma^*)^2(2\sigma)^2(2\sigma^*)^2(1\pi)^2$. However, Hund's rule states that we should place one electron in each of the two 1π orbitals with unpaired spins. Thus there are two unpaired electrons in B_2 and so B_2 is paramagnetic.

12-50. Using Figure 12-27 and Equation (12-1), we find that the ground-state electron configurations and bond orders of F_2 and F_2^+ are

	ground-state electron configuration	bond order
F_2	$(1\sigma)^2(1\sigma^*)^2(2\sigma)^2(2\sigma^*)^2(1\pi)^4(3\sigma)^2(1\pi^*)^4$	1
F_2^+	$(1\sigma)^2(1\sigma^*)^2(2\sigma)^2(2\sigma^*)^2(1\pi)^4(3\sigma)^2(1\pi^*)^3$	$1\frac{1}{2}$

Because the bond order of F_2^+ is greater than that of F_2, we predict that F_2^+ has a larger bond energy and a shorter bond length than F_2.

12-52. Using Figure 12-27 and Equation (12-1), we find that the ground-state electron configurations and bond orders of NF, NF^+ and NF^- are

	ground-state electron configuration	bond order
NF	$(1\sigma)^2(1\sigma^*)^2(2\sigma)^2(2\sigma^*)^2(1\pi)^4(3\sigma)^2(1\pi^*)^2$	2
NF^+	$(1\sigma)^2(1\sigma^*)^2(2\sigma)^2(2\sigma^*)^2(1\pi)^4(3\sigma)^2(1\pi^*)^1$	$2\frac{1}{2}$
NF^-	$(1\sigma)^2(1\sigma^*)^2(2\sigma)^2(2\sigma^*)^2(1\pi)^4(3\sigma)^2(1\pi^*)^3$	$1\frac{1}{2}$

According to Hund's rule, two electrons in NF occupy different $1\pi^*$ orbitals with unpaired spins. Thus NF is paramagnetic. Both NF^+ and NF^- contain an odd number of electrons and, therefore, an unpaired electron. Thus both NF^+ and NF^- are paramagnetic.

12-54. The ground-state electron configuration of the cyanide ion, CN^- is $(1\sigma)^2(1\sigma^*)^2(2\sigma)^2(2\sigma^*)^2(1\pi)^4(3\sigma)^2$. The bond order of CN^- is 3. The Lewis formula for CN^- is $\ominus$:C≡N: . Both molecular orbital theory and the Lewis formula predict that there is a triple bond in CN^-.

CHAPTER 13

SOLUTIONS TO THE EVEN-NUMBERED PROBLEMS

13-2. The amount of heat q_p required to vaporize 100 μg of einsteinium is

$$q_p = n\Delta\overline{H}_{vap}$$

$$= (100\ \mu g)\left(\frac{1\ g}{10^6\ \mu g}\right)\left(\frac{1\ mol}{252\ g}\right)(128\ kJ\cdot mol^{-1})$$

$$= 5.08 \times 10^{-5}\ kJ = 5.08 \times 10^{-2}\ J$$

13-4. The number of moles in 100.0 g of ethyl alcohol is

$$n = (100.0\ g)\left(\frac{1\ mol}{46.07\ g}\right) = 2.171\ mol$$

If 87.9 kJ are required to vaporize 2.171 mol, then

$$\Delta\overline{H}_{vap} = \frac{87.9\ kJ}{2.171\ mol} = 40.5\ kJ\cdot mol^{-1}$$

13-6. The number of moles of water in 16 one-ounce ice cubes is

$$n = (16\ \text{ice cubes})\left(\frac{1\ oz}{1\ \text{ice cube}}\right)\left(\frac{28\ g}{1\ oz}\right)\left(\frac{1\ mol}{18.02\ g}\right)$$

$$= 24.86\ mol$$

The heat *released* when the water is cooled from 18°C to 0°C is

$$q_n = n\overline{C}_p\Delta T \qquad (13\text{-}1)$$

$$= (24.86\ mol)(75.3\ J\cdot K^{-1}\cdot mol^{-1})(-18\ K)$$

$$= -33700\ J = -33.7\ kJ$$

where $\overline{C}_p$ for water is 75.3 $J\cdot K^{-1}\cdot mol^{-1}$ (Example 13-1). The heat *released* when the water is converted into ice is

$$q_p = n(-\Delta H_{fus})$$

$$= (24.86\ mol)(-6.01\ kJ\cdot mol^{-1})$$

$$= -149\ kJ$$

The total heat released is

$$q_p = 33.7 \text{ kJ} + 149 \text{ kJ} = 183 \text{ kJ}$$

The mass of Freon-12 that is vaporized by absorbing 183 kJ is found by

$$183 \text{ kJ} = (155 \text{ J}\cdot\text{g}^{-1})\left(\frac{1 \text{ kJ}}{1000 \text{ J}}\right) \text{(mass of Freon)}$$

$$183 \text{ kJ} = (0.155 \text{ kJ}\cdot\text{g}^{-1})\text{(mass of Freon)}$$

$$\text{mass of Freon} = \frac{183 \text{ kJ}}{0.155 \text{ kJ}\cdot\text{g}^{-1}} = 1180 \text{ g} = 1.18 \text{ kg}$$

13-8. The number of moles in 7.0 g of ice is

$$n = (7.0 \text{ g})\left(\frac{1 \text{ mol}}{18.02 \text{ g}}\right) = 0.388 \text{ mol}$$

The heat required to raise the temperature of ice from $-10^{\circ}C$ to $0^{\circ}C$ is

$$\begin{aligned} q_p &= n\overline{C}_P(T_2 - T_1) \\ &= (0.388 \text{ mol})(37.7 \text{ J}\cdot\text{K}^{-1}\cdot\text{mol}^{-1})(10 \text{ K}) \\ &= 150 \text{ J} \end{aligned}$$

where $\overline{C}_P$ for ice is $37.7 \text{ J}\cdot\text{K}^{-1}\cdot\text{mol}^{-1}$ (Example 13-1).

The heat absorbed when the ice melts is

$$\begin{aligned} q_p &= n\Delta\overline{H}_{fus} \\ &= (0.388 \text{ mol})(6.01 \text{ kJ}\cdot\text{mol}^{-1}) \\ &= 2.3 \text{ kJ} = 230 \text{ J} \end{aligned}$$

The heat required to raise the temperature of water from $0^{\circ}C$ to $52^{\circ}C$ is

$$q_p = n\bar{C}_p(T_2 - T_1)$$
$$= (0.388 \text{ mol})(75.3 \text{ J·K}^{-1}\text{·mol}^{-1})(52 \text{ K})$$
$$= 1520 \text{ J}$$

The total heat required is

$$q_p = 150 \text{ J} + 2300 \text{ J} + 1520 \text{ J}$$
$$= 3970 \text{ J} = 4.0 \text{ kJ}$$

13-10. Your body uses energy to melt the snow. It requires

$$q_p = n\Delta\bar{H}_{fus}$$
$$= (1.00 \text{ g})\left(\frac{1 \text{ mol}}{18.02 \text{ g}}\right)(6.01 \text{ kJ·mol}^{-1})$$
$$= 0.334 \text{ kJ}$$

to melt one gram of snow if the temperature of the snow is $0°C$.

13-12. The number of moles in 10.0 g of water is

$$(10.0 \text{ g})\left(\frac{1 \text{ mol}}{18.02}\right) = 0.555 \text{ mol}$$

The heat required to raise the temperature of 10.0g of water from 50°C to 100°C is given by

$$q_p = n\bar{C}_p\Delta T$$
$$= (0.555 \text{ mol})(75.3 \text{ J·K}^{-1}\text{·mol}^{-1})(50 \text{ K})$$
$$= 2090 \text{ J}$$

The heat required to vaporize 10.0 g of water is given by

$$q_p = n\Delta\bar{H}_{vap}$$
$$= (0.555 \text{ mol})(40.7 \text{ kJ·mol}^{-1})$$
$$= 22.6 \text{ kJ} = 22{,}600 \text{ J}$$

It would take longer to vaporize the water.

13-14. The boiling point of acetone is 56.2°C, and the temperature remains constant during the vaporization of the acetone. The heat required to vaporize the acetone is

$$q_p = (500.0\ \text{J}\cdot\text{min}^{-1})(55.0\ \text{min}) = 27{,}500\ \text{J}$$

The number of moles in 50.0 g of acetone is

$$n = (50.0\ \text{g})\left(\frac{1\ \text{mol}}{58.08\ \text{g}}\right) = 0.861\ \text{mol}$$

and so

$$\Delta\overline{H}_{vap} = \frac{27500\ \text{J}}{0.861\ \text{mol}} = 31900\ \text{J}\cdot\text{mol}^{-1} = 31.9\ \text{kJ}\cdot\text{mol}^{-1}$$

13-16. The only ionic compound is NaCl. Thus we predict that NaCl has the highest boiling point. Ammonia, NH_3, is a polar molecule and so it is hydrogen bonded. Both Ar and Kr are nonpolar; thus we predict that the boiling point of NH_3 is greater than those of Ar and Kr. The noble gas, Kr, has a greater mass and a greater number of electrons than Ar. Therefore, the predicted order is

$$T_b[\text{Ar}] < T_b[\text{Kr}] < T_b[\text{NH}_3] < T_b[\text{NaCl}]$$

13-18. The molecule, H_2, is nonpolar. The molecule, HF, is polar, and there is hydrogen bonding between HF molecules. The molecule, CH_4, is nonpolar. The molecule, CH_3OH, is polar, and there is hydrogen-bonding between molecules. The molecules HF and CH_3OH involve hydrogen bonding.

13-20. All four molecules are nonpolar molecules. The molar enthalpy of vaporization of non-polar molecules depends upon the molecular mass. The value of $\Delta\overline{H}_{vap}$ increases with increasing molecular mass. Therefore, the order of the molar enthalpies of vaporization is

$$\Delta\overline{H}_{vap}[\text{CH}_4] < \Delta\overline{H}_{vap}[\text{CCl}_4] < \Delta\overline{H}_{vap}[\text{SiCl}_4] < \Delta\overline{H}_{vap}[\text{SiBr}_4]$$

13-22. $$\Delta\overline{H}_{vap} = (85\ J\cdot K^{-1}\cdot mol^{-1})T_b$$

$$= (85\ J\cdot K^{-1}\cdot mol^{-1})(373\ K)$$

$$= 31700\ J\cdot mol^{-1} = 31.7\ kJ\cdot mol^{-1}$$

The actual value of $\Delta\overline{H}_{vap}$ for water is 40.7 $kJ\cdot mol^{-1}$. Water is strongly hydrogen-bonded, and so has strong specific intermolecular interactions. Therefore, Trouton's rule does not apply to water.

13-24. The heat absorbed by the sublimation of 1.00 mole of $CO_2(s)$ is

$$q_p = (1.00\ mol)(25.2\ kJ\cdot mol^{-1})$$

The heat absorbed by the sublimation of $CO_2(s)$ is the heat evolved by the freezing of water. The heat evolved by the freezing of n moles of water is

$$q_p = (n)(6.01 kJ\cdot mol^{-1}) = 25.2\ kJ$$

or solving for n

$$n = \frac{25.2\ kJ}{6.01 kJ\cdot mol^{-1}} = 4.19\ mol$$

13-26. The energy absorbed by an area 1.0m^2 in one day is 8.1 MJ = 8100 kJ. The enthalpy of sublimation of ice is 6.01 $kJ\cdot mol^{-1}$ + 40.7 $kJ\cdot mol^{-1}$ = 46.7 $kJ\cdot mol^{-1}$. The amount of ice that sublimes is given by

$$q_p = (n)(46.7\ kJ\cdot mol^{-1}) = 8100\ kJ$$

or

$$n = \frac{8100\ kJ}{46.7\ kJ\cdot mol^{-1}} = 173\ mol$$

The mass of water in 173 mol is

$$m = (173\ \text{mol})\left(\frac{18.02\ \text{g}}{1\ \text{mol}}\right) = 3120\ \text{g}$$

The volume occupied by 3120 g of ice is obtained from

$$d = \frac{m}{V} \qquad \text{or} \qquad V = \frac{m}{d}$$

$$V = \frac{3120\ \text{g}}{0.917\ \text{g}\cdot\text{cm}^{-3}} = 3400\ \text{cm}^3$$

The depth of ice melted in an area 1m^2 is

$$\text{depth} = \left(\frac{3400\ \text{cm}^3}{1\text{m}^2}\right)\left(\frac{1\text{m}^2}{10^4\text{cm}^2}\right)$$

$$= 0.340\ \text{cm}$$

13-28. From Figure 13-18, the boiling point of ethyl alcohol at 2 atm is around 370 K.

13-30. From Table 13-2, the equilibrium vapor pressure of water at 25^0C is 23.8 torr or 0.0313 atm. We can find the concentration of water using the ideal gas law

$$\frac{n}{V} = \frac{P}{RT}$$

$$M = \frac{n}{V} = \frac{0.0313\ \text{atm}}{(0.08206\ \text{L}\cdot\text{atm}\cdot\text{K}^{-1}\cdot\text{mol}^{-1})(298\ \text{K})}$$

$$= 0.00128\ \text{mol}\cdot\text{L}^{-1}$$

13-32. Using Figure 13-18, the equilibrium vapor pressure of water at 93°C is 0.74 atm. The normal atmospheric pressure in Mexico City is 0.74 atm or 560 torr.

13-34. The boiling point of water was used to determine the normal atmospheric pressure of Lhasa. Plots of normal atmospheric pressure versus altitude were used to find the altitude corresponding to the atmospheric pressure of Lhasa.

13-36. The equilibrium vapor pressure of water at 10°C is 9.2 torr or 0.0121 atm (Table 13-2). We can find the partial pressure of water from the definition of relative humidity

$$\text{relative humidity} = \frac{P_{H_2O}}{P^0_{H_2O}} \times 100$$

$$P_{H_2O} = \frac{(50)(9.2\text{ torr})}{100} = 4.6\text{ torr}$$

The equilibrium vapor pressure of water at 25°C is 23.8 torr. The partial pressure of water is 4.6 torr, and so the relative humidity at 25°C is

$$\text{relative humidity} = \frac{4.6\text{ torr}}{238\text{ torr}} \times 100 = 19\%$$

13-38. The air near the surface of the glass is colder than the air in the room, and so is able to sustain less water vapor.

13-40. The surfactant lowers the surface tension of water, allowing the water to spread out, or "sheet".

13-42. The volume is maintained when a drop is dispersed into smaller drops. The volume of a drop of 3.0 mm radius is

$$V = \frac{4\pi R^3}{3} = \frac{4}{3}\pi(3.0 \text{ mm})^3 = 113 \text{ mm}^3$$

The volume of a drop 3.0×10^{-3}mm radius is

$$V = \frac{4}{3}\pi(3.0 \times 10^{-3} \text{ mm})^3 = 113 \times 10^{-9} \text{ mm}^3$$

We can now find how many smaller drops are formed from the large drop.

V(large drop) = V(total small drops) = (number of drops) (V of small drops)

$$113 \text{ mm}^3 = (\text{number of drops})(113 \times 10^{-9} \text{ mm}^3)$$

$$\text{number of drops} = \frac{113 \text{ mm}^3}{113 \times 10^{-9} \text{ mm}^3} = 1.00 \times 10^9 \text{ drops}$$

The surface area of the large drop is

$$A = 4\pi R^2 = 4\pi(3.0 \text{ mm})^2 = 113 \text{ mm}^2$$

The energy of the surface is given by

$$\text{surface tension} = \frac{\text{energy}}{\text{area}}$$

$$\text{energy} = \text{surface tension} \times \text{area}$$

$$= (72 \text{ mJ}\cdot\text{m}^{-2})(113 \text{ mm}^2)\left(\frac{1\text{m}^2}{10^6 \text{ mm}^2}\right)$$

$$= 8.1 \times 10^{-3} \text{ mJ}$$

The surface area of one small drop is

$$A = 4\pi R^2 = 4\pi(3.0 \times 10^{-3} \text{ mm})^2$$

$$= 1.13 \times 10^{-4} \text{ mm}^2$$

The total surface area of the small drops is

$$A = (\text{number of drops})(\text{area of 1 drop})$$

$$A = (1.00 \times 10^{9} \text{ drops})(1.13 \times 10^{-4} \text{ mm}^2)$$

$$= 1.13 \times 10^{5} \text{ mm}^2$$

The energy of the total surface is

$$\text{energy} = (72 \text{ mJ·m}^{-2})(1.13 \times 10^{5} \text{ mm}^2)\left(\frac{1 \text{ m}^2}{10^6 \text{ mm}^2}\right)$$

$$= 8.1 \text{ mJ}$$

The energy required to disperse the drop into smaller drops is

$$\text{energy} = 8.1 \text{ mJ} - 8.1 \times 10^{-3} \text{ mJ} = 8.1 \text{ mJ}$$

13-44. The molar volume of tantalum is given by

$$V = \frac{\text{molar mass}}{\text{density}} = \frac{180.9 \text{ g·mol}^{-1}}{16.69 \text{ g·cm}^{-3}} = 10.84 \text{ cm}^3\text{·mol}^{-1}$$

There are two tantalum atoms per unit cell (Example 13-8 and Figure 13-30b) and so the number of unit cells per mole of tantalum is

$$\text{unit cells per mole} = \frac{6.022 \times 10^{23} \text{atom·mol}^{-1}}{\frac{2 \text{ atom}}{\text{unit cell}}} = 3.011 \times 10^{23} \frac{\text{unit cell}}{\text{mol}}$$

The volume of a unit cell is

$$v = \frac{10.84 \text{ cm}^3\text{·mol}^{-1}}{3.011 \times 10^{23} \frac{\text{unit cell}}{\text{mol}}} = 3.600 \times 10^{-23} \frac{\text{cm}^3}{\text{unit cell}}$$

The length of an edge of a unit cell is

$$l = (3.600 \times 10^{-23} \text{cm}^3)^{1/3} = 3.302 \times 10^{-8} \text{ cm}$$

$$= 330.2 \text{ pm}$$

13-46. The volume of a unit cell of lithium is

$$v = (351\ \text{pm})^3 = (3.51 \times 10^{-8}\text{cm})^3 = 4.324 \times 10^{-23}\ \text{cm}^3$$

There are two lithium atoms per unit cell (Problem 13-44) and the mass of a unit cell is

$$m = \left(\frac{2\ \text{atom}}{\text{unit cell}}\right)\left(\frac{6.941\ \text{g}\cdot\text{mol}^{-1}}{6.022 \times 10^{23}\ \text{atom}\cdot\text{mol}^{-1}}\right) = 2.305 \times 10^{-23}\ \frac{\text{g}}{\text{unit cell}}$$

The density is

$$d = \frac{m}{v} = \frac{2.305 \times 10^{-23}\ \frac{\text{g}}{\text{unit cell}}}{4.324 \times 10^{-23}\ \frac{\text{cm}^3}{\text{unit cell}}} = 0.533\ \text{g}\cdot\text{cm}^{-3}$$

13-48. The molar volume of chromium is

$$\overline{V} = \frac{\text{molar mass}}{\text{density}} = \frac{52.00\ \text{g}\cdot\text{mol}^{-1}}{720\ \text{g}\cdot\text{cm}^{-3}} = 7.22\ \text{cm}^3\cdot\text{mol}^{-1}$$

The volume of a unit cell is

$$v = (288.4\ \text{pm})^3 = (2.884 \times 10^{-8}\text{cm})^3 = 2.399 \times 10^{-23}\ \text{cm}^3$$

The number of unit cells in a molar volume of chromium is

$$\text{unit cell per mole} = \frac{7.222\ \text{cm}^3\cdot\text{mol}^{-1}}{2.399 \times 10^{-23}\ \frac{\text{cm}^3}{\text{unit cell}}} = 3.01 \times 10^{23}\ \frac{\text{unit cell}}{\text{mole}}$$

There are two chromium atoms per unit cell, and so

$$\text{Avogadro's number} = \left(\frac{2\ \text{atom}}{\text{unit cell}}\right)\left(\frac{3.01 \times 10^{23}\ \text{unit cell}}{\text{mol}}\right)$$

$$= 6.02 \times 10^{23}\ \text{atom}\cdot\text{mol}^{-1}$$

13-50. The molar volume of CsBr is

$$\overline{V} = \frac{\text{molar mass}}{\text{density}} = \frac{212.8\ \text{g}\cdot\text{mol}^{-1}}{4.44\ \text{g}\cdot\text{cm}^{-3}} = 47.9\ \text{cm}^3\cdot\text{mol}^{-1}$$

There is one CsBr formula unit per unit cell (Figure 13-34), so there are 6.022×10^{23} unit cells per mole. The volume of a unit cell is

$$v = \frac{47.93\,\text{cm}^3\cdot\text{mol}^{-1}}{6.022 \times 10^{23}\ \frac{\text{unit cell}}{\text{mol}}} = 7.95 \times 10^{-23}\ \frac{\text{cm}^3}{\text{unit cell}}$$

The length of an edge of a unit cell is

$$l = (7.959 \times 10^{-23}\ \text{cm}^3)^{1/3} = 4.30 \times 10^{-8}\ \text{cm}$$

From Figure 13-33 or 13-34 we see that the nearest-neighbor distance is one half of the main diagonal of the unit cell. The length of the main diagonal is $\sqrt{3}\,l$, and so the nearest neighbor distance is $\sqrt{3}\,l/2$, or 3.72×10^{-8} cm, or 372 pm.

13-52. The molar volume of KBr is

$$\overline{V} = \frac{\text{molar mass}}{\text{density}} = \frac{119.00\ \text{g}\cdot\text{mol}^{-1}}{2.75\ \text{g}\cdot\text{cm}^{-3}} = 43.3\ \text{cm}^3\cdot\text{mol}^{-1}$$

The volume of a unit cell is

$$v = (654\ \text{pm})^3 = (6.54 \times 10^{-8}\ \text{cm})^3 = 2.80 \times 10^{-22}\ \text{cm}^3$$

The number of unit cells per mole is

$$\frac{\overline{V}}{v} = \frac{43.3\ \text{cm}^3\cdot\text{mol}^{-1}}{2.80 \times 10^{-22}\ \frac{\text{cm}^3}{\text{unit cell}}} = 1.55 \times 10^{23}\ \frac{\text{unit cell}}{\text{mol}}$$

This result says that there are four formula units of KBr in a unit cell, and so the unit cell must be the NaCl-type.

13-54. The molecular volume of sodium chloride is

$$\overline{V} = \frac{\text{molar mass}}{\text{density}} = \frac{58.44\ \text{g}\cdot\text{mol}^{-1}}{2.163\ \text{g}\cdot\text{cm}^{-3}} = 27.02\ \text{cm}^3\cdot\text{mol}^{-1}$$

From Figure 13-33 a, we can see that the length of one side of the unit cell is 282 pm + 282 pm = 564 pm. The volume of the unit cell is

$$v = (564\ \text{pm})^3 = (5.64 \times 10^{-8}\text{cm})^3 = 1.794 \times 10^{-22}\ \text{cm}^3$$

The number of unit cells per mole is

$$\text{unit cells per mole} = \frac{\overline{V}}{v} = \frac{27.02\ \text{cm}^3\cdot\text{mol}^{-1}}{1.794 \times 10^{-22}\ \frac{\text{cm}^3}{\text{unit cell}}}$$

$$= 1.506 \times 10^{23}\ \text{unit cell}\cdot\text{mol}^{-1}$$

From Example 13-11,we see that there are four sodium ions and four chloride ions in a unit cell of sodium chloride. Thus the number of formula units per mole (Avogadro's number) of sodium chloride is given by

$$\text{Avogadro's number} = \left(1.506 \times 10^{23}\ \frac{\text{unit cell}}{\text{mol}}\right)\left(\frac{4\ \text{formula units}}{\text{unit cell}}\right)$$

$$= 6.02 \times 10^{23}\ \text{formula units}\cdot\text{mol}^{-1}$$

13-56. a) gas

b) liquid

c) gas

d) solid

13-58. The phase diagram (not to scale) of oxygen is

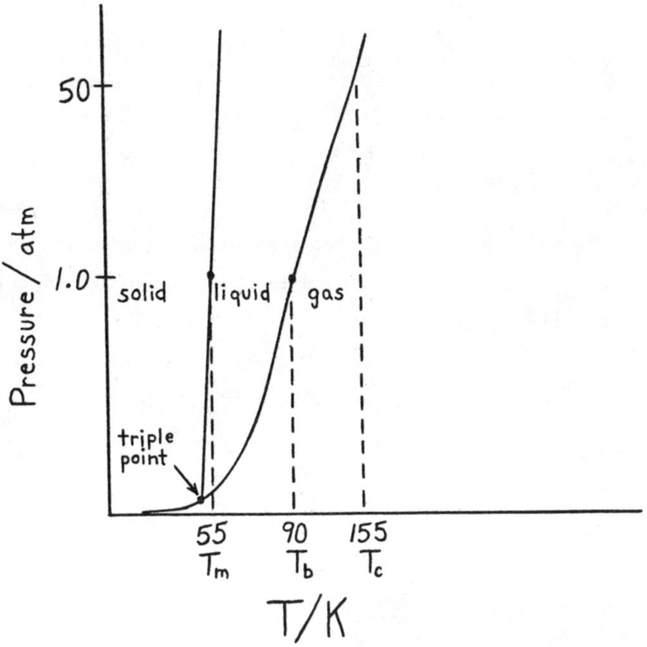

The melting point curve slopes to the right so that the melting point increases with pressure. Thus solid oxygen does not melt under an applied pressure.

13-60. The Clausius-Clapeyron equation is

$$\log \frac{P_2}{P_1} = \frac{\Delta\overline{H}_{vap}}{2.3\ R}\left(\frac{1}{T_1} - \frac{1}{T_2}\right)$$

We let P_2 = 92.5 torr, P_1 = 31.8 torr, T_2 = 323 K, and T_1 = 303 K, and write

$$\log\left(\frac{92.5\ \text{torr}}{31.8\ \text{torr}}\right) = \frac{\Delta\overline{H}_{vap}}{(2.3)(8.314\ \text{J}\cdot\text{K}^{-1}\cdot\text{mol}^{-1})}\left(\frac{1}{303\ \text{K}} - \frac{1}{323\ \text{K}}\right)$$

$$\log 2.91 = \frac{1.07 \times 10^{-5}\ \text{J}^{-1}\cdot\text{mol}\Delta\overline{H}_{vap}}{19.12\ \text{J}\cdot\text{K}^{-1}\cdot\text{mol}^{-1}}$$

$$0.464 = (1.07 \times 10^{-5}\ J^{-1}\cdot mol)\Delta\overline{H}_{vap}$$

$$\Delta\overline{H}_{vap} = \frac{0.464}{1.07 \times 10^{-5}\ J^{-1}\cdot mol} = 4.34 \times 10^{4}\ J\cdot mol^{-1}$$

$$= 43.4\ kJ\cdot mol^{-1}$$

The value of $\Delta\overline{H}_{vap}$ varies some with temperature. The value of $\Delta\overline{H}_{vap}$ (40.7 $kJ\cdot mol^{-1}$) used in the chapter is an average over the liquid range.

13-62. The vapor pressure of the solid is equal to the vapor pressure of the liquid at the triple point. Thus we write

$$10.560 + \frac{-1640}{T_t} = 7.769 + \frac{-1159}{T_t}$$

$$2.791 = \frac{481}{T_t}$$

$$T_t = \frac{481}{2.791} = 172\ K$$

The triple point pressure is given by

$$\log P_t = 7.769 + \frac{-1159}{172}$$

$$= 7.769 + (-6.738) = 1.03$$

or

$$\log P_t = 10.560 + \frac{-1640}{172} = 1.03$$

and so

$$P_t = 10^{1.03} = 10.7\ torr$$

CHAPTER 14

SOLUTIONS TO THE EVEN-NUMBERED PROBLEMS

14-2. In 100 g of solution, the mass of formaldehyde is 40 g, the mass of methyl alcohol is 10 g, and the mass of water is 50 g. The mole fractions are given by

$$x_i = \frac{n_i}{n_1 + n_2 + n_3} \qquad \text{where } i = 1, 2, \text{ or } 3$$

$$x_{H_2CO} = \frac{(40\text{ g})\left(\dfrac{1\text{ mol }H_2CO}{30.03\text{ g }H_2CO}\right)}{(40\text{ g})\left(\dfrac{1\text{ mol }H_2CO}{30.03\text{ g }H_2CO}\right) + (10\text{ g})\left(\dfrac{1\text{ mol }CH_3OH}{32.04\text{ g }CH_3OH}\right) + (50\text{ g})\left(\dfrac{1\text{ mol }H_2O}{18.02\text{ g }H_2O}\right)}$$

$$= \frac{1.332\text{ mol}}{1.332\text{ mol} + 0.312\text{ mol} + 2.775\text{ mol}} = \frac{1.332\text{ mol}}{4.419\text{ mol}}$$

$$= 0.30$$

$$x_{CH_3OH} = \frac{(10\text{ g})\left(\dfrac{1\text{ mol }CH_3OH}{32.04\text{ g }CH_3OH}\right)}{4.419\text{ mol}}$$

$$= 0.071$$

$$x_{H_2O} = \frac{(50\text{ g})\left(\dfrac{1\text{ mol }H_2O}{18.02\text{ g }H_2O}\right)}{4.419\text{ mol}}$$

$$= 0.63$$

14-4. The mole fraction of formaldehyde is given by

$$x_{H_2CO} = \frac{n_{H_2CO}}{n_{H_2CO} + n_{H_2O}} = 0.186$$

The mole fraction of water is given by

$$X_{H_2O} = \frac{n_{H_2O}}{n_{H_2CO} + n_{H_2O}}$$

but

$$X_{H_2CO} + X_{H_2O} = \frac{n_{H_2CO} + n_{H_2O}}{n_{H_2CO} + n_{H_2O}} = 1$$

thus

$$X_{H_2O} = 1.000 - X_{H_2CO} = 1.000 - 0.186 = 0.814$$

14-6. The mass of water is

$$m = dV = (0.993\ g \cdot mL^{-1})(100.0\ mL) = 99.3\ g$$

The mole fraction of water in the solution is

$$X_{H_2O} = \frac{n_{H_2O}}{n_{H_2O}+n_{glycerin}} = \frac{(99.3\ g)\left(\frac{1\ mol\ H_2O}{18.02\ g\ H_2O}\right)}{(99.3\ g)\left(\frac{1\ mol\ H_2O}{18.02\ g\ H_2O}\right)+(50.0\ g)\left(\frac{1\ mol\ glycerin}{92.09\ g\ glycerin}\right)}$$

$$= 0.910$$

Raoult's law is

$$P_{H_2O} = X_{H_2O}P^0_{H_2O} = (0.910)(47.1\ torr) = 42.9\ torr$$

The vapor pressure lowering is

$$P^0_{H_2O} - P_{H_2O} = 47.1\ torr - 42.9\ torr = 4.2\ torr$$

14-8. The mole fraction of water in the solution is

$$X_{H_2O} = \frac{n_{H_2O}}{n_{H_2O} + n_{C_2H_6O_2}} = \frac{(100\text{ g})\left(\frac{1\text{ mol }H_2O}{18.02\text{ g }H_2O}\right)}{(100\text{ g})\left(\frac{1\text{ mol }H_2O}{18.02\text{ g }H_2O}\right)(50.0\text{ g})\left(\frac{1\text{ mol }C_2H_6O_2}{62.07\text{ g }C_2H_6O_2}\right)}$$

$$= 0.873$$

From Raoult's law we compute

$$P_{H_2O} = X_{H_2O}P^0_{H_2O} = (0.873)(1.00\text{ atm}) = 0.873\text{ atm}$$

The vapor pressure lowering is

$$P^0_{H_2O} - P_{H_2O} = 1.00\text{ atm} - 0.873\text{ atm} = 0.13\text{ atm}$$

14-10. When $X_2 = 1$, the mole fraction of the solvent is zero. Therefore, the vapor pressure, P_1 is zero. We have

$$P_1^0 - P_1 = P_1^0 - 0 = P_1^0 = cX_2 = (c)(1)$$

We have then that

$$P_1^0 = c$$

14-12. The mole fraction of benzene is

$$X_B = \frac{n_{benzene}}{n_{benzene} + n_{toluene}} = \frac{(50.0\text{ g})\left(\frac{1\text{ mol benzene}}{78.11\text{ g benzene}}\right)}{(50.0\text{ g})\left(\frac{1\text{ mol benzene}}{78.11\text{ g benzene}}\right) + (50.0\text{ g})\left(\frac{1\text{ mol toluene}}{92.13\text{ g toluene}}\right)}$$

$$= 0.541$$

The partial pressure of benzene is

$$P_B = X_BP_B^0 = (0.541)(380\text{ torr}) = 206\text{ torr}$$

The mole fraction of toluene is

$$X_T = \frac{n_{toluene}}{n_{toluene} + n_{benzene}} = \frac{(50.0\text{ g})\left(\dfrac{1\text{ mol toluene}}{92.13\text{ g toluene}}\right)}{(50.0\text{ g})\left(\dfrac{1\text{ mol benzene}}{78.11\text{ g benzene}}\right) + (50.0\text{ g})\left(\dfrac{1\text{ mol toluene}}{92.13\text{ toluene}}\right)}$$

$$= 0.459$$

The partial pressure of toluene is

$$P_T = X_T P_T^0 = (0.459)(140\text{ torr}) = 64.3\text{ torr}$$

The total pressure is given by (Problem 14-11)

$$P_{total} = P_2^0 + X_1(P_1^0 - P_2^0)$$

Let 2 = toluene (T) and 1 = benzene (B), then we have

$$P_{total} = P_T^0 + X_B(P_B^0 - P_T^0)$$

$$= 140\text{ torr} + (0.541)(380\text{ torr} - 140\text{ torr})$$

$$= 270\text{ torr}$$

Also

$$P_{total} = P_T + P_B = 206\text{ torr} + 64.3\text{ torr} = 270\text{ torr}$$

14-14. A 1.75 m solution contains 1.75 mol of $Ba(NO_3)_2$ in a 1000 g of water. The mass of 1.75 mol of $Ba(NO_3)_2$ is

$$m = (1.75\text{ mol})\left(\frac{261.3\text{ g }Ba(NO_3)_2}{1\text{ mol }Ba(NO_3)_2}\right) = 457\text{ g}$$

Dissolve 457 g of $Ba(NO_3)_2$ in 1000 g of water.

14-16. The number of moles of oxalic acid in 18.0 g is

$$n = (18.0\text{ g})\left(\frac{1\text{ mol } H_2C_2O_4}{90.04\text{ g } H_2C_2O_4}\right) = 0.200\text{ mol}$$

Molality is given by

$$\text{molality} = \frac{\text{moles of solute}}{\text{kilogram of solvent}}$$

$$0.050\text{ m} = \frac{0.200\text{ mol } H_2C_2O_4}{\text{kilograms of } H_2O}$$

$$\text{kilograms of } H_2O = \frac{0.200\text{ mol}}{0.050\text{ mol}\cdot\text{kg}^{-1}} = 4.0\text{ kg}$$

14-18. a) There is one solute particle per CH_3OH formula unit because CH_3OH does not dissociate in water. Thus the colligative molality is 1.0 m_c.

b) There are four ions per $Al(NO_3)_3$ formula unit because $Al(NO_3)_3$ dissociates into Al^{3+} and $3NO_3^-$ in water. Thus the colligative molality is 4.0 m_c.

c) There are three ions per $Fe(NO_3)_2$ formula unit because $Fe(NO_3)_2$ dissociates into Fe^{2+} and 2 NO_3^- in water. Thus the colligative molality is 3.0 m_c.

14-20. The mole fraction of water in the solution is

$$X_{H_2O} = \frac{n_{H_2O}}{n_{H_2O} + n_{solute}} = \frac{(1000\text{ g})\left(\frac{1\text{ mol } H_2O}{18.02\text{ g } H_2O}\right)}{(1000\text{ g})\left(\frac{1\text{ mol } H_2O}{18.02\text{ g } H_2O}\right) + 0.30\text{ mol}}$$

$$= 0.9946$$

The vapor pressure of water is

$$P_{H_2O} = X_{H_2O}P^0_{H_2O} = (0.9946)(47.1\text{ torr}) = 46.8\text{ torr}$$

14-22. Take 100 g of solution. The solution contains 24.0 g of K_2CrO_4 and 76.0 g of H_2O. The number of moles of K_2CrO_4 in 24.0 g is

$$n = (24.0\ g)\left(\frac{1\ mol\ K_2CrO_4}{194.20\ g\ K_2CrO_4}\right) = 0.1236\ mol$$

The molality of the solution is

$$\text{molality} = \frac{\text{moles of solute}}{\text{kilograms of solvent}}$$

$$= \frac{0.1236\ mol}{0.0760\ kg} = 1.63\ m$$

The volume of the solution is

$$V = \frac{m}{d} = \frac{100\ g}{1.21\ g\cdot mL^{-1}} = 82.6\ mL = 0.0826\ L$$

The molarity is

$$\text{molarity} = \frac{\text{moles of solute}}{\text{volume of solution}}$$

$$= \frac{0.1236\ mol}{0.0826\ L} = 1.50\ M$$

14-24. The colligative molality of a solute that raises the boiling point of water by $1^{\circ}C$ is (Table 14-4)

$$T_b - T_b^0 = 1.0\ K = K_b m_c = (0.52\ K\cdot m_c^{-1})(m_c)$$

$$m_c = \frac{1.0\ K}{0.52\ K\cdot m_c^{-1}} = 1.92\ m_c$$

The molality of the NaCl(aq) solution is $\frac{1.92}{2} = 0.96\ m$. There are 0.96 moles of NaCl dissolved in 1000 g of water. The mass of NaCl is

$$\text{mass} = (0.96\ mol)\left(\frac{58.44\ g\ NaCl}{1\ mol\ NaCl}\right)$$

$$= 56\ g$$

14-26. The boiling point elevation is (Table 14-4)

$$T_b - T_b^0 = K_b m_c = (0.52\ \text{K}\cdot m_c^{-1})(1.10\ m_c)$$

$$= 0.57\ \text{K} = 0.57^\circ\text{C}$$

The boiling point is

$$T_b = 100.00^\circ\text{C} + 0.57^\circ\text{C} = 100.57^\circ\text{C}$$

The mole fraction of water in a 1.10 m solution is

$$X_{H_2O} = \frac{n_{H_2O}}{n_{H_2O} + n_{solute}} = \frac{(1000\ \text{g})\left(\frac{1\ \text{mol}\ H_2O}{18.02\ \text{g}\ H_2O}\right)}{(1000\ \text{g})\left(\frac{1\ \text{mol}\ H_2O}{18.02\ \text{g}\ H_2O}\right) + 1.10\ \text{mol}}$$

$$= 0.9806$$

The vapor pressure is

$$P_{H_2O} = X_{H_2O} P^0_{H_2O} = (0.9806)(12.79\ \text{torr})$$

$$= 12.5\ \text{torr}$$

14-28. The molality of the solution is

$$\text{molality} = \frac{\text{moles of solute}}{\text{kilograms of solvent}}$$

$$= \frac{(50.0\ \text{g})\left(\frac{1\ \text{mol ethylene glycol}}{62.07\ \text{g ethylene glycol}}\right)}{0.0500\ \text{kg water}}$$

$$= 16.1\ \text{m}$$

$$m_c = 16.1\ m_c$$

The boiling point elevation is

$$T_b - T_b^0 = K_b m_c = (0.52\ K\cdot m_c^{-1})(16.1\ m_c)$$

$$= 8.4\ K = 8.4^\circ C$$

The boiling point is

$$T_b = 100.00^\circ C + 8.4^\circ C = 108.4^\circ C$$

14-30. The colligative molality of the solution is

$$m_c = 2\ m = (2)(0.15\ m) = 0.30\ m_c$$

The freezing point depression is

$$T_f^0 - T_f = K_f m_c = (1.86\ K\cdot m_c^{-1})(0.30\ m_c)$$

$$= 0.56\ K = 0.56^\circ C$$

The freezing point is

$$T_f = 0.00^\circ C - 0.56^\circ C = -0.56^\circ C$$

14-32. From the freezing-point-depression equation

$$T_f^0 - T_f = K_f m_c$$

we have

$$6.23\ K = (20.2\ K\cdot m_c^{-1})m_c$$

Thus

$$m_c = \frac{6.23\ K}{20.2\ K\cdot m_c^{-1}} = 0.308\ m_c$$

Because the mass is given, we have the correspondance

$$0.308\ \text{mol}\cdot\text{kg}^{-1} \approx \frac{1.00\ \text{g quinine}}{0.0100\ \text{kg cyclohexane}} = 100\ \text{g}\cdot\text{kg}^{-1}$$

and therefore

$$0.308\ \text{mol} \approx 100\ \text{g}$$

Dividing both sides by 0.308, we have that

$$1.00\ \text{mol} \approx 325\ \text{g}$$

Therefore, the molecular mass of quinine is 325.

14-34. We can find the molality of the solution from the freezing point depression

$$T_f^0 - T_f = K_f m_c$$

$$1.22\ \text{K} = (40.0\ \text{K}\cdot m_c^{-1})m_c$$

$$m_c = \frac{1.22\ \text{K}}{40.0\ \text{K}\cdot m_c^{-1}} = 0.0305\ m_c$$

We can calculate the solute mass per kilogram of solvent from the molality because the mass of PCB is given. We have the correspondance

$$0.0305\ \text{mol}\cdot\text{kg}^{-1} \approx \frac{0.100\ \text{g PCB}}{0.0100\ \text{kg camphor}} = 10.0\ \text{g}\cdot\text{kg}^{-1}$$

and therefore

$$0.0305\ \text{mol} \approx 10.0\ \text{g}$$

Dividing both sides by 0.0305 mol, we have that

$$1.00\ \text{mol} \approx 328\ \text{g}$$

The molecular mass of the PCB is 328.

14-36. Take 100 mL of wine. There are 12 mL of ethyl alcohol and 88 mL of water. The molality of ethyl alcohol in wine is

$$m = \frac{(12\ \text{mL})(0.79\ \text{g}\cdot\text{mL}^{-1})\left(\frac{1\ \text{mol ethyl alcohol}}{46.07\ \text{g ethyl alcohol}}\right)}{(88\ \text{mL})(1.00\,\text{g}\cdot\text{mL}^{-1})\left(\frac{1\ \text{kg}}{1000\ \text{g}}\right)}$$

$$= 2.34\ \text{m}$$

The freezing point depression is

$$T_f^0 - T_f = K_f m_c = (1.86\ \text{K}\cdot m_c^{-1})(2.34\ m_c)$$

$$= 4.4\ \text{K} = 4.4^\circ\text{C}$$

The freezing point is

$$T_f = 0.00^\circ\text{C} - 4.4^\circ\text{C} = -4.4^\circ\text{C}$$

Take 100 mL of vodka. There are 40 mL of ethyl alcohol and 60 mL of water. The molality of the solution is

$$m = \frac{(40\ \text{mL})(0.79\ \text{g}\cdot\text{mL}^{-1})\left(\frac{1\ \text{mol}}{46.07\ \text{g}}\right)}{(60\ \text{mL})(1.00\text{g}\cdot\text{mL}^{-1})\left(\frac{1\ \text{kg}}{1000\ \text{g}}\right)}$$

$$= 11.4\ \text{m}$$

The freezing point depression is

$$T_f^0 - T_f = K_f m_c = (1.86\ \text{K}\cdot m_c^{-1})(11.4\ m_c)$$

$$= 21\ \text{K} = 21^\circ\text{C}$$

The freezing point is

$$T_f = 0.00^\circ\text{C} - 21^\circ\text{C} = -21^\circ\text{C}$$

14-38. We can find the molality of the solution from the freezing-point depression

$$T_f^0 - T_f = K_f m_c$$

$$6.5^\circ C - (-1.95^\circ C) = (20.2\ K \cdot m_c^{-1}) m_c$$

$$m_c = \frac{8.5\ K}{20.2\ K \cdot m_c^{-1}} = 0.421\ m_c$$

We can calculate the mass of menthol in 1.00 kilogram of cyclohexane. We have the correspondance

$$0.421\ mol \cdot kg^{-1} \approx \frac{6.54\ g\ menthol}{0.100\ kg\ cyclohexane} = 65.4\ g \cdot kg^{-1}$$

and therefore

$$0.421\ mol \approx 65.4\ g$$

Dividing both sides by 0.421, we have that

$$1.00\ mol \approx 155\ g$$

The formula mass of menthol is 155.

14-40. We can find the colligative molality from the freezing-point depression

$$T_f^0 - T_f = K_f m_c$$

$$1.41\ K = (1.86\ K \cdot m_c^{-1}) m_c$$

$$m_c = \frac{1.41\ K}{1.86\ K \cdot m_c^{-1}} = 0.758\ m_c$$

The colligative molality is 3 times the molality (3 x 0.25 m = 0.75 m_c). Thus K_2HgI_4 in water yields 3 ions per formula unit

$$K_2HgI_4(s) \xrightarrow[H_2O(l)]{} 2K^+(aq) + HgI_4^{2-}(aq)$$

14-42. The osmotic pressure is given by

$$\pi = RTM_c$$

Thus

$$\pi = (0.0821\ L\cdot atm\cdot K^{-1}\cdot mol^{-1})(310\ K)(1.10\ mol\cdot L^{-1})$$

$$= 28.0\ atm$$

14-44. The concentration of pepsin in the aqueous solution is

$$M_c = \frac{\pi}{RT} = \frac{(0.213\ atm)}{(0.0821\ L\cdot atm\cdot K^{-1}\cdot mol^{-1})(298\ K)}$$

$$= 8.71 \times 10^{-3}\ mol\cdot L^{-1}$$

The molecular mass can be calculated from the concentration. We have the correspondance

$$8.71 \times 10^{-3}\ mol\cdot L^{-1} \approx \frac{3.00\ g}{0.0100\ L} = 3.00 \times 10^2\ g\cdot L^{-1}$$

Thus

$$8.71 \times 10^{-3}\ mol \approx 3.00 \times 10^2\ g$$

Dividing both sides by 8.71×10^{-3}, we have that

$$1.00\ mol \approx 3.44 \times 10^4\ g$$

The molecular mass of pepsin is about 34,400.

14-46. The concentration of the polymer in the aqueous solution is

$$M_c = \frac{\pi}{RT} = \frac{2.0 \times 10^{-2}\ \text{atm}}{(0.0821\ \text{L}\cdot\text{atm}\cdot\text{K}^{-1}\cdot\text{mol}^{-1})(293\ \text{K})}$$

$$= 8.31 \times 10^{-4}\ \text{mol}\cdot\text{L}^{-1}$$

We have that

$$8.31 \times 10^{-4}\ \text{mol}\cdot\text{L}^{-1} \mathrel{\hat{=}} \frac{2.0\ \text{g}}{0.0100\ \text{L}} = 200\ \text{g}\cdot\text{L}^{-1}$$

and thus

$$8.31 \times 10^{-4}\ \text{mol} \mathrel{\hat{=}} 200\ \text{g}$$

and therefore

$$1.00\ \text{mol} \mathrel{\hat{=}} 2.41 \times 10^{5}\ \text{g}$$

The formula mass is 2.41×10^5. The empirical formula is $CH_2C(CH_3)_2$. The formula mass per unit is 56.10. The number of units in the polymer is given by

$$x(56.10) = 2.41 \times 10^{5}$$

thus

$$x = \frac{2.41 \times 10^{5}}{56.10} = 4.30 \times 10^{3}$$

Therefore the number of subunits in the polymer is 4300.

14-48. The total colligative molarity is

$$M_c = M_c(\text{NaCl}) + M_c(\text{MgSO}_4)$$

$$= (2)(0.15\ \text{M}) + (2)(0.015\ \text{M})$$

$$= 0.33\ M_c$$

The osmotic pressure of the solution is

$$\pi = RTM_c = (0.0821\ \text{L·atm·K}^{-1}\text{·mol}^{-1})(298\ \text{K})(0.33\ \text{mol·L}^{-1})$$

$$= 8.1\ \text{atm}$$

Thus the minimum pressure is 8.1 atm.

14-50. The Henry's law constants are 1.6×10^3 atm·M^{-1} for N_2, 7.8×10^2 atm·M^{-1} for O_2, and 29 atm·M^{-1} for CO_2 at 25°C (Table 14-5). The concentration of N_2 is given by

$$M_{N_2} = \frac{P_{N_2}}{k_h} = \frac{1.0\ \text{atm}}{1.6 \times 10^3\ \text{atm·M}^{-1}} = 6.3 \times 10^{-4}\ \text{M}$$

The concentration of O_2 is

$$M_{O_2} = \frac{1.0\ \text{atm}}{7.8 \times 10^2\ \text{atm·M}^{-1}} = 1.3 \times 10^{-3}\ \text{M}$$

The concentration of CO_2 is

$$M_{CO_2} = \frac{1.0\ \text{atm}}{29\ \text{atm·M}^{-1}} = 3.4 \times 10^{-2}\ \text{M}$$

The CO_2 at 1.0 atm pressure has the highest concentration in water.

14-52. The partial pressure of O_2 is 0.20 atm. The concentration of O_2 is (Table 14-5)

$$M_{O_2} = \frac{0.20\ \text{atm}}{7.8 \times 10^2\ \text{atm·M}^{-1}} = 2.6 \times 10^{-4}\ \text{M}$$

The partial pressure of N_2 is 0.79 atm. The concentration of N_2 is

$$M_{N_2} = \frac{0.79\ \text{atm}}{1.6 \times 10^3\ \text{atm·M}^{-1}} = 4.9 \times 10^{-4}\ \text{M}$$

14-54. The increase in air pressure at 90 ft is

$$(90\ \text{ft})\left(\frac{1\ \text{atm}}{33\ \text{ft}}\right) = 2.7\ \text{atm}$$

The total pressure in the lungs at 90 ft is 3.7 atm. The partial pressure of O_2 is

$$P_{O_2} = (\text{fraction by volume of } O_2)P_{\text{total}}$$

$$0.20\ \text{atm} = (\text{fraction } O_2)(3.7\ \text{atm})$$

$$\text{fraction } O_2 \text{ by volume} = \frac{0.20\ \text{atm}}{3.7\ \text{atm}} = 5.4 \times 10^{-2}$$

The air in a diver's tank should be 5.4% oxygen by volume.

CHAPTER 15

SOLUTIONS TO THE EVEN-NUMBERED PROBLEMS

15-2. a) Each product concentration factor, raised to a power equal to its balancing coefficient, appears in the numerator of the K_c expression, and each reactant concentration factor, raised to a power equal to its balancing coefficient, appears in the denominator of the K_c expression. Pure solids and liquids do not appear in the K_c expression.

$$K_c = \frac{[CH_3OH]}{[CO][H_2]^2}$$

b) $K_c = [CO_2][H_2O]$

15-4. a)

$$K_c = \frac{[NO]^2}{[N_2][O_2]}$$

b)

$$K_c = [O_2]$$

15-6. a) $K_c = [CO_2][NH_3]^2$

b)

$$K_c = \frac{[CO]^2}{[CO_2]}$$

c)

$$K_c = \frac{[N_2O_4]}{[N_2][O_2]^2}$$

15-8. a) $K_p = P^2_{NH_3} P_{CO_2}$

b) $K_p = \dfrac{P^2_{CO}}{P_{CO_2}}$

c) $K_p = \dfrac{P_{N_2O_4}}{P_{N_2} P_{O_2}}$

15-10. $H_2SO_4(l) + 2NaCl(s) \longrightarrow 2HCl(g) + Na_2SO_4(s)$

$$K_c = [HCl]^2 \qquad K_p = P^2_{HCl}$$

15-12. We can write the equilibrium constant expression by applying the Law of Concentration Action to the reaction

$$K_c = \frac{[PCl_3][Cl_2]}{[PCl_5]}$$

We now set up a table of initial concentrations and equilibrium concentration.

Because PCl_5 has decomposed to PCl_3 and Cl_2, the concentration of PCl_3 and the concentration of Cl_2 are equal to the decrease in the concentration of PCl_5. Substituting in the values of the equilibrium concentrations in the K_c expression, we have

	$PCl_5(g)$ $\rightleftharpoons$	$PCl_3(g)$ +	$Cl_2(g)$
Initial Concentration	1.10 M	0	0
Equilibrium Concentration	0.33 M	1.10M-0.33M = 0.77M	1.10M-).33M=0.77M

$$K_c = \frac{(0.77\ M)(0.77\ M)}{(0.33\ M)}$$

$$K_c = 1.8\ M \text{ at } 250^\circ C$$

15-14. The K_p expression for the reaction is

$$K_p = \frac{P_{CO}P^3_{H_2}}{P_{CH_4}P_{H_2O}}$$

Substituting the values of the equilibrium pressures in the K_p expression, we find that

$$K_p = \frac{(0.57\ \text{atm})(2.26\ \text{atm})^3}{(0.31\ \text{atm})(0.83\ \text{atm})}$$

$$K_p = 26\ \text{atm}^2 \quad \text{at } 1000^\circ\text{C}$$

15-16. The K_p expression for the reaction is

$$K_p = \frac{P_{NO}^2 P_{O_2}}{P_{NO_2}^2}$$

The total pressure, P_{tot}, is the sum of the partial pressure of the gases.

$$P_{tot} = P_{NO} + P_{O_2} + P_{NO_2}$$

Because we started with pure NO_2, at equilibrium, we have from the reaction stoichiometry, $P_{NO} = 2P_{O_2}$ (two NO molecules are produced for each O_2 molecule produced), and $P_{NO_2} =$ 0.500 atm - P_{NO} (for every NO_2 molecule that decomposes, one NO molecule is produced). We now write the expression for the total pressure in terms of P_{NO}

$$P_{tot} = P_{NO} + \frac{1}{2} P_{NO} + 0.500\ \text{atm} - P_{NO} = 0.732\ \text{atm}$$

thus

$$\frac{1}{2} P_{NO} = 0.732\ \text{atm} - 0.500\ \text{atm} = 0.232\ \text{atm}$$

$$P_{NO} = 0.464\ \text{atm}$$

Further

$$P_{O_2} = \frac{1}{2} P_{NO} = \frac{1}{2}(0.464\ \text{atm}) = 0.232\ \text{atm}$$

and

$$P_{NO_2} = 0.500 \text{ atm} - 0.464 \text{ atm} = 0.036 \text{ atm}$$

Substituting the values of the equilibrium pressures in the K_p expression, we find that

$$K_p = \frac{(0.464 \text{ atm})^2(0.232 \text{ atm})}{(0.036 \text{ atm})^2}$$

$$K_p = 39 \text{ atm at } 1000 \text{ K}$$

15-18. The K_c expression for the reaction is

$$K_c = \frac{[NO_2]^2}{[N_2O_4]}$$

The relation between K_c and K_p is obtained using the relation

$P = [gas]RT$, or $[gas] = P/RT$, thus

$$K_c = \frac{\left(\frac{P_{NO_2}}{RT}\right)^2}{\left(\frac{P_{N_2O_4}}{RT}\right)} = \frac{1}{RT}\frac{P^2_{NO_2}}{P_{N_2O_4}} = \frac{1}{RT} K_p$$

$$K_p = RTK_c$$

Substituting the values of R, T and K_c into the above equation, we have

$$K_p = (0.0821 \text{ L·atm·K}^{-1}\text{·mol}^{-1})(373 \text{ K})(0.20 \text{ mol·L}^{-1})$$

$$K_p = 6.1 \text{ atm at } 100^\circ\text{C}$$

15-20. We can write the equilibrium constant expression for the reaction by applying the Law of Concentration Action

$$K_c = \frac{[S_2][H_2]^2}{[H_2S]^2}$$

Substituting the values of the equilibrium concentrations into the K_c expression we find that

$$K_c = \frac{(0.075\ M)(0.027\ M)^2}{(0.36\ M)^2}$$

$$K_c = 4.2 \times 10^{-4}\ M \text{ at } 1200^{\circ}C$$

The relation between K_c and K_p is

$$K_c = \frac{\left(\frac{P_{S_2}}{RT}\right)\left(\frac{P_{H_2}}{RT}\right)^2}{\left(\frac{P_{H_2S}}{RT}\right)^2} = \left(\frac{1}{RT}\right)\left(\frac{P_{S_2}P_{H_2}^2}{P_{H_2S}^2}\right) = \frac{1}{RT} K_p$$

and

$$K_p = RTK_c$$

Thus at $1200^{\circ}C$ we have

$$K_p = (0.0821\ L\cdot atm\cdot K^{-1}\cdot mol^{-1})(1473\ K)(4.2 \times 10^{-4} mol\cdot L^{-1})$$

$$K_p = 5.1 \times 10^{-2} \text{ atm at } 1200^{\circ}C$$

15-22. From the Law of Concentration Action, we have

$$K_p = \frac{P_{CO}^2}{P_{CO_2}}$$

Substituting the values of the equilibrium pressure of CO and of K_p in the K_p expression, we have that

$$1.90 \text{ atm} = \frac{(1.50 \text{ atm})^2}{P_{CO_2}}$$

Thus

$$P_{CO_2} = \frac{(1.50 \text{ atm})^2}{1.90 \text{ atm}} = 1.18 \text{ atm at equilibrium at 1000 K}$$

15-24. From the Law of Concentration Action we have

$$K_c = \frac{[CS_2]}{[S_2]}$$

We set up a table of initial concentrations and equilibrium concentrations.

Let X be the number of moles per liter of CS_2 that is produced by the reaction of S_2 with C(s). From the reaction stoichiometry, we see that at equilibrium $[S_2]$ = 2.00 M-X.

	S_2(g)	+ C(s) $\rightleftharpoons$	CS_2(g)
Initial Concentration	$\frac{10.0 \text{ mol}}{5.00 \text{ L}} = 2.00$ M	—	0
Equilibrium Concentration	2.00 M-X	—	X

Substituting the equilibrium concentration expressions in the K_c expression, we have

$$K_c = \frac{X}{2.00 \text{ M} - X} = 9.40$$

Thus

$$X = (9.40)(2.00\text{ M}-X) = 18.80\text{M} - 9.40\text{ X}$$

or

$$10.40\text{ X} = 18.8\text{ M}$$

and

$$X = \frac{18.8\text{ M}}{10.40} = 1.808\text{ M}$$

The number of moles of CS_2 prepared is

$$(1.808\text{ mol}\cdot\text{L}^{-1})(5.00\text{ L}) = 9.04\text{ mol}$$

The mass of CS_2 prepared is

$$(9.04\text{ mol})\left(\frac{76.13\text{ g}}{1\text{ mol}}\right) = 688\text{ g of } CS_2$$

15-26. From the Law of Concentration Action we have

$$K_p = \frac{P_{CH_4}}{P_{H_2}^2} = 0.263\text{ atm}^{-1}$$

From the initial concentrations and the reaction stoichiometry we have

	C(s) +	$2H_2(g)$ $\rightleftharpoons$	$CH_4(g)$
Initial Concentration	—	0	$\frac{0.250\text{ mol}}{4.00\text{ L}} = 0.0625\text{ M}$
Equilibrium Concentration	—	2X	0.0625 M-X

where X is the number of moles per liter of CH_4 that reacts. Thus 2X is the number of moles per liter of H_2 that is produced.

The equilibrium gas pressures are given by the relation

$$P = [gas]RT$$

Thus we have from the K_p expression

$$K_p = \frac{P_{CH_4}}{P^2_{H_2}} = \frac{(RT)[CH_4]}{(RT)^2[H_2]^2} = \frac{(0.0625\ M-X)(RT)}{(2X)^2(RT)^2} = 0.263\ atm^{-1}$$

and

$$0.0625\ M-X = (4X^2)(0.263\ atm^{-1})(RT)$$

Thus at $1000^\circ C$

$$0.0625\ M-X = 4(0.263\ atm^{-1})(0.0821\ L\cdot atm\cdot K^{-1}\cdot mol^{-1})(1273\ K)X^2$$

$$0.0625\ M-X = (110\ M^{-1})X^2$$

or

$$110X^2 + 1\ MX - 0.0625\ M^2 = 0$$

Using the quadratic formula we obtain

$$X = \frac{-1M \pm \sqrt{1M^2 - (4)(110)(-0.0625\ M^2)}}{(2)(110)}$$

and

$$X = 0.0197\ M$$

Thus

$$[CH_4] = 0.0625\ M-X = 0.0625\ M - 0.0197\ M = 0.0428\ M$$

The value of P_{CH_4} at equilibrium is

$$P_{CH_4} = [CH_4](RT)$$

$$= (0.0428\ mol\cdot L^{-1})(0.0821\ L\cdot atm\cdot K^{-1}\cdot mol^{-1})(1273\ K) = 4.47\ atm$$

15-28. From the Law of Concentration Action we have

$$K_p = P_{CO_2} P_{H_2O}$$

Let X be the equilibrium pressure of CO_2. Since we started with only $NaHCO_3$ the reaction stoichiometry, at equilibrium $P_{CO_2} = P_{H_2O}$. We have then that

$$K_p = (X)(X) = X^2 = 0.25 \text{ atm}^2$$

Taking the square root of both sides, we have

$$X = 0.50 \text{ atm}$$

At equilibrium $P_{CO_2} = P_{H_2O} = 0.50$ atm.

15-30. From the Law of Concentration Action, we have

$$K_c = \frac{[COCl_2]}{[CO][Cl_2]} = 4.0 \text{ M}^{-1}$$

We set up a table of initial concentrations and equilibrium concentrations.

Let X be the number of moles per liter of CO and of Cl_2 that react. We see from the reaction stoichiometry that the number of moles per liter of $COCl_2$ produced is also X.

	CO(g) +	Cl_2(g) ⇌	$COCl_2$(g)
Initial Concentration	$\frac{5.00\text{ mol}}{10.0\text{ L}} = 0.500$M	$\frac{2.50\text{ mol}}{10.0\text{ L}} = 0.250$M	0
Equilibrium Concentration	0.500M−X	0.250M−X	X

Substituting the equilibrium concentration expression in the K_c expression, yields

$$\frac{X}{(0.500M-X)(0.250M-X)} = \frac{X}{0.125\ M^2-0.750MX+X^2} = 4.0\ M^{-1}$$

or

$$X = (4.0\ M^{-1})(0.125\ M^2 - 0.750MX + X^2)$$

$$X = 0.50\ M - 3.0\ X + 4.0\ M^{-1}\ X^2$$

Rearranging to the standard form of a quadratic equation, we have

$$4.0\ M^{-1}\ X^2 - 4.0\ X + 0.50\ M = 0$$

The solution to this equation is given by the quadratic formula

$$X = \frac{4.0 \pm \sqrt{16.0 - (4)(4.0\ M^{-1})(0.50\ M)}}{(2)(4.0\ M^{-1})}$$

$$X = \frac{4.0 \pm 2.8}{8.0\ M^{-1}}$$

$$= 0.85\ M \text{ and } 0.15\ M$$

We can eliminate the result X = 0.85 M because we started with only 0.25M $Cl_2(g)$ and thus the value of $[COCl_2]$ at equilibrium cannot exceed 0.25 M. At equilibrium

$$[COCl_2] = 0.15\ M$$

$$[Cl_2] = 0.25\ M - 0.15\ M = 0.10\ M$$

$$[CO] = 0.50\ M - 0.15\ M = 0.35\ M$$

15-32. From the Law of Concentration Action we have

$$K_c = \frac{[H_2O]^2}{[H_2]^2}$$

We can find the value of K_c by substituting the values of the equilibrium concentrations

$$K_c = \frac{(0.25\ M)^2}{(0.25\ M)^2} = 1.0$$

We set up a table of initial concentrations and equilibrium concentrations. Let X be the number of moles per liter of H_2O produced by the reaction of SnO_2 and H_2. One mole of H_2 reacts to produce one mole of H_2O.

	$SnO_2(s)$ +	$2H_2(g)$ ⇌	$Sn(s)$ +	$2H_2O(g)$
Initial Concentration	—	0.50 M	—	0.25 M
Equilibrium Concentration	—	0.50 M-X	—	0.25 M+X

Substituting the equilibrium concentration expressions in the K_c expression, we have

$$K_c = \frac{(0.25\ M+X)^2}{(0.50\ M-X)^2} = 1.0$$

Taking the square root of both sides, we have that

$$\frac{0.25\ M+X}{0.50\ M-X} = 1.0$$

or

$$0.25\ M+X = (1.0)(0.50\ M-X) = 0.50\ M-X$$

$$2X = 0.25\ M$$

$$X = \frac{0.25\ M}{2} = 0.13\ M$$

Thus at the new equilibrium state we have

$$[H_2O] = 0.25\ M + 0.13\ M = 0.38\ M$$

$$[H_2] = [H_2O] = 0.38\ M$$

15-34. From the Law of Concentration Action we have

$$K_p = \frac{P^2_{HI}}{P_{H_2}} = 8.6\ atm$$

The total pressure is equal to the sum of the partial pressures of HI and H_2

$$P_{tot} = P_{HI} + P_{H_2} = 4.5\ atm$$

Solving for P_{HI}, we have

$$P_{HI} = 4.5\ atm - P_{H_2}$$

Substituting the expression for P_{HI} into the K_p expression, we have

$$\frac{(4.5\ atm - P_{H_2})^2}{P_{H_2}} = \frac{20.25\ atm^2 - 9.0\ atm\ P_{H_2} + P^2_{H_2}}{P_{H_2}} = 8.6\ atm$$

or

$$20.25\ atm^2 - 9.0\ atm\ P_{H_2} + P^2_{H_2} = 8.6\ atm\ P_{H_2}$$

We now rearrange to the standard form of the quadratic equation to get

$$P^2_{H_2} - 17.6\ atm\ P_{H_2} + 20.25\ atm^2 = 0$$

The solution is given by the quadratic formula

$$P_{H_2} = \frac{17.6 \text{ atm} \pm \sqrt{309.76 \text{ atm}^2 - (4)(1)(20.25 \text{ atm}^2)}}{2}$$

$$= \frac{17.6 \text{ atm} \pm 15.1 \text{ atm}}{2} = 16.4 \text{ atm and } 1.3 \text{ atm}$$

We reject the solution P_{H_2} = 16.4 atm as physically impossible because P_{tot} = 4.5 atm. Thus at equilibrium P_{H_2} = 1.3 atm and P_{HI} = 4.5 atm - 1.3 atm = 3.2 atm.

15-36. From the Law of Concentration Action we have

$$K_p = \frac{P_{CH_4}}{P_{H_2}^2} = 0.263 \text{ atm}^{-1}$$

The total pressure is the sum of the partial pressures of CH_4 and H_2

$$P_{tot} = P_{CH_4} + P_{H_2} = 2.11 \text{ atm}$$

Solving for P_{CH_4}, we have that

$$P_{CH_4} = 2.11 \text{ atm} - P_{H_2}$$

Substituting the expression for P_{CH_4} in the K_p expression, we have

$$\frac{2.11 \text{ atm} - P_{H_2}}{P_{H_2}^2} = 0.263 \text{ atm}^{-1}$$

or

$$2.11 \text{ atm} - P_{H_2} = 0.263 \text{ atm}^{-1} P_{H_2}^2$$

Rearranging this equation to the standard quadratic form yields

$$0.263 \text{ atm}^{-1} P_{H_2}^2 + P_{H_2} - 2.11 \text{ atm} = 0$$

We can solve this quadratic equation using the quadratic formula

$$P_{H_2} = \frac{-1 \pm \sqrt{(1)^2 - (4)(0.263 \text{ atm}^{-1})(-2.11 \text{ atm})}}{(2)(0.263 \text{ atm}^{-1})}$$

$$= \frac{-1 \pm 1.794}{0.526 \text{ atm}^{-1}}$$

$$= 1.51 \text{ atm}$$

where we have rejected the physically impossible solution -5.31 atm. Thus at equilibrium $P_{H_2} = 1.51$ atm and $P_{CH_4} = 2.11$ atm $- 1.51$ atm $= 0.60$ atm.

15-38. From the Law of Concentration Action we have

$$K_p = \frac{P_{NH_3}^2}{P_{N_2} P_{H_2}^3} = 0.10 \text{ atm}^{-2}$$

a) Substituting in the values of P_{N_2} and P_{H_2} in the K_p expression, we have

$$\frac{P_{NH_3}^2}{(1.00 \text{ atm})(3.00 \text{ atm})^3} = 0.10 \text{ atm}^{-2}$$

$$P_{NH_3}^2 = (0.10 \text{ atm}^{-2})(27.0 \text{ atm}^4) = 2.7 \text{ atm}^2$$

Taking the square root we obtain

$$P_{NH_3} = 1.6 \text{ atm}$$

b) Recall that the mole fraction is defined as $X_a = \frac{n_a}{n_{tot}}$ where n_a is the number of moles of component a in a mixture and n_{tot} is the total number of moles in the mixture. The partial pressure of a gas in a gas mixture is equal to the mole fraction of that gas in the mixture times the total gas pressure. Thus

$$P_{N_2} = X_{N_2}P_{tot} = X_{N_2}(2.00\ atm)$$

$$P_{H_2} = X_{H_2}P_{tot} = (0.20)(2.00\ atm) = 0.40\ atm$$

$$P_{NH_3} = X_{NH_3}P_{tot} = X_{NH_3}(2.00\ atm)$$

Using the relationship

$$X_{N_2} + X_{H_2} + X_{NH_3} = 1$$

together with $X_{H_2} = 0.20$, we have

$$X_{N_2} + 0.20 + X_{NH_3} = 1$$

Solving for X_{N_2}, we obtain

$$X_{N_2} = 0.80 - X_{NH_3}$$

Thus

$$P_{N_2} = (0.80 - X_{NH_3})(2.00\ atm) = 1.60\ atm - 2.00\ atm\ X_{NH_3}$$

Substituting the partial pressure expressions in the K_p expression, we have

$$\frac{(2.00\ atm\ X_{NH_3})^2}{(1.60\ atm - 2.00\ atm\ X_{NH_3})(0.40\ atm)^3} = 0.10\ atm^{-2}$$

Thus

$$\frac{4.00\ atm^2 X_{NH_3}^2}{0.102\ atm^4 - 0.128\ atm^4\ X_{NH_3}} = 0.10\ atm^{-2}$$

and therefore

$$4.00\ atm^2 X_{NH_3}^2 = 0.0102\ atm^2 - 0.0128\ atm^2 X_{NH_3}$$

or

$$4.00 X_{NH_3}^2 = 0.0102 - 0.0128\ X_{NH_3}$$

Rearranging to the standard quadratic form, we have

$$4.00X_{NH_3}^2 + 0.0128X_{NH_3} - 0.0102 = 0$$

We can solve this equation using the quadratic formula

$$X_{NH_3} = \frac{-0.0128 \pm \sqrt{1.64 \times 10^{-4} - (4)(4.00)(-0.0102)}}{(2)(4.00)}$$

$$= \frac{-0.0128 \pm 0.404}{8.00}$$

$$= 0.0489$$

15-40. Increasing the reaction volume has no effect on the reaction equilibrium because there are the same number of moles of gas on both sides of the reaction.

15-42. The temperature should be decreased to increase the extent of conversion to NOBr. The reaction is exothermic so that a lowering of the temperature favors the evolution of heat.

15-44. a) $\longleftarrow$; the reaction is exothermic and thus a higher temperature causes the reaction to shift to the left.

b) $\longleftarrow$; an increase in the reaction volume is partially offset by a shift in the reaction equilibrium toward the side with the larger number of moles of gas.

c) $\longrightarrow$; a shift to the right will partially restore the decrease in $[Ni(CO)_4]$.

d) No change; the addition of Ni(s) has no effect on the concentration of Ni(s) and thus has no effect on the equilibrium concentrations.

15-46. a) Shift to the right, $\rightarrow$; the reaction is endothermic.

b) Shift to the right, $\rightarrow$; there are more moles of gas on the right.

c) Shift to the left, ←; addition of $H_2O(l)$ decreases the concentration of $N_2(aq)$.

d) Shift to the left, ←; a shift to the left decreases the concentration of $N_2(g)$.

15-48. a) An increase in the total volume will lead to an increase in the percentage conversion of H_2O to CO and H_2. There are more moles on the right side and thus an increase in total volume leads to a shift to the right.

b) The reaction is endothermic and thus a decrease in temperature shifts the equilibrium to the left, leading to a decrease in the percentage conversion of H_2O to CO and H_2.

15-50. Because the reaction involving the formation of $CH_3OH(g)$ from $CO(g)$ and $H_2(g)$ is exothermic and $\Delta n_{gas} = -2$, the use of Le Châtelier's Principle suggests that low temperature and high pressure (small reaction volume) favors the formation of methanol.

15-52. The K_p expression is

$$K_p = \frac{P_{CH_4}}{P^2_{H_2}} = 2.69 \times 10^3 \text{ atm}^{-1}$$

The Q_p expression is

$$Q_p = \frac{(P_{CH_4})_0}{(P_{H_2})^2_0}$$

From the data given we compute

$$Q_p = \frac{3.0 \text{ atm}}{(0.20 \text{ atm})^2} = 75 \text{ atm}^{-1}$$

Because Q_p does not equal K_p, the reaction is not at equilibrium when P_{H_2} = 0.20 atm and P_{CH_4} = 3.0 atm.

$$\frac{Q_p}{K_p} = \frac{75\ \text{atm}^{-1}}{2.69 \times 10^3\ \text{atm}^{-1}} = 2.8 \times 10^{-2}$$

When the value of Q/K is less than 1, reaction proceeds left to right toward equilibrium.

15-54. The Q_p expression for the reaction is

$$Q_p = \frac{\left(P_{CS_2}\right)_0}{\left(P_{S_2}\right)_0}$$

Thus

$$Q_p = \frac{0.794\ \text{atm}}{1.78\ \text{atm}} = 0.446$$

The reaction is not at equilibrium because Q_p does not equal K_p

$$\frac{Q_p}{K_p} = \frac{0.446}{9.40} = 0.0475$$

Because Q/K < 1, the reaction proceeds from left to right toward equilibrium.

15-56. The Q_p expression for the reaction is

$$Q_p = \frac{\left(P_{CH_3OH}\right)_0}{\left(P_{H_2}\right)_0^2\left(P_{CO}\right)_0}$$

Thus

$$Q_p = \frac{10.0\ \text{atm}}{(0.010\ \text{atm})^2(0.0050\ \text{atm})} = 2.0 \times 10^7\ \text{atm}^{-2}$$

The value of Q_p/K_p is

$$\frac{Q_p}{K_p} = \frac{2.0 \times 10^7\ \text{atm}^{-2}}{2.25 \times 10^4\ \text{atm}^{-2}} = 8.9 \times 10^2$$

Thus the reaction proceeds right to left toward equilibrium.

15-58. For the reaction

(1) $2NO_2(g) \rightleftharpoons N_2O_4(g)$

$$K_{c1} = \frac{1}{0.20\ M} = 5.0\ M^{-1}$$

where

$$K_{c1} = \frac{[N_2O_4]}{[NO_2]^2}$$

For the reaction

(2) $NO_2(g) \rightleftharpoons \frac{1}{2} N_2O_4(g)$

$$K_{c2} = \frac{[N_2O_4]^{1/2}}{[NO_2]}$$

We see that

$$K_{c2} = K_{c1}^{1/2}$$

$$= (5.0\ M^{-1})^{1/2} = 2.2\ M^{-1/2}$$

The relation between K_{c2} and K_{p2} is

$$K_{c2} = \frac{\left(\frac{P_{N_2O_4}}{RT}\right)^{\frac{1}{2}}}{\left(\frac{P_{NO_2}}{RT}\right)} = (RT)^{\frac{1}{2}} \frac{P^{\frac{1}{2}}_{N_2O_4}}{P_{NO_2}} = (RT)^{\frac{1}{2}} K_{p2}$$

Thus

$$K_{p2} = \frac{K_{c2}}{(RT)^{\frac{1}{2}}} = \frac{2.2\ M^{-\frac{1}{2}}}{(0.0821\ L{\cdot}atm{\cdot}K^{-1}{\cdot}mol^{-1})^{\frac{1}{2}}(373\ K)^{\frac{1}{2}}}$$

$$= 0.40\ atm^{-\frac{1}{2}}$$

15-60. The equation

$$(3) \qquad CaCO_3(s) + C(s) \rightleftharpoons CaO(s) + 2CO(g)$$

is the sum of the two given equations. Thus we have

$$K_3 = K_1K_2 = (0.039\ atm)(1.9\ atm)$$

$$= 0.074\ atm^2$$

15-62.

$$\log \frac{K_2}{K_1} = \frac{\Delta H^0_{rxn}}{2.30\ R}\left(\frac{T_2 - T_1}{T_2T_1}\right)$$

Substituting in the values of K_1, T_1, T_2 and ΔH_{rxn} into the van't Hoff equation we have

$$\log\left(\frac{K_2}{617}\right) = \frac{(-10.2\ kJ{\cdot}mol^{-1})(1000\ J{\cdot}kJ^{-1})(373\ K-298\ K)}{(2.30)(8.31\ \ J{\cdot}mol^{-1}{\cdot}K^{-1})(373\ K)(298\ K)}$$

$$= -0.360$$

Taking the antilogarithm of both sides, we have

$$\frac{K_2}{617} = 0.436$$

Thus

$$K_2 = (0.436)(617) = 269\ M$$

15-64. The van't Hoff equation is

$$\log \frac{K_2}{K_1} = \frac{\Delta H^0_{rxn}}{2.30\ R}\left(\frac{T_2 - T_1}{T_1 T_2}\right)$$

Substituting in the values of K_1, T_1, T_2 and ΔH^0_{rxn} yields

$$\log \frac{K_2}{4 \times 10^{-4}} = \frac{(181\ kJ\cdot mol^{-1})(1000\ J\cdot kJ^{-1})(298\ K - 2273\ K)}{(2.30)(8.31\ J\cdot mol^{-1}\cdot K^{-1})(2273\ K)(298\ K)}$$

$$= -27.61$$

Taking the antilogarithm of both sides, we have

$$\frac{K_2}{4 \times 10^{-4}} = 2.44 \times 10^{-28}$$

$$K_2 = (2.44 \times 10^{-28})(4 \times 10^{-4})$$

$$= 1 \times 10^{-31}$$

15-66.

$$\log \frac{K_2}{K_1} = \frac{\Delta H^0_{rxn}}{2.30\ R}\left(\frac{T_2 - T_1}{T_1 T_2}\right)$$

Substituting in the values of K_1, T_1, T_2 and ΔH^0_{rxn} we obtain

$$\log\left(\frac{K_2}{2.25 \times 10^4\ atm^{-2}}\right) = \frac{(-128\ kJ\cdot mol^{-1})(1000\ J\cdot kJ^{-1})(573\ K - 298}{(2.30)(8.31\ J\cdot mol^{-1}\cdot K^{-1})(298\ K)(573\ K)}$$

$$= -10.786$$

Thus

$$\frac{K_2}{2.25 \times 10^4 \text{ atm}^{-2}} = 1.64 \times 10^{-11}$$

$$K_2 = (1.64 \times 10^{-11})(2.25 \times 10^4 \text{ atm}^{-2})$$

$$K_2 = 3.69 \times 10^{-7} \text{atm}^{-2}$$

CHAPTER 16

SOLUTIONS TO THE EVEN-NUMBERED PROBLEMS

16-2. a) The rate law for a zero-order reaction is

$$\text{rate} = k$$

Thus the units of k are the same as the units of the reaction rate, that is, $M \cdot s^{-1}$.

b) The rate law for a third-order reaction is

$$\text{rate} = k[A]^3$$

Thus

$$k = \frac{\text{rate}}{[A]^3}$$

The units are

$$k = \frac{M \cdot s^{-1}}{M^3} = M^{-2} \cdot s^{-1}$$

16-4. a) As the reaction proceeds, oxygen gas is produced. The pressure of oxygen increases as the reaction proceeds. You can measure the rate of the reaction by measuring the rate of increase of the pressure of oxygen.

b) You cannot measure the rate of this reaction by measuring pressure because there will be no change in the pressure. However, bromine is reddish-brown in color whereas HBr and H_2 are colorless. We can measure the rate of the reaction by measuring the rate of the increase in the intensity of the color of the gaseous mixture.

16-6. We have that (see Problem 16-5)

$$\text{rate} = (3.0 \times 10^{6}\ M^{-1}\cdot s^{-1})[O_3][NO]$$

Thus the initial rate is

$$\text{rate} = (3.0 \times 10^{6}\ M^{-1}\cdot s^{-1})(3.0 \times 10^{-4} M)(6.3 \times 10^{-3} M)$$

$$= 5.7\ M\cdot s^{-1}$$

16-8. From the reaction stoichiometry we have

$$\frac{\Delta[C_{10}H_{12}]}{\Delta t} = -\left(\frac{1}{2}\right)\frac{\Delta[C_5H_6]}{\Delta t}$$

because one molecule of $C_{10}H_{12}$ is produced when two molecules of C_5H_6 react. Thus

$$\frac{\Delta[C_{10}H_{12}]}{\Delta t} = \left(\frac{1}{2}\right)(2.3\ \text{torr}\cdot s^{-1})$$

$$= 1.2\ \text{torr}\cdot s^{-1}$$

16-10. The average rate of disappearance of $H^{+}(aq)$ is

$$\text{rate} = \frac{-\Delta[H^{+}]}{\Delta t} = \frac{-([H^{+}]_2 - [H^{+}]_1)}{t_2 - t_1} = \frac{-[H^{+}]_2 + [H^{+}]_1}{t_2 - t_1}$$

For the interval 0 to 31 min we have

$$\text{rate} = \frac{-1.90\ M + 2.12\ M}{31\ \text{min} - 0\ \text{min}} = 7.1 \times 10^{-3}\ M\cdot \text{min}^{-1}$$

For the interval 31 min to 61 min we have

$$\text{rate} = \frac{-1.78\ M + 1.90\ M}{61\ \text{min} - 31\ \text{min}} = 4.0 \times 10^{-3}\ M\cdot \text{min}^{-1}$$

For the interval 61 min to 121 min we have

$$\text{rate} = \frac{-1.61\ M + 1.78\ M}{121\ \text{min} - 61\ \text{min}} = 2.8 \times 10^{-3}\ M\cdot \text{min}^{-1}$$

From the reaction stoichiometry we conclude that the average rate of disappearance of $CH_3OH(aq)$ is equal to the average rate of disappearance of $H^+(aq)$

$$\text{rate} = \frac{-\Delta[CH_3OH]}{\Delta t} = \frac{-\Delta[H^+]}{\Delta t}$$

Similarly we find that the average rate of appearance of $CH_3Cl(aq)$ is

$$\text{rate} = \frac{\Delta[CH_3Cl]}{\Delta t} = \frac{-\Delta[H^+]}{\Delta t}$$

16-12. The reaction rate is given by

$$\text{rate} = \frac{-\Delta P_{SO_2Cl_2}}{\Delta t} = \frac{-P_{SO_2Cl_2,2} + P_{SO_2Cl_2,1}}{t_2 - t_1}$$

Thus the rate over the interval 0 to 5000 s is

$$\text{rate} = \frac{-680 \text{ torr} + 760 \text{ torr}}{5000 \text{ s} - 0 \text{ s}} = 1.6 \times 10^{-2} \text{ torr}\cdot\text{s}^{-1}$$

The rate over the interval 5000 s to 10,000 s is

$$\text{rate} = \frac{-610 \text{ torr} + 680 \text{ torr}}{10{,}000 \text{ s} - 5000 \text{ s}} = 1.4 \times 10^{-2} \text{ torr}\cdot\text{s}^{-1}$$

16-14. The data indicates that the rate of the reaction is proportional to $[NOBr]^2$ because when $[NOBr]$ is doubled the rate quadruples. We have a second-order rate law.

$$\text{rate} = k[NOBr]^2$$

or

$$k = \frac{\text{rate}}{[NOBr]^2}$$

Let's show that the ratio of the rate to $[NOBr]^2$ has a constant value

$$\frac{\text{rate}}{[NOBr]^2} = \frac{0.80\ M\cdot s^{-1}}{(0.20\ M)^2} = 20\ M^{-1}\cdot s^{-1}$$

$$= \frac{3.20\ M\cdot s^{-1}}{(0.40\ M)^2} = 20\ M^{-1}\cdot s^{-1}$$

$$= \frac{7.20\ M\cdot s^{-1}}{(0.60\ M)^2} = 20\ M^{-1}\cdot s^{-1}$$

$$\frac{12.80\ M\cdot s^{-1}}{(0.80\ M)^2} = 20\ M^{-1}\cdot s^{-1}$$

16-16. The rate law is

$$\text{rate} = k[C_5H_6]^x$$

When $[P_{C_5H_6}]_0$ is doubled, the initial rate quadruples. Thus the rate law is second-order in $P_{C_5H_6}$

$$\text{rate} = kP_{C_5H_6}$$

The value of the rate constant is

$$k = \frac{\text{rate}}{P^2_{C_5H_6}}$$

We can calculate k using the data for any one of the three runs

$$k = \frac{5.76\ \text{torr}\cdot s^{-1}}{(200\ \text{torr})^2} = 1.44 \times 10^{-4}\ \text{torr}^{-1}\cdot s^{-1}$$

$$k = \frac{23.04\ \text{torr}\cdot s^{-1}}{(400\ \text{torr})^2} = 1.44 \times 10^{-4}\ \text{torr}^{-1}\cdot s^{-1}$$

$$k = \frac{92.2\ \text{torr}\cdot s^{-1}}{(800\ \text{torr})^2} = 1.44 \times 10^{-4}\ \text{torr}^{-1}\cdot s^{-1}$$

16-18. When $P_{N_2O_3}$ is increased by a factor of 1.4/0.91 = 1.5, the initial rate is increased by a factor of 5.5/8.4 = 1.5. Thus the rate law is first-order in $P_{N_2O_3}$. The rate law for the reaction is

$$\text{rate} = kP_{N_2O_3}$$

We can calculate the rate constant from the data of any of the three runs. The value of the rate constant is given by

$$k = \frac{\text{rate}}{P_{N_2O_3}}$$

$$= \frac{5.5 \text{ torr}\cdot\text{s}^{-1}}{0.91 \text{ torr}} = 6.0 \text{ s}^{-1}$$

$$= \frac{8.4 \text{ torr}\cdot\text{s}^{-1}}{1.4 \text{ torr}} = 6.0 \text{ s}^{-1}$$

$$= \frac{13 \text{ torr}\cdot\text{s}^{-1}}{2.1 \text{ torr}} = 6.2 \text{ s}^{-1}$$

16-20. When $[CoBr(NH_3)_5^{2+}]_0$ is doubled and $[OH^-]_0$ remains the same, the initial rate doubles. Thus the rate is first-order in $[CoBr(NH_3)_5^{2+}]$. When $[OH^-]_0$ is tripled and $[CoBr(NH_3)_5^{2+}]_0$ remains the same, the initial rate triples. Thus the rate is first-order in $[OH^-]$. The rate law is

$$\text{rate} = k[CoBr(NH_3)_5^{2+}][OH^-]$$

The rate law is second-order overall. We can calculate the rate constant using the data from any of the four runs. The value of the rate constant is given by

$$k = \frac{\text{rate}}{[CoBr(NH_3)_5^{2+}][OH^-]}$$

$$k = \frac{1.37 \times 10^{-3}\ M\cdot s^{-1}}{(0.030\ M)(0.030\ M)} = 1.5\ M^{-1}\cdot s^{-1}$$

$$= \frac{2.74 \times 10^{-3} M\cdot s^{-1}}{(0.06\ M)(0.030\ M)} = 1.5\ M^{-1}\cdot s^{-1}$$

$$= \frac{4.11 \times 10^{-3} M\cdot s^{-1}}{(0.030\ M)(0.090\ M)} = 1.5\ M^{-1}\cdot s^{-1}$$

$$= \frac{1.23 \times 10^{-2}\ M\cdot s^{-1}}{(0.090\ M)(0.090\ M)} = 1.5\ M^{-1}\cdot s^{-1}$$

16-22. When $[BrO_3^-]_0$ is doubled, and $[I^-]_0$ and $[H^+]_0$ remain the same, the initial rate doubles. Thus the rate is first-order in $[BrO_3^-]$ or

$$\text{rate} = k[BrO_3^-][I^-]^x[H^+]^y$$

when $[I^-]_0$ is doubled and $[BrO_3^-]_0$ and $[H^+]_0$ remain the same, the initial rate doubles. Thus the rate is first-order in $[I^-]$ or

$$\text{rate} = k[BrO_3^-][I^-][H^+]^y$$

When $[H^+]_0$ and $[I^-]_0$ are both doubled and $[BrO_3^-]_0$ remains the same, the initial rate doubles. Because we know that the rate law is first order in $[I^-]_0$, we conclude that the rate law does not depend upon $[H^+]$. Thus the rate law is

$$\text{rate} = k[BrO_3^-][I^-]$$

16-24. From the rate law we compute the initial rate for run (1)

$$\text{rate} = (5.0 \times 10^{-3}\ M^{-1}\cdot s^{-1})(0.20\ M)(0.20\ M)$$

$$= 2.0 \times 10^{-4}\ M\cdot s^{-1}$$

For run (2) we have

$$4.0 \times 10^{-4}\ M\cdot s^{-1} = (5.0 \times 10^{-3}\ M^{-1}\cdot s^{-1})(0.20\ M)[I^-]_0$$

Thus

$$[I^-]_0 = \frac{4.0 \times 10^{-4}\ M\cdot s^{-1}}{(5.0 \times 10^{-3}\ M^{-1}\cdot s^{-1})(0.20\ M)} = 0.40\ M$$

For run (3) we have

$$8.0 \times 10^{-4}\ M\cdot s^{-1} = (5.0 \times 10^{-3} M^{-1}\cdot s^{-1})[C_2H_4Br_2]_0(0.20\ M)$$

Thus

$$[C_2H_4Br_2]_0 = \frac{8.0 \times 10^{-4}\ M\cdot s^{-1}}{(5.0 \times 10^{-3}\ M^{-1}\cdot s^{-1})(0.20\ M} = 0.80\ M$$

16-26. When $[CH_3COCH_3]_0$ is doubled and $[Br_2]_0$ and $[H^+]_0$ remain the same, the initial rate doubles, thus the rate law is first order in $[CH_3COCH_3]$. When $[Br_2]_0$ is doubled and $[CH_3COCH_3]_0$ and $[H^+]_0$ remain the same, the initial rate remains the same, thus the rate law is independent of, or zero-order in $[Br_2]$. When $[H^+]_0$ and $[Br_2]_0$ are doubled, and $[CH_3COCH_3]_0$ remains the same, the initial rate doubles, thus the rate is first-order in $[H^+]$. The rate law is

$$\text{rate} = k[CH_3COCH_3][H^+]$$

The value of the rate constant is given by

$$k = \frac{\text{rate}}{[CH_3COCH_3][H^+]}$$

Thus

$$k = \frac{4.0 \times 10^{-3}\ M\cdot s^{-1}}{(1.00\ M)(1.00\ M)} = 4.0 \times 10^{-3}\ M^{-1}\cdot s^{-1}$$

$$k = \frac{8.0 \times 10^{-3}\ M\cdot s^{-1}}{(2.00\ M)(1.00\ M)} = 4.0 \times 10^{-3}\ M^{-1}\cdot s^{-1}$$

$$k = \frac{8.0 \times 10^{-3}\ M\cdot s^{-1}}{(1.00\ M)(2.00\ M)} = 4.0 \times 10^{-3}\ M^{-1}\cdot s^{-1}$$

16-28. For a first order reaction

$$\log \frac{[A]}{[A]_0} = \frac{-kt}{2.30}$$

We are given that $[A]_0 - [A] = 0.10[A]_0$, thus

$$\frac{[A]}{[A]_0} = \frac{[A]_0 - 0.10[A]_0}{[A]_0} = 1 - 0.10 = 0.90$$

Thus

$$\log 0.90 = \frac{-10\ s^{-1}\ t}{2.30}$$

$$-0.0458 = -4.35\ s^{-1}\ t$$

Solving for t, we have

$$t = \frac{0.0458}{4.35\ s^{-1}} = 0.011\ s$$

16-30. The value of the half-life is given by

$$t_{1/2} = \frac{0.693}{k}$$

Thus

$$t_{1/2} = \frac{0.693}{5.5 \times 10^{-4} s^{-1}} = 1.26 \times 10^3\ s$$

$$= (1.26 \times 10^3\ s)\left(\frac{1\ min}{60\ s}\right)\left(\frac{1\ hr}{60\ min}\right) = 0.35\ hr$$

The number of half-lives in 2.0 hrs is

$$\frac{2.0\ hr}{0.35\ hr/half\text{-}life} = 5.71\ half\text{-}lives$$

The concentration of a reactant that remains unreacted after n half-lives is given by

$$[A] = [A]_0 \left(\frac{1}{2}\right)^n$$

$$[\text{cyclopropane}] = (1.00 \times 10^{-3}\ M)\left(\frac{1}{2}\right)^{5.71}$$

$$= 1.9 \times 10^{-5}\ M$$

We can also use the integrated form of the rate law to compute the concentration of cyclopropane after 2.0 hr

$$\log [A] = \log [A]_0 - \frac{kt}{2.30}$$

$$\log[\text{cyclopropane}] = \log(1.00\times10^{-3}) - \frac{(5.5\times10^{-4}\,s^{-1})(2.0\,hr)\left(\frac{60\ min}{1\ hr}\right)\left(\frac{60\ s}{1\ min}\right)}{2.30}$$

$$= -3.00-1.72 = -4.72$$

$$[\text{cyclopropane}] = 1.9 \times 10^{-5}\ M$$

16-32. The production of bacteria is a first-order reaction with a certain doubling-time (see Problem 16-31). Thus the number of bacteria present after n doubling-times is given by

$$\text{number of bacteria} = (\text{number of bacteria})_0 (2)^n$$

The number of doubling-times in 10 days is

$$\frac{(10\ d)\left(\frac{24\ hr}{1\ d}\right)}{39\ hr/\text{doubling-time}} = 6.15$$

Thus if there are 20,000 bacteria per mL present initially, then after 10 days at 40^0F

$$\text{number of bacteria} = (20{,}000)(2)^{6.15}$$

$$= 1.4 \times 10^6\ \text{bacteria/mL}$$

16-34. The fraction of a reactant remaining after time t is given by

$$\log \frac{[A]}{[A]_0} = \frac{-kt}{2.30}$$

Thus

$$\log \frac{[A]}{[A]_0} = \frac{-(1.5 \times 10^{-4}\ s^{-1})(1\ ms)\left(\frac{1\ s}{10^3\ ms}\right)}{2.30}$$

$$= -6.52 \times 10^{-8}$$

Taking antilogarithms, we have

$$\frac{[A]}{[A]_0} = \text{fraction remaining} = 1.00$$

In other words, very little CH_3I decomposes in 1 ms.

16-36. The fraction of a reactant remaining after time t is given by

$$\log \frac{[A]}{[A]_0} = \frac{-kt}{2.30}$$

The time for 99.99% of the reaction to take place is given by

$$\log 0.0001 = \frac{-kt_1}{2.30}$$

$$\frac{kt_1}{2.30} = 4.00$$

where $\frac{[A]}{[A]_0} = 1 - 0.9999 = 0.0001$. The time for 99.0% of the reaction to take place is given by

$$\log 0.010 = \frac{-kt_2}{2.30}$$

$$\frac{kt_2}{2.30} = 2.00$$

where $\frac{[A]}{[A]_0} = 1 - 0.990 = 0.010.$

Thus

$$\frac{t_1}{t_2} = \frac{4.00}{2.00} = 2$$

16-38. The total pressure is given by

$$P_{tot} = P_{H_2C_2O_4} + P_{CO_2} + P_{HCHO_2}$$

From the reaction stoichiometry we have

$$P_{H_2C_2O_4} = \left(P_{H_2C_2O_4}\right)_0 - P_{CO_2}$$

and

$$P_{CO_2} = P_{HCHO_2}$$

Thus

$$P_{tot} = P_{H_2C_2O_4} + 2P_{CO_2}$$

Using the fact that $P_{CO_2} = (P_{H_2C_2O_4})_0 - P_{H_2C_2O_4}$, we write

$$= P_{H_2C_2O_4} + 2\left[(P_{H_2C_2O_4})_0 - P_{H_2C_2O_4}\right]$$

$$= 2(P_{H_2C_2O_4})_0 - P_{H_2C_2O_4}$$

Thus

$$P_{H_2C_2O_4} = 2(P_{H_2C_2O_4})_0 - P_{tot}$$

Application of the above equation to the data yields

Run #	$(P_{H_2C_2O_4})_0$ /torr	$P_{H_2C_2O_4}$ at 2.00 x 10^4 s/torr
1	5.0	2.8
2	7.0	4.0
3	8.4	4.8

We have that

$$\frac{[A]}{[A]_0} = \left(\frac{1}{2}\right)^n$$

for a first-order reaction. The ratio of $P_{H_2C_2O_4}$ at 2.00 x 10^4 s to $(P_{H_2C_2O_4})_0$ should be the same for the three runs if the reaction is first order.

Run 1 $$\frac{[A]}{[A]_0} = \frac{P_{H_2C_2O_4}}{(P_{H_2C_2O_4})_0} = \frac{2.8 \text{ torr}}{5.0 \text{ torr}} = 0.56$$

Run 2 $$= \frac{4.0 \text{ torr}}{7.0 \text{ torr}} = 0.57$$

Run 3 $$= \frac{4.8 \text{ torr}}{8.4 \text{ torr}} = 0.57$$

Thus the rate law is first-order in $P_{H_2C_2O_4}$

$$\text{rate} = kP_{H_2C_2O_4}$$

We can calculate the rate constant k from the integrated form of the rate law

$$\log[A] = \log[A]_0 - \frac{kt}{2.30}$$

Choosing data from Run 2 yields

$$\log 4.0 = \log 7.0 - \frac{k(2.00\text{x } 10^4 \text{ s})}{2.30}$$

$$0.602 = 0.845 - \frac{(2.00 \text{ x } 10^4 \text{ s})k}{2.30}$$

$$k = \frac{(2.30)(0.243)}{2.00 \text{ x } 10^4 \text{ s}} = 2.8 \text{ x } 10^{-5} \text{ s}^{-1}$$

16-40. The rate of loss of neurons is given as constant, thus the rate law is

$$\text{Rate} = k = 2 \text{ x } 10^5 \text{ neurons}\cdot\text{day}^{-1}$$

The number of days required for 1 x 10^{10} neurons to be lost is

$$\left(2 \text{ x } 10^5 \frac{\text{neurons}}{\text{day}}\right)(\text{number of days}) = 1 \text{ x } 10^{10} \text{ neurons}$$

Thus, the number of days is

$$\frac{1 \text{ x } 10^{10} \text{ neurons}}{2 \text{ x } 10^5 \text{ neurons}\cdot\text{day}^{-1}} = 5 \text{ x } 10^4 \text{ day}$$

$$(5 \text{ x } 10^4 \text{ day})\left(\frac{1 \text{ yr}}{365 \text{ day}}\right) = 140 \text{ yr}$$

Thus the age in years at which the number of neurons is one half the original value is

$$140 \text{ yr} + 30 \text{ yr} = 170 \text{ years}$$

16-42. The concentration of a reactant in a second-order reaction at a time t is given by

$$\frac{1}{[A]} = \frac{1}{[A]_0} + kt$$

Thus

$$\frac{1}{[NO_2]} = \frac{1}{1.25\ M} + (0.54\ M^{-1}\cdot s^{-1})(2\ min)\left(\frac{60\ s}{1\ min}\right)$$

$$= 0.800\ M^{-1} + 64.8\ M^{-1} = 65.6\ M^{-1}$$

and

$$[NO_2] = \frac{1}{65.6\ M^{-1}} = 0.015\ M$$

16-44. The half-life of a second-order reaction is given by

$$t_{\frac{1}{2}} = \frac{1}{k[A]_0}$$

Since $[OH^-]_0 = [H^+]_0$, we can use either OH^- or H^+ as A. Thus

$$t_{\frac{1}{2}} = \frac{1}{(1.3 \times 10^{11}\ M^{-1}\cdot s^{-1})(1.0 \times 10^{-3}\ M)} = 7.7 \times 10^{-9}\ s$$

16-46. a) The rate law is second-order in [A].

b) The rate law is second-order in [A].

c) The rate law is first-order in [A]. The half-life is independent of $[A]_0$.

16-48. a) rate = $k[K][HCl]$

b) rate = $k[Cl][ICl]$

c) rate = $k[NO_3][CO]$

16-50. The overall equation is the sum of the two equations

$$NO_2(g) + NO_2(g) + NO_3(g) \longrightarrow NO(g) + NO_3(g) + NO(g) + O_2(g)$$

Thus after combining terms

$$2NO_2(g) \longrightarrow 2NO(g) + O_2(g)$$

16-52. The rate law derived from mechanism 1 is determined by the slow elementary step

$$\text{rate} = k[N_2O_5]$$

The rate law derived from mechanism 2 is

$$\text{rate} = k[NO_3][N_2O_5]$$

Because NO_3 is an intermediate, we must eliminate NO_3 from the rate law. The equilibrium reaction is fast. Thus at any instant, we have

$$K = \frac{[NO_2][NO_3]}{[N_2O_5]}$$

and therefore

$$[NO_3] = \frac{K[N_2O_5]}{[NO_2]}$$

The corresponding rate law is

$$\text{rate} = kK[N_2O_5]^2/[NO_2]$$

Only the rate law derived from mechanism 1 is consistent with the observed rate law.

16-54. If we multiply the third equation by 2 and add the first two equations, then we obtain

$$H_2O_2(aq) + I^-(aq) + HOI(aq) + I^-(aq) + 2OH^-(aq) + 2H^+(aq)$$

$$\longrightarrow HOI(aq) + OH^-(aq) + I_2(aq) + OH^-(aq) + 2H_2O(l)$$

The overall reaction is

$$H_2O_2(aq) + 2I^-(aq) + 2H^+(aq) \longrightarrow I_2(aq) + 2H_2O(l)$$

The rate law is given by the slow elementary step

$$\text{rate} = k[H_2O_2][I^-]$$

16-56. The rate law of the slow elementary step is

$$\text{rate} = k[COCl_2][Cl]$$

The species Cl is an intermediate and its concentration is not easily measured. We can eliminate [Cl] because the first reaction is in rapid equilibrium. At any instant we have

$$K = \frac{[Cl]^2}{[Cl_2]}$$

Thus

$$[Cl] = (K[Cl_2])^{1/2} = K^{1/2}[Cl_2]^{1/2}$$

Therefore we have

$$\text{rate} = k[COCl_2](K^{1/2})[Cl_2]^{1/2}$$

If we let $kK^{1/2}$ be a new rate constant k_1, then we have

$$\text{rate} = k_1[COCl_2][Cl_2]^{1/2}$$

which is the observed rate law.

16-58. The rate of the reaction is equal to the rate of the slow rate-determining step

$$\text{rate} = k_1[Hg][Tl^{3+}]$$

Step (1) is a rapid equilibrium, thus

$$K_1 = \frac{[Hg][Hg^{2+}]}{[Hg_2^{2+}]}$$

and

$$[Hg] = \frac{K_1[Hg_2^{2+}]}{[Hg^{2+}]}$$

Substitution of this expression for [Hg] into the rate expression yields

$$\text{rate} = \frac{k_1K_1[Hg_2^{2+}][Tl^{3+}]}{[Hg^{2+}]} = k\frac{[Hg_2^{2+}][Tl^{3+}]}{[Hg^{2+}]}$$

16-60. The rate law of the elementary slow step is

$$\text{rate} = k[N_2O_5^*]$$

The species $N_2O_5^*$ is an intermediate. We can eliminate $[N_2O_5^*]$ from the rate law using the equilibrium constant expression of Step (1).

$$K = \frac{[N_2O_5^*][N_2O_5]}{[N_2O_5]^2} = \frac{[N_2O_5^*]}{[N_2O_5]}$$

$$[N_2O_5^*] = K[N_2O_5]$$

Substituting $[N_2O_5^*]$ into the rate law, we have

$$\text{rate} = \underbrace{kK}_{k'}[N_2O_5]$$

$$= k'[N_2O_5]$$

16-62. The Arrhenius equation is

$$\log\left(\frac{k_2}{k_1}\right) = \frac{E_a}{2.30\ R}\left(\frac{T_2 - T_1}{T_1T_2}\right)$$

Thus

$$\log\left(\frac{0.750\ M^{-1}\cdot s^{-1}}{0.0234\ M^{-1}\cdot s^{-1}}\right) = \frac{E_a}{(2.30)(8.31\ J\cdot K^{-1}\cdot mol^{-1})}\frac{(773\ K - 673\ K)}{(673\ K)(773\ K)}$$

$$1.506 = (1.01 \times 10^{-5}\ J^{-1}\cdot mol)E_a$$

$$E_a = \frac{1.506}{1.01 \times 10^{-5}\ J^{-1}\cdot mol} = 1.49 \times 10^5\ J\cdot mol^{-1}$$

$$= 149\ kJ\cdot mol^{-1}$$

16-64. The Arrhenius equation is

$$\log\left(\frac{k_2}{k_1}\right) = \frac{E_a}{2.30\ R}\left(\frac{T_2 - T_1}{T_1T_2}\right)$$

Thus

$$\log\left(\frac{k_2}{5.0 \times 10^{-4}\ s^{-1}}\right) = \frac{102 \times 10^3\ J\cdot mol^{-1}}{(2.30)(8.31\ J\cdot mol^{-1}\cdot K^{-1})}\frac{(338\ K - 318\ K)}{(318\ K)(338\ K)}$$

$$= 0.993$$

$$\frac{k_2}{5.0 \times 10^{-4}\ s^{-1}} = 9.84$$

$$k_2 = (9.84)(5.0 \times 10^{-4}\ s^{-1})$$

$$= 4.9 \times 10^{-3}\ s^{-1}$$

16-66. The Arrhenius equation is

$$\log\left(\frac{k_2}{k_1}\right) = \frac{E_a}{2.30\ R}\left(\frac{T_2 - T_1}{T_1T_2}\right)$$

We shall consider the pulse rate to be the rate constant k, thus

$$\log\left(\frac{k_2}{75\ \text{beats}\cdot\text{min}^{-1}}\right) = \frac{(30 \times 10^3\ \text{J}\cdot\text{mol}^{-1})(295\ \text{K} - 310\ \text{K})}{(2.30)(8.31\ \ \text{J}\cdot\text{K}^{-1}\cdot\text{mol}^{-1})(295\ \text{K})(310\ \text{K})}$$

$$= -0.2575$$

See the inside back cover for the conversion factor of ^{0}F to ^{0}C.

Hence

$$\frac{k_2}{75\ \text{beats}\cdot\text{min}^{-1}} = 0.553$$

and

$$k_2 = (0.553)(75\ \text{beats}\cdot\text{min}^{-1})$$

$$= 41\ \text{beats}\cdot\text{min}^{-1}$$

16-68. We first must calculate log k and 1/T

$(1/T)/10^{-3}K^{-1}$	log k
3.66	-6.10
3.36	-4.46
3.14	-3.30
2.96	-2.31

A plot of log k vs. 1/T is a straight line. To estimate k at 50^0C, we read the value of log k at $1/323\ \text{K} = 3.10 \times 10^{-3}\text{K}^{-1}$ from the plot. We see that

$$\log k = -3.2$$

$$k = 10^{-3.2} = 6 \times 10^{-4}\text{s}^{-1}$$

We can calculate E_a using the Arrhenius equation and any pair of data

$$\log\left(\frac{k_2}{k_1}\right) = \frac{E_a}{2.30R}\left(\frac{T_2 - T_1}{T_1T_2}\right)$$

$$\log\left(\frac{49.8 \times 10^{-5}\text{s}^{-1}}{3.46 \times 10^{-5}\text{s}^{-1}}\right) = \frac{E_a(318\text{ K} - 298\text{ K})}{(2.30)(8.31\text{ J}\cdot\text{K}^{-}\cdot\text{mol}^{-})(318\text{ K})(298\text{ K})}$$

$$1.158 = (1.104 \times 10^{-5}\text{J}^{-1}\cdot\text{mol})E_a$$

$$E_a = \frac{1.158}{1.104 \times 10^{-5}\text{J}^{-1}\cdot\text{mol}} = 1.05 \times 10^5\text{ J}\cdot\text{mol}^{-1}$$

$$= 105\text{ kJ}\cdot\text{mol}^{-1}$$

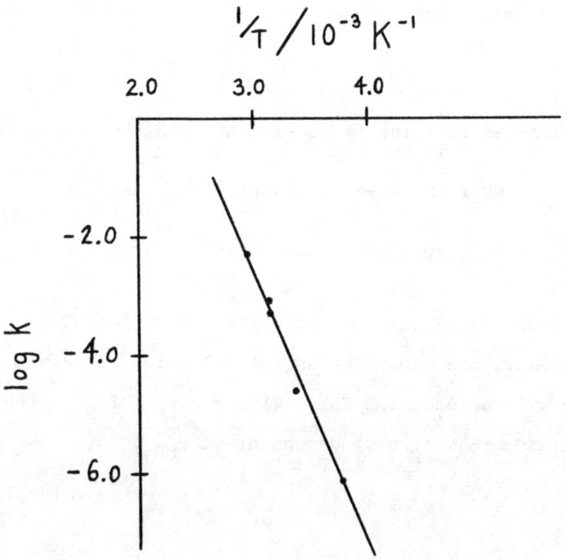

16-70. A catalyst lowers the activation energy thereby speeding up the reaction rates in both directions.

16-72. When $[H_2O_2]_0$ is doubled and $[I^-]_0$ and $[H^+]_0$ remain the same, the initial rate doubles. The reaction is first-order in $[H_2O_2]$. When $[I^-]_0$ is doubled and $[H^+]_0$ and $[H_2O_2]_0$ remain the same, the initial rate doubles. The reaction is first-order in $[I^-]$. When $[H^+]_0$ and $[I^-]_0$ are doubled and $[H_2O_2]_0$ remains the same, the initial rate increases by a factor of 8. Because the reaction is first-order in $[I^-]$, the reaction is second order in $[H^+]$. The rate law is

$$\text{rate} = k[H_2O_2][I^-][H^+]^2$$

The catalysts are $H^+(aq)$ and $I^-(aq)$ because they do not appear in the reaction as reactants.

16-74. Run the reaction with all initial concentration held constant in the dark and in the light. If the reaction rate differs in the light, then the reaction is influenced by light.

16-76. The observed rate law is consistent with the machanism

$$NH_3(g) \longrightarrow NH_3(\text{surface}) \qquad \text{fast}$$

$$2NH_3(\text{surface}) \longrightarrow N_2(g) + 3H_2(g) \qquad \text{slow}$$

If the number of surface sites for adsorption of NH_3 is small compared to the number of $NH_3(g)$ molecules, then all surface sites will be occupied for a wide range of P_{NH_3} values and the reaction rate will be independent of P_{NH_3}.

16-78. The equilibrium constant is related to the rate constants by

$$K = \frac{k_f}{k_r}$$

Thus

$$k_r = \frac{k_f}{K}$$

$$= \frac{8.4 \times 10^3\ M^{-1}\cdot s^{-1}}{8.4 \times 10^7\ M^{-1}} = 1.0 \times 10^{-4}\ s^{-1}$$

16-80. At equilibrium

$$\text{rate}_f = \text{rate}_r$$

Thus

$$\text{rate}_r = k_f[HNO_2][OH^-]$$

We want to find the rate law in the reverse direction in terms of k_r. We have that

$$K = \frac{k_f}{k_r}$$

or

$$k_f = k_r K$$

The equilibrium constant expression is

$$K = \frac{[NO_2^-]}{[HNO_2][OH^-]}$$

Thus

$$k_f = k_r \frac{[NO_2^-]}{[HNO_2][OH^-]}$$

Substituting the expression for k_f into the rate law, we have

$$\text{rate}_r = k_r \frac{[NO_2^-][HNO_2][OH^-]}{[HNO_2][OH^-]}$$

$$= k_r[NO_2^-]$$

16-82. The equilibrium constant for the reaction is

$$K = \frac{[H^+][NO_3^-][NO]^2}{[HNO_2]^3}$$

The forward reaction rate law is

$$\text{rate} = \frac{k_f[HNO_2]^4}{[NO]^2}$$

The backward reaction rate law is

$$\text{rate} = k_b X$$

The two rates are equal at equilibrium

$$kf \frac{[HNO_2]^4}{[NO]^2} = k_b X$$

But

$$K = \frac{k_f}{k_b}$$

and so

$$\frac{k_f}{k_b} = \frac{[NO]^2 X}{[HNO_2]^4} = K = \frac{[H^+][NO_3^-][NO]^2}{[HNO_2]^3}$$

Thus we see that

$$X = [H^+][NO_3^-][HNO_2]$$

and that the backward reaction rate law is

$$\text{rate} = k_b[H^+][NO_3^-][HNO_2]$$

CHAPTER 17

SOLUTIONS TO THE EVEN-NUMBERED PROBLEMS

17-2. We can calculate the value of $[OH^-]$ from the K_w expression

$$K_w = [H_3O^+][OH^-] = 1.00 \times 10^{-14}\ M^2$$

If we solve for $[OH^-]$, then we have

$$[OH^-] = \frac{1.00 \times 10^{-14}\ M^2}{[H_3O^+]} = \frac{1.00 \times 10^{-14}\ M^2}{2.5 \times 10^{-4}\ M} = 4.0 \times 10^{-11}\ M$$

Because $[H_3O^+] > [OH^-]$, the solution is acidic.

17-4. Because KOH is a strong base in water, it is completely dissociated and thus

$$[OH^-] = 0.25\ M \quad \text{and} \quad [K^+] = 0.25\ M$$

We can calculate $[H_3O^+]$ from the K_w expression

$$K_w = [H_3O^+][OH^-] = 1.00 \times 10^{-14}\ M^2$$

or

$$[H_3O^+] = \frac{1.00 \times 10^{-14}\ M^2}{[OH^-]} = \frac{1.00 \times 10^{-14}\ M^2}{0.25\ M} = 4.0 \times 10^{-14}\ M$$

Because $[OH^-] > [H_3O^+]$, the solution is basic.

17-6. We first find the number of moles in 0.60 g of $Ca(OH)_2$

$$n = (0.60\ g)\left(\frac{1\ mol}{74.10\ g}\right) = 8.10 \times 10^{-3}\ mol$$

The molarity of the solution is

$$\text{molarity} = \frac{\text{moles of solute}}{\text{volume of solution}} = \frac{8.10 \times 10^{-3}\ mol}{1.500\ L} = 5.40 \times 10^{-3}\ M$$

The strong base, $Ca(OH)_2$, is completely dissociated in water and yields two OH^- per mole $Ca(OH)_2$.

$$[OH^-] = (2)(5.40 \times 10^{-3} \text{ M}) = 0.0108 \text{ M}$$

$$[Ca^+] = 5.40 \times 10^{-3} \text{ M}$$

We can calculate $[H_3O^+]$ from the K_w expression

$$K_w = [H_3O^+][OH^-] = 1.00 \times 10^{-14} \text{ M}^2$$

$$[H_3O^+] = \frac{1.00 \times 10^{-14} \text{ M}^2}{[OH^-]} = \frac{1.00 \times 10^{-14} \text{ M}^2}{0.0108 \text{ M}} = 9.3 \times 10^{-13} \text{ M}$$

17-8. Because CsOH is a strong base in water, it is completely dissociated and thus

$$[OH^-] = 0.10 \text{ M}$$

We can find the value of $[H_3O^+]$ from the K_w expression

$$[H_3O^+] = \frac{1.00 \times 10^{-14} \text{ M}^2}{[OH^-]} = \frac{1.00 \times 10^{-14} \text{ M}^2}{0.10 \text{ M}} = 1.0 \times 10^{-13} \text{ M}$$

The pH of the solution is

$$pH = -\log[H_3O^+] = -\log(1.0 \times 10^{-13}) = 13.00$$

Because pH > 7, the solution is basic.

17-10. Because $Ba(OH)_2$ is a strong base in water, it is completely dissociated to yield two $OH^-(aq)$ per mole of $Ba(OH)_2(aq)$.

$$[OH^-] = (2)(0.0010 \text{ M}) = 0.0020 \text{ M} = 2.0 \times 10^{-3} \text{ M}$$

We can find the value of $[H_3O^+]$ from the K_w expression

$$[H_3O^+] = \frac{1.00 \times 10^{-14} \text{M}^2}{[OH^-]} = \frac{1.00 \times 10^{-14} \text{M}^2}{2.0 \times 10^{-3} \text{ M}} = 5.0 \times 10^{-12} \text{M}$$

The pH of the solution is

$$pH = -\log[H_3O^+] = -\log(5.0 \times 10^{-12}) = 11.30$$

The solution is basic.

17-12. We first must calculate the number of moles in 6.25 g of NaOH

$$n = (6.25\ g)\left(\frac{1\ mol}{40.00\ g}\right) = 0.156\ mol$$

The molarity of the solution is

$$\text{molarity} = \frac{\text{moles of solute}}{\text{volume of solution}} = \frac{0.156\ mol}{0.100\ L} = 1.56\ M$$

Because NaOH is a strong base, it is completely dissociated in water and thus

$$[OH^-] = 1.56\ M$$

We can calculate $[H_3O^+]$ from the K_w expression

$$[H_3O^+] = \frac{1.00 \times 10^{-14}\ M^2}{[OH^-]} = \frac{1.00 \times 10^{-14}\ M^2}{1.56\ M} = 6.41 \times 10^{-15}\ M$$

The pH of the solution is

$$pH = -\log[H_3O^+] = -\log(6.41 \times 10^{-15}) = 14.19$$

17-14. The ion-product constant for water at 37°C is

$$K_w = [H_3O^+][OH^-] = 2.40 \times 10^{-14}\ M^2$$

In a neutral aqueous solution

$$[H_3O^+] = [OH^-]$$

Therefore

$$K_w = [H_3O^+][H_3O^+] = [H_3O^+]^2 = 2.40 \times 10^{-14}\ M^2$$

Taking the square root of both sides, we get

$$[H_3O^+] = 1.55 \times 10^{-7}\ M$$

The pH of the solution is

$$pH = -\log[H_3O^+] = -\log(1.55 \times 10^{-7}) = 6.81$$

A solution of pH = 7.00 is basic at 37°C , because the pH of the solution is greater than 6.81, the pH of a neutral solution at 37°C.

17-16. From the definition of pH

$$pH = -\log[H_3O^+]$$

we write

$$[H_3O^+] = 10^{-pH}$$

$$[H_3O^+] = 10^{-12} = 1 \times 10^{-12}\ M$$

We can calculate $[OH^-]$ from the K_w expression

$$[OH^-] = \frac{1.00 \times 10^{-14}\ M^2}{[H_3O^+]} = \frac{1.00 \times 10^{-14}\ M^2}{1 \times 10^{-12}\ M} = 1 \times 10^{-2}\ M$$

The reaction is

$$NH_3(aq) + H_2O(l) \rightleftharpoons NH_4^+(aq) + OH^-(aq)$$

17-18. From the definition of pH

$$pH = -\log[H_3O^+]$$

we write

$$[H_3O^+] = 10^{-pH}$$

For normal rain water the pH is 5.6, and so

$$[H_3O^+] = 10^{-5.6}$$

$$= 2.51 \times 10^{-6}\ M$$

For acid rain water the pH is 3.0 and so

$$[H_3O^+] = 10^{-3.0}$$

$$= 1.00 \times 10^{-3}\ M$$

The ratio of the concentration of $H_3O^+(aq)$ in acid rain to the concentration of $H_3O^+(aq)$ in normal rain is

$$\text{ratio} = \frac{[H_3O^+]_{\text{acid rain}}}{[H_3O^+]_{\text{normal rain}}} = \frac{1.00 \times 10^{-3}\ M}{2.51 \times 10^{-6}\ M}$$

$$= 400$$

The acidity of normal rain is due to dissolved CO_2.

$$CO_2(aq) + 2H_2O(l) \rightleftharpoons H_3O^+(aq) + HCO_3^-(aq)$$

17-20. From the definition of pH

$$pH = -\log[H_3O^+]$$

we find

$$[H_3O^+] = 10^{-pH}$$

The pH of the world's oceans is 8.15 and so

$$[H_3O^+] = 10^{-8.15} = 7.08 \times 10^{-9}\ M$$

We can calculate $[OH^-]$ from the K_w expression

$$[OH^-] = \frac{1.00 \times 10^{-14}\ M^2}{[H_3O^+]} = \frac{1.00 \times 10^{-14}\ M^2}{7.08 \times 10^{-9}\ M} = 1.41 \times 10^{-6}\ M$$

17-22. From the definition of pH

$$pH = -\log[H_3O^+]$$

we write

$$[H_3O^+] = 10^{-pH} = 10^{-10.52} = 3.02 \times 10^{-11}\ M$$

We can calculate $[OH^-]$ from the K_w expression

$$[OH^-] = \frac{1.00 \times 10^{-14}\ M^2}{[H_3O^+]} = \frac{1.00 \times 10^{-14}\ M^2}{3.02 \times 10^{-11}\ M} = 3.31 \times 10^{-4}\ M$$

Because one mole of $Mg(OH)_2$ yields two moles of OH^- in water, the concentration of $Mg(OH)_2$ is

$$[Mg(OH)_2] = \frac{1}{2}[OH^-] = 1.656 \times 10^{-4}\ M$$

Solubility often is expressed as the number of grams per 100 mL of solution. The number of moles in 100 mL of solution is

$$n = \text{molarity} \times \text{volume} = (1.656 \times 10^{-4}\ mol \cdot L^{-1})(0.100\ L)$$

$$= 1.656 \times 10^{-5}\ mol$$

The mass in 1.656×10^{-5} mol of $Mg(OH)_2$ is

$$\text{mass} = (1.656 \times 10^{-5}\ mol)\left(\frac{58.33\ g}{1\ mol}\right) = 9.66 \times 10^{-4}\ g$$

The solubility of $Mg(OH)_2$ is 9.66×10^{-4} g per 100 mL of solution.

17-24. The reaction is

$$HOCN(aq) + H_2O(l) \rightleftharpoons H_3O^+(aq) + OCN^-(aq)$$

The acid-dissociation constant expression is

$$K_a = \frac{[H_3O^+][OCN^-]}{[HOCN]}$$

We can find the value of $[H_3O^+]$ from the pH

$$[H_3O^+] = 10^{-pH} = 10^{-2.19} = 6.46 \times 10^{-3}\ M$$

Set up a table of initial concentrations and equilibrium concentrations.

	$HOCN(aq)$	$+\ H_2O(l) \rightleftharpoons$	$H_3O^+(aq)$	$+\ OCN^-(aq)$
Initial Concentration	0.20 M		0 $[H_3O^+]$	0 $[IO_4^-] = [H_3O^+]$
Equilibrium Concentration	$0.20\ M - [H_3O^+]$ $0.20\ M - 6.46 \times 10^{-3}$ 0.19 M		$[H_3O^+]$ $6.46 \times 10^{-3}\ M$	$[OCN^-] = [H_3O^+]$ $6.46 \times 10^{-3}\ M$

Substituting the equilibrium concentrations into the K_a expression, we have

$$K_a = \frac{(6.46 \times 10^{-3}\ M)(6.46 \times 10^{-3}\ M)}{0.19\ M}$$

$$K_a = 2.2 \times 10^{-4}\ M$$

17-26. The reaction is

$$HCHO_2(aq) + H_2O(l) \rightleftharpoons H_3O^+(aq) + CHO_2^-(aq)$$

The acid-dissociation constant expression is

$$K_a = \frac{[H_3O^+][CHO_2^-]}{[HCHO_2]}$$

We can find the value of $[H_3O^+]$ from the pH of the solution.

$$[H_3O^+] = 10^{-pH} = 10^{-2.38} = 4.17 \times 10^{-3}\ M$$

Set up a table of initial concentrations and equilibrium concentrations.

	$HCHO_2(aq)$	$+ H_2O(l) \rightleftharpoons$	$H_3O^+(aq)$	$+ CHO_2^-(aq)$
Initial Concentration	0.10 M		0	0
Equilibrium Concentration	0.10 M-$[H_3O^+]$ 0.10 M-4.17x10^{-3}M 0.096 M		$[H_3O^+]$ 4.17x10^{-3}M	$[CHO_2^-] = [H_3O^+]$ 4.17x10^{-3}M

Substituting the values of the equilibrium concentrations in the K_a expression, we have

$$K_a = \frac{(4.17 \times 10^{-3}\ M)(4.17 \times 10^{-3}\ M)}{0.096\ M}$$

$$= 1.8 \times 10^{-4}\ M$$

17-28. The reaction is

$$HC_2H_2ClO_2(aq) + H_2O(l) \rightleftharpoons H_3O^+(aq) + C_2H_2ClO_2^-(aq)$$

The acid dissociation constant expression is

$$K_a = \frac{[H_3O^+][C_2H_2ClO_2^-]}{[HC_2H_2ClO_2]} = 1.35 \times 10^{-3}\ M$$

and so

$$\frac{[C_2H_2ClO_2^-]}{[HC_2H_2ClO_2]} = \frac{K_a}{[H_3O^+]} = \frac{1.35 \times 10^{-3}\ M}{[H_3O^+]}$$

We can find the value of $[H_3O^+]$ from the pH

$$[H_3O^+] = 10^{-pH} = 10^{-4.61} = 2.45 \times 10^{-5}\ M$$

$$\frac{[C_2H_2ClO_2^-]}{[HC_2H_2ClO_2]} = \frac{1.35 \times 10^{-3}\ M}{2.45 \times 10^{-5}\ M} = 55.1$$

17-30. The reaction is

$$\text{uric acid(aq)} + H_2O(l) \rightleftharpoons H_3O^+(aq) + \text{urate ion(aq)}$$

The acid dissociation constant expression is

$$K_a = \frac{[H_3O^+][\text{urate ion}]}{[\text{uric acid}]} = 4.0 \times 10^{-6}\ M$$

The ratio of urate ion to uric acid is

$$\frac{[\text{urate ion}]}{[\text{uric acid}]} = \frac{K_a}{[H_3O^+]} = \frac{4.0 \times 10^{-6}\ M}{[H_3O^+]}$$

The concentration of $H_3O^+(aq)$ is

$$[H_3O^+] = 10^{-pH} = 10^{-6.0} = 1.0 \times 10^{-6}\ M$$

$$\frac{[\text{urate ion}]}{[\text{uric acid}]} = \frac{4.0 \times 10^{-6}\ M}{1.0 \times 10^{-6}\ M} = 4.0$$

17-32. The reaction is

$$\underset{\text{benzoic acid}}{HC_7H_5O_2(aq)} + H_2O(l) \rightleftharpoons H_3O^+(aq) + \underset{\text{benzoate}}{C_7H_5O_2^-(aq)}$$

The acid dissociation constant expression is

$$K_a = \frac{[H_3O^+][C_7H_5O_2^-]}{[HC_7H_5O_2]} = \frac{[H_3O^+][\text{benzoate}]}{[\text{benzoic acid}]}$$

$$\frac{[HC_7H_5O_2]}{[C_7H_5O_2^-]} = \frac{[\text{benzoic acid}]}{[\text{benzoate}]} = \frac{[H_3O^+]}{K_a} = \frac{[H_3O^+]}{6.46 \times 10^{-5}\ M}$$

The concentration of H_3O^+ (aq) is

$$[H_3O^+] = 10^{-pH} = 10^{-3.0} = 1.0 \times 10^{-3}\ M$$

The ratio is

$$\frac{[\text{benzoic acid}]}{[\text{benzoate}^-]} = \frac{1.0 \times 10^{-3}\ M}{6.46 \times 10^{-5}\ M} = 15$$

17-34. The reaction is

$$HClO(aq) + H_2O(l) \rightleftharpoons H_3O^+(aq) + ClO^-(aq)$$

We can set up a table of initial and equilibrium concentrations.

	$HClO(aq)$	$+ H_2O(l) \rightleftharpoons$	$H_3O^+(aq)$	$+ ClO^-(aq)$
Initial Concentration	0.15 M		0	0
Equilibrium Concentration	$0.15\ M - [H_3O^+]$		$[H_3O^+]$	$[ClO^-] = [H_3O^+]$

Substituting the equilibrium concentration expressions in the K_a expression, we have

$$K_a = \frac{[H_3O^+][ClO^-]}{[HClO]} = \frac{[H_3O^+]^2}{0.15\ M - [H_3O^+]} = 2.95 \times 10^{-8}\ M$$

The value of K_a is very small so that we expect that $[H_3O^+]$ shall also be very small. We ignore $[H_3O^+]$ compared to 0.15 M, and obtain

$$\frac{[H_3O^+]^2}{0.15\ M} = 2.95 \times 10^{-8}\ M$$

$$[H_3O^+]^2 = 4.43 \times 10^{-9}\ M^2$$

$$[H_3O^+] = 6.66 \times 10^{-5}\ M$$

We see that $[H_3O^+]$ is indeed very small compared to 0.15 M. The pH of the solution is

$$pH = -\log[H_3O^+] = -\log(6.66 \times 10^{-5}) = 4.18$$

17-36. The reaction is

$$HC_2H_2ClO_2(aq) + H_2O(l) \rightleftharpoons H_3O^+(aq) + C_2H_2ClO_2^-(aq)$$

To find $[H_3O^+]$ in the solution, we set up a table of initial and equilibrium concentrations

	$HC_2H_2ClO_2(aq)$	$+\ H_2O(l)$	$\rightleftharpoons\ H_3O^+(aq)$	$+\ C_2H_2ClO_2^-(aq)$
Initial Concentration	0.10 M		0	0
Equilibrium Concentration	$0.10\ M - [H_3O^+]$		$[H_3O^+]$	$[C_2H_2ClO_2^-] = [H_3O^+]$

Substituting the equilibrium concentration expressions in the K_a expression, we have

$$K_a = \frac{[H_3O^+][C_2H_2ClO_2^-]}{[HC_2H_2ClO_2]} = \frac{[H_3O^+]^2}{0.10\ M - [H_3O^+]} = 1.35 \times 10^{-3}\ M$$

or

$$[H_3O^+]^2 + 1.35 \times 10^{-3}\ M[H_3O^+] - 1.35 \times 10^{-4}\ M^2 = 0$$

The quadratic formula gives

$$[H_3O^+] = \frac{-1.35 \times 10^{-3}\ M \pm \sqrt{1.82 \times 10^{-6}\ M^2 - (4)(1)(-1.35 \times 10^{-4}\ M^2)}}{(2)(1)}$$

$$= \frac{-1.35 \times 10^{-3}\ M \pm 2.34 \times 10^{-2}\ M}{2}$$

$$= 1.10 \times 10^{-2}\ M \text{ and } -1.24 \times 10^{-2}\ M$$

We reject the negative root because concentrations must be positive. Thus

$$[H_3O^+] = 1.10 \times 10^{-2}\ M = 0.0110\ M$$

and

$$pH = -\log[H_3O^+] = -\log(0.0110) = 1.96$$

17-38. The number of moles in two 5 grain aspirin tablets is

$$n = (2)(324\ mg)\left(\frac{1\ g}{1000\ mg}\right)\left(\frac{1\ mol}{180.15\ g}\right) = 0.00360\ mol$$

The molarity of the solution is

$$\text{molarity} = \frac{\text{moles of solute}}{\text{volume of solution}} = \frac{0.00360\ mol}{0.500\ L} = 0.00720\ M$$

We can set up a table of initial and equilibrium concentrations.

	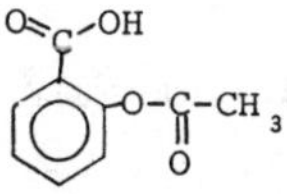 (aq)	+ $H_2O(l)$ ⇌	$H_3O^+(aq)$ +	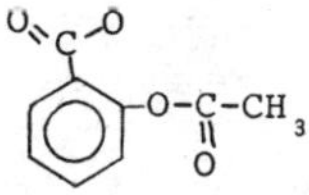 (aq)
Initial Concentration	0.00720 M		0	0
Equilibrium Concentration	0.00720 M − $[H_3O^+]$		$[H_3O^+]$	$[H_3O^+]$

Substituting the equilibrium concentration expression in the K_a expression, we have

$$K_a = \frac{[H_3O^+]^2}{0.00720\ M - [H_3O^+]} = 2.75 \times 10^{-5}\ M$$

Write this equation in the form of a quadratic equation

$$[H_3O^+]^2 + 2.75 \times 10^{-5}\ M[H_3O^+] - 1.98 \times 10^{-7}\ M^2 = 0$$

The quadratic formula gives

$$[H_3O^+] = \frac{-2.75 \times 10^{-5}\text{ M} \pm \sqrt{7.56 \times 10^{-10}\text{ M}^2 - (4)(1)(-1.98 \times 10^{-7}\text{ M}^2)}}{(2)(1)}$$

$$= \frac{-2.75 \times 10^{-5}\text{ M} \pm 8.90 \times 10^{-4}\text{ M}}{2}$$

$$= 4.31 \times 10^{-4}\text{ M} \quad \text{and} \quad -4.59 \times 10^{-4}\text{ M}$$

We reject the negative root and calculate the pH of the solution by

$$\text{pH} = -\log[H_3O^+] = -\log(4.31 \times 10^{-4}) = 3.37$$

17-40. The number of moles in 10.0 g of $NaHSO_4$ is

$$n = (10.0\text{ g})\left(\frac{1\text{ mol}}{120.06\text{ g}}\right) = 0.0833\text{ mol}$$

The molarity of the solution is

$$\text{molarity} = \frac{\text{moles of solute}}{\text{volume of solution}} = \frac{0.0833\text{ mol}}{0.100\text{ L}} = 0.833\text{ M}$$

We can set up a table of initial and equilibrium concentrations.

	$HSO_4^-(aq)$	+ $H_2O(l)$ $\rightleftharpoons$	$H_3O^+(aq)$	+ $SO_4^{2-}(aq)$
Initial Concentration	0.833 M		0	0
Equilibrium Concentration	0.833 M − $[H_3O^+]$		$[H_3O^+]$	$[SO_4^{2-}] = [H_3O^+]$

Substituting the equilibrium concentration expressions in the K_a expression, we have

$$K_a = \frac{[SO_4^{2-}][H_3O^+]}{[HSO_4^-]} = \frac{[H_3O^+]^2}{0.833\text{ M} - [H_3O^+]} = 0.012\text{ M}$$

or

$$[H_3O^+]^2 + 0.012\ M[H_3O^+] - 0.0100\ M^2 = 0$$

The quadratic formula gives

$$[H_3O^+] = \frac{-0.012\ M \pm \sqrt{0.000144 M^2 - (4)(1)(-0.0100\ M^2)}}{(2)(1)}$$

$$= \frac{-0.012\ M \pm 0.200\ M}{2}$$

$$= 0.094\ M$$

The pH of the solution is

$$pH = -\log[H_3O^+] = -\log(0.094) = 1.03$$

17-42. The reaction is

$$C_2H_5NH_2(aq) + H_2O(l) \rightleftharpoons C_2H_5NH_3^+(aq) + OH^-(aq)$$

We can find the concentration of $OH^-(aq)$ from the pH and the ion-

$$[H_3O^+] = 10^{-pH} = 10^{-12.20} = 6.31 \times 10^{-13}\ M$$

$$[OH^-] = \frac{1.00 \times 10^{-14}\ M^2}{[H_3O^+]} = \frac{1.00 \times 10^{-14}\ M^2}{6.31 \times 10^{-13}\ M} = 1.58 \times 10^{-2}\ M$$

We can set up a table of initial and equilibrium concentrations.

	$C_2H_5NH_2(aq)$	+ $H_2O(l) \rightleftharpoons$	$C_2H_5NH_3^+(aq)$	+ $OH^-(aq)$
Initial Concentration	0.50 M		0	0
Equilibrium Concentration	0.50 M − $[OH^-]$ 0.48 M		$[C_2H_5NH_3^+] = [OH^-]$ 1.58×10^{-2} M	$[OH^-]$ 1.58×10^{-2} M

The base protonation constant expression is

$$K_b = \frac{[C_2H_5NH_3^+][OH^-]}{[C_2H_5NH_2]} = \frac{(1.58 \times 10^{-2}\ M)(1.58 \times 10^{-2}\ M)}{(0.48\ M)}$$

$$= 5.2 \times 10^{-4}\ M$$

17-44. The reaction is

$$HONH_2(aq) + H_2O(l) \rightleftharpoons HONH_3^+(aq) + OH^-(aq)$$

We can set up a table of initial and equilibrium concentrations.

	$HONH_2(aq)$	$+ H_2O(l) \rightleftharpoons$	$HONH_3^+(aq)$	$+ OH^-(aq)$
Initial Concentration	0.125 M		0	0
Equilibrium Concentration	$0.125\ M - [OH^-]$		$[HONH_3^+] = [OH^-]$	$[OH^-]$

The expression for K_b is

$$K_b = \frac{[HONH_3^+][OH^-]}{[HONH_2]} = \frac{[OH^-]^2}{0.125\ M - [OH^-]} = 1.07 \times 10^{-8}\ M$$

Because K_b is so small, we expect that $[OH^-]$ will be small and thus negligible compared to 0.125 M. The K_b expression becomes

$$K_b = \frac{[OH^-]^2}{0.125\ M} = 1.07 \times 10^{-8}\ M$$

$$[OH^-]^2 = 1.34 \times 10^{-9}\ M^2$$

$$[OH^-] = 3.66 \times 10^{-5}\ M$$

$$[H_3O^+] = \frac{1.00 \times 10^{-14}\ M^2}{[OH^-]} = \frac{1.00 \times 10^{-14}\ M^2}{3.66 \times 10^{-5}\ M} = 2.73 \times 10^{-10}\ M$$

The pH of the solution is

$$pH = -\log[H_3O^+] = -\log(2.73 \times 10^{-10}) = 9.56$$

17-46. We can set up a table of initial and equilibrium concentrations.

	$NH_3(aq)$ + $H_2O(l)$		$\rightleftharpoons$ $NH_4^+(aq)$ +	$OH^-(aq)$
Initial Concentration	0.20 M		0	0
Equilibrium Concentration	$0.20\ M - [OH^-]$		$[NH_4^+] = [OH^-]$	$[OH^-]$

The expression for K_b is

$$K_b = \frac{[NH_4^+][OH^-]}{[NH_3]} = \frac{[OH^-]^2}{0.20\ M - [OH^-]} = 1.8 \times 10^{-5}\ M$$

The value of K_b may not be small enough to ignore the $[OH^-]$ in the denominator and so we shall use the quadratic formula

$$[OH^-]^2 + 1.8 \times 10^{-5}\ M[OH^-] - 3.6 \times 10^{-6}\ M^2 = 0$$

The quadratic formula gives

$$[OH^-] = \frac{-1.8 \times 10^{-5}\ M + \sqrt{3.24 \times 10^{-10}\ M^2 - (4)(1)(-3.6 \times 10^{-6}\ M^2)}}{(2)(1)}$$

$$[OH^-] = \frac{-1.8 \times 10^{-5}\ M \pm 3.8 \times 10^{-3}\ M}{2}$$

$$[OH^-] = 1.9 \times 10^{-3}\ M$$

$$[H_3O^+] = \frac{1.00 \times 10^{-14}\ M^2}{[OH^-]} = \frac{1.00 \times 10^{-14}\ M^2}{1.9 \times 10^{-3}\ M} = 5.3 \times 10^{-12}\ M$$

The pH of the solution is

$$pH = -\log[H_3O^+] = -\log(5.3 \times 10^{-12}) = 11.28$$

17-48. a) The equilibrium is shifted toward $NH_4^+(aq)$. The added HCl decreases the concentration of $OH^-(aq)$.

b) The equilibrium is shifted toward $NH_3(aq)$. The concentration of $OH^-(aq)$ has been increased.

c) The equilibrium is shifted toward $HCHO_2(aq)$. The concentration of $H_3O^+(aq)$ has been increased.

d) The equilibrium is shifted away from $HCHO_2(aq)$. The added $OH^-(aq)$ reacts with the $H_3O^+(aq)$ to produce H_2O, thereby decreasing the concentration of $H_3O^+(aq)$.

17-50. a) The equilibrium is shifted from left to right.

b) The equilibrium is shifted from right to left because $[H_3O^+]$ has increased.

17-52. a) The equilibrium is shifted from right to left because $\Delta H_{rxn} < 0$.

b) The equilibrium is shifted from right to left because $NO_2^-(aq)$ is added.

c) The equilibrium is shifted from left to right. The $OH^-(aq)$ added reacts with the $H_3O^+(aq)$ so that the concentration of $H_3O^+(aq)$ is decreased.

d) The equilibrium is shifted from left to right.

17-54. a) $NH_3(l) + NH_3(l) \rightleftharpoons NH_4^+(am) + NH_2^-(am)$

conjugate acid-base pair (NH_3, NH_4^+)

conjugate acid-base pair (NH_3, NH_2^-)

b) $HNO_2(aq) + H_2O(l) \rightleftharpoons H_3O^+(aq) + NO_2^-(aq)$

conjugate acid-base pair (H_2O, H_3O^+)

conjugate acid-base pair (HNO_2, NO_2^-)

c) $C_5H_5N(aq) + H_2O(l) \rightleftharpoons C_5H_5NH^+(aq) + OH^-(aq)$

conjugate acid-base pair (H_2O, OH^-)

conjugate acid-base pair (C_5H_5N, $C_5H_5NH^+$)

17-56. a) $NO_3^-(aq)$ c) $C_6H_5O^-(aq)$

b) $CHO_2^-(aq)$ d) $CH_3NH_2(aq)$

17-58. a) $HC_2H_2ClO_2$ is an acid; $C_2H_2ClO_2^-$ is its conjugate base.

b) NH_3 is a base; NH_4^+ is its conjugate acid.

c) ClO^- is a base; HClO is its conjugate acid.

d) CHO_2^- is a base; $HCHO_2$ is its conjugate acid.

e) HN_3 is an acid; N_3^- is its conjugate base.

f) NO_2^- is a base; HNO_2 is its conjugate acid.

17-60. We have that $K_a = \frac{K_w}{K_b}$

a) $K_a = \frac{1.00 \times 10^{-14}\ M^2}{1.5 \times 10^{-9}\ M} = 6.7 \times 10^{-6}\ M$ for $C_5H_5NH^+$

b) $K_a = \frac{1.00 \times 10^{-14}\ M^2}{2.1 \times 10^{-5}\ M} = 4.8 \times 10^{-10}\ M$ for HCN

c) $K_a = \frac{1.00 \times 10^{-14}\ M^2}{4.6 \times 10^{-11}\ M} = 2.2 \times 10^{-4}\ M$ for HCNO

d) $K_a = \frac{1.00 \times 10^{-14}\ M^2}{1.1 \times 10^{-7}\ M} = 9.1 \times 10^{-8}\ M$ for H_2S

17-62. a) The equation is the sum of two equations

(1) $C_2H_3O_2^-(aq) + H_2O(l) \rightleftharpoons HC_2H_3O_2(aq) + OH^-(aq)$

$K_1 = K_b = 5.9 \times 10^{-10}\ M$

(2) $NH_4^+(aq) + OH^-(aq) \rightleftharpoons NH_3(aq) + H_2O(l)$

$K_2 = \frac{1}{K_b} = \frac{1}{1.7 \times 10^{-5}\ M}$

The equilibrium constant for a reaction that is the sum of two equations is the product of their equilibrium constants.

$$K = K_1K_2 = \frac{5.9 \times 10^{-10}\ M}{1.7 \times 10^{-5}\ M} = 3.5 \times 10^{-5}\ M$$

b) The equation is the sum of the two equations

1) $C_6H_5O^-(aq) + H_2O(l) \rightleftharpoons HC_6H_5O(aq) + OH^-(aq)$

$$K_1 = K_b = 1.0 \times 10^{-4}\ M$$

2) $C_5H_5NH^+(aq) + OH^-(aq) \rightleftharpoons C_5H_5N(aq) + H_2O(l)$

$$K_2 = \frac{1}{K_b} = \frac{1}{1.5 \times 10^{-9}\ M}$$

The equilibrium constant for the reaction is

$$K = K_1K_2 = \frac{1.0 \times 10^{-4}\ M}{1.5 \times 10^{-9}\ M} = 6.7 \times 10^4$$

17-64. The value of pK_a is given by

$$pK_a = -\log K_a$$

a) The value of K_a for $NH_4^+(aq)$ is 5.6×10^{-10} M, Thus

$$pK_a = -\log(5.6 \times 10^{-10}) = 9.25$$

b) The value of K_a for $HSO_4^-(aq)$ is 1.2×10^{-2} M, thus

$$pK_a = -\log(1.2 \times 10^{-2}) = 1.92$$

17-66. At 25°C, we have that

$$K_aK_b = 1.00 \times 10^{-14}$$

If we take the logarithm of both sides, then we have

$$\log(K_aK_b) = \log(1.00 \times 10^{-14}) = -14.00$$

Recall that

$$\log ab = \log a + \log b$$

Therefore, we have

$$\log(K_a K_b) = \log K_a + \log K_b = -14.00$$

If we multiply both sides by -1, then we have

$$-\log K_a - \log K_b = 14.00$$

By definition

$$-\log K_a = pK_a$$

and

$$-\log K_b = pK_b$$

Thus

$$pK_a + pK_b = 14.00$$

17-68. a) Acidic cation, neutral anion; acidic solution.

b) Neutral cation, neutral anion; neutral solution.

c) Neutral cation, acidic anion; acidic solution.

d) Neutral cation, basic anion; basic solution.

e) Acidic cation, basic anion; cannot predict without calculations.

17-70. a) Acidic cation, basic anion; cannot predict without calculations.

b) Neutral cation, basic anion; basic solution.

c) Acidic cation, neutral anion; acidic solution.

d) Neutral cation, acidic anion; acidic solution.

e) Neutral cation, neutral anion; neutral solution.

17-72. The aluminum salts produce acidic solutions because $Al^{3+}(aq)$ is an acidic cation and $SO_4^{2-}(aq)$ is a neutral anion.

17-74. The reaction is

$$C_3H_7O_2^-(aq) + H_2O(l) \rightleftharpoons HC_3H_7O_2(aq) + OH^-(aq)$$

The equilibrium constant is

$$K_b = \frac{K_w}{K_a} = \frac{1.00 \times 10^{-14}\ M^2}{1.34 \times 10^{-5}\ M} = 7.46 \times 10^{-10}\ M$$

We can set up a table of initial and equilibrium concentrations.

	$C_3H_7O_2^-(aq)$	+ $H_2O(l)$ ⇌	$HC_3H_7O_2(aq)$ +	$OH^-(aq)$
Initial Concentration	0.20 M		0	0
Equilibrium Concentration	0.20 M − $[OH^-]$		$[HC_3H_7O_2] = [OH^-]$	$[OH^-]$

The expression for K_b is

$$K_b = \frac{[HC_3H_7O_2][OH^-]}{[C_3H_7O_2^-]} = \frac{[OH^-]^2}{0.20\ M - [OH^-]} = 7.46 \times 10^{-10}\ M$$

We can neglect $[OH^-]$ relative to 0.20 M because K_b is very small. Thus we have

$$\frac{[OH^-]^2}{0.20\ M} = 7.46 \times 10^{-10}\ M$$

$$[OH^-] = 1.22 \times 10^{-5}\ M$$

Using the ion-product constant of water, we obtain

$$[H_3O^+] = \frac{1.00 \times 10^{-14}\ M^2}{[OH^-]} = \frac{1.00 \times 10^{-14}\ M^2}{1.22 \times 10^{-5}\ M} = 8.20 \times 10^{-10}\ M$$

The pH of the solution is

$$pH = -\log[H_3O^+] = -\log(8.20 \times 10^{-10}) = 9.09$$

17-76. The reaction is

$$NO_2^-(aq) + H_2O(l) \rightleftharpoons HNO_2(aq) + OH^-(aq)$$

The equilibrium constant is (Table 17-2)

$$K_b = \frac{K_w}{K_a} = \frac{1.00 \times 10^{-14}\ M}{4.47 \times 10^{-4}\ M} = 2.24 \times 10^{-11}\ M$$

The table of initial and equilibrium concentrations is

	$NO_2^-(aq)$ +	$H_2O(l)$ $\rightleftharpoons$	$HNO_2(aq)$ +	$OH^-(aq)$
Initial Concentration	0.25 M		0	0
Equilibrium Concentration	$0.25\ M - [OH^-]$		$[HNO_2] = [OH^-]$	$[OH^-]$

The K_b expression is

$$K_b = \frac{[HNO_2][OH^-]}{[NO_2^-]} = \frac{[OH^-]^2}{0.25\ M - [OH^-]} = 2.24 \times 10^{-11}\ M$$

Neglecting $[OH^-]$ with respect to 0.25 M, we have

$$\frac{[OH^-]^2}{0.25\ M} = 2.24 \times 10^{-11}\ M$$

$$[OH^-] = 2.37 \times 10^{-6}\ M$$

and

$$[HNO_2] = [OH^-] = 2.37 \times 10^{-6}\ M$$

and

$$[NO_2^-] = 0.25\ M - [OH^-] = 0.25\ M - 2.37 \times 10^{-6}\ M = 0.25\ M$$

Using the ion-product constant of water, we obtain

$$[H_3O^+] = \frac{1.00 \times 10^{-14}\ M^2}{[OH^-]} = \frac{1.00 \times 10^{-14}\ M^2}{2.37 \times 10^{-6}\ M} = 4.22 \times 10^{-9}\ M$$

$$pH = -\log[H_3O^+] = -\log(4.22 \times 10^{-9}) = 8.37$$

17-78. The reaction is

$$CN^-(aq) + H_2O(l) \rightleftharpoons HCN(aq) + OH^-(aq)$$

A table of initial and equilibrium concentrations is

	$CN^-(aq)$	$+\ H_2O(l)$ $\rightleftharpoons$	$HCN(aq)$ +	$OH^-(aq)$
Initial Concentration	0.15 M		0	0
Equilibrium Concentration	$0.15\ M - [OH^-]$		$[HCN] = [OH^-]$	$[OH^-]$

The K_b expression is

$$K_b = \frac{[HCN][OH^-]}{[CN^-]} = \frac{[OH^-]^2}{0.15\ M - [OH^-]} = 2.1 \times 10^{-5}\ M$$

(Table 17-4)

Neglecting $[OH^-]$ relative to 0.15 M, we have

$$\frac{[OH^-]^2}{0.15\ M} = 2.1 \times 10^{-5}\ M$$

$$[OH^-] = 1.8 \times 10^{-3}\ M$$

and

$$[H_3O^+] = \frac{1.00 \times 10^{-14}\ M^2}{[OH^-]} = \frac{1.00 \times 10^{-14}\ M^2}{1.8 \times 10^{-3}\ M} = 5.6 \times 10^{-12}\ M$$

The pH of the solution is

$$pH = -\log[H_3O^+] = -\log(5.6 \times 10^{-12}) = 11.25$$

17-80. The number of moles in 25.0 g of barium acetate, $Ba(C_2H_3O_2)_2$, is

$$n = (25.0\ g)\left(\frac{1\ mol}{255.4\ g}\right) = 0.0979\ mol$$

The molarity of the barium acetate solution is

$$\text{molarity} = \frac{n}{V} = \frac{0.0979\ mol}{1.00\ L} = 0.0979\ M$$

The reaction is

$$C_2H_3O_2^-(aq) + H_2O(l) \rightleftharpoons HC_2H_3O_2(aq) + OH^-(aq)$$

From Table 17-4 we find that

$$K_b = 5.9 \times 10^{-10}\ M$$

We have

	$C_2H_3O_2^-(aq)$	+ $H_2O(l)$ $\rightleftharpoons$	$HC_2H_3O_2(aq)$	+ $OH^-(aq)$
Initial Concentration	0.196 M		0	0
Equilibrium Concentration	0.196 M − $[OH^-]$		$[OH^-]$	$[OH^-]$

Notice that $[C_2H_3O_2^-]$ is 2 x $[Ba(C_2H_3O_2)_2]$. The K_b expression is

$$K_b = \frac{[HC_2H_3O_2][OH^-]}{[C_2H_3O_2^-]} = \frac{[OH^-]^2}{0.196\ M - [OH^-]} = 5.9 \times 10^{-10}\ M$$

Neglecting $[OH^-]$ relative to 0.196 M, we have

$$\frac{[OH^-]^2}{0.196\ M} = 5.9 \times 10^{-10}\ M$$

$$[OH^-] = 1.08 \times 10^{-5} \text{ M}$$

$$[H_3O^+] = \frac{1.00 \times 10^{-14} \text{ M}^2}{[OH^-]}$$

$$= \frac{1.00 \times 10^{-14} \text{ M}^2}{1.08 \times 10^{-5} \text{ M}} = 9.26 \times 10^{-10} \text{ M}$$

The pH of the solution is

$$pH = -\log[H_3O^+] = -\log(9.26 \times 10^{-10}) = 9.03$$

17-82. In solution Tl^{3+} exists as $Tl(H_2O)_6^{3+}(aq)$.

	$Tl(H_2O)_6^{3+}(aq)$	$+ \; H_2O(l) \rightleftharpoons$	$H_3O^+(aq)$	$+ \; Tl(OH)(H_2O)_5^{2+}(aq)$
Initial Concentration	0.10 M		0	0
Equilibrium Concentration	$0.10 \text{ M} - [H_3O^+]$		$[H_3O^+]$	$[H_3O^+]$

The K_a expression is

$$K_a = \frac{[H_3O^+][Tl(OH)(H_2O)_5^{2+}]}{[Tl(H_2O)_6^{3+}]} = \frac{[H_3O^+]^2}{0.10 \text{ M} - [H_3O^+]} = 6 \times 10^{-2} \text{ M}$$

The value of K_a is not small enough to neglect $[H_3O^+]$ in the denominator, and so we must use the quadratic equation

$$[H_3O^+]^2 + 6 \times 10^{-2} \text{ M}[H_3O^+] - 6 \times 10^{-3} \text{ M}^2 = 0$$

The quadratic formula gives

$$[H_3O^+] = \frac{-6 \times 10^{-2}\ M \pm \sqrt{3.6 \times 10^{-3}\ M^2 - (4)(1)(-6 \times 10^{-3}\ M^2)}}{(2)(1)}$$

$$= \frac{-6 \times 10^{-2}\ M \pm 1.7 \times 10^{-1}\ M}{2}$$

$$= 5.5 \times 10^{-2}\ M$$

The pH of the solution is

$$pH = -\log[H_3O^+] = -\log(5.5 \times 10^{-2}) = 1.3$$

17-84. a) NH_3 yields $OH^-(aq)$ in aqueous solution and thus is an Arrhenius base. NH_3 is a proton acceptor and thus is a Brønsted-Lowry base. NH_3 can act as an electron-pair donor and thus is a Lewis base.

b) Br^- is not an Arrhenius base nor a Bronsted-Lowry base. Br^- can act as an electron-pair donor and thus is a Lewis base.

c) NaOH yields $OH^-(aq)$ and thus is an Arrhenius base. NaOH can act as a proton acceptor and thus is a Brønsted-Lowry base. NaOH can act as an electron-pair donor in aqueous solution and thus is a Lewis base.

17-86. a) $CH_3-\ddot{N}(CH_3)-CH_3$ has a lone pair of electrons and can act as an electron-pair donor. Thus it is a Lewis base.

b) BCl_3 is an electron deficient species and thus is a Lewis acid.

c) BeF_2 is an electron deficient species and thus is a Lewis acid.

CHAPTER 18

SOLUTIONS TO THE EVEN-NUMBERED PROBLEMS

18-2. Using Figure 18-2, we see that the pH at which methyl orange is yellow is greater than 4.5 and the pH at which bromcresol purple is yellow is less than 5. The pH of the solution is between 4.5 and 5.

18-4. Using Figure 18-2, we see that the pH at which o-cresol red is yellow is greater than 2, the pH at which methyl orange is orange is between 3 and 4.5, and the pH at which methyl red is red is less than 4.5. Therefore, the pH of the solution is between 3 and 4.5. To obtain a better estimate we should use an indicator that changes color at a pH less than 4, such as bromcresol green.

18-6. We see from Figure 18-2 that the transition color range of neutral red is between pH 7 and 8. When neutral red is added to the nutrient broth, an orange color indicates that the pH of the broth is between 7 and 8. When the broth is red, the pH is below 7 and when the broth is yellow, the pH is above 8.

18-8. At the middle of the color change $[HI_n] \approx [In^-]$, therefore

$$K_{ai} \approx [H_3O^+]$$

The middle of the color change of Nile blue A occurs at pH = 10.5

$$[H_3O^+] = 10^{-pH} = 10^{-10.5} = 10^{0.5} \times 10^{-11} = 3.2 \times 10^{-11}\ M$$

$$K_{ai} \approx 3.2 \times 10^{-11}\ M$$

18-10. The number of moles of $OH^-(aq)$ present in 50.0 mL of 0.100 M NaOH(aq) is

$$\text{moles of } OH^-(aq) = MV = (0.100\ mol\cdot L^{-1})(0.0500\ L) = 5.00 \times 10^{-3} mol$$

The number of moles of $H_3O^+(aq)$ present in 20.0 mL of 0.100 M HCl(aq) is

$$\text{moles of } H_3O^+(aq) = MV = (0.100\ \text{mol}\cdot\text{L}^{-1})(0.0200\ \text{L}) = 2.00 \times 10^{-3}\text{mol}$$

We have not added enough $H_3O^+(aq)$ to react with all the $OH^-(aq)$. The number of moles of $OH^-(aq)$ remaining is

$$\text{moles of } OH^-(aq) \text{ remaining} = 5.00 \times 10^{-3}\ \text{mol} - 2.00 \times 10^{-3}\ \text{mol}$$

$$= 3.00 \times 10^{-3}\ \text{mol}$$

The volume of the solution after adding the HCl solution is 50.0 mL + 20.0 mL = 70.0 mL. The concentration of the remaining $OH^-(aq)$ is

$$[OH^-] = \frac{3.00 \times 10^{-3}\ \text{mol}}{0.0700\ \text{L}} = 4.29 \times 10^{-2}\ \text{M}$$

The concentration of H_3O^+ is

$$[H_3O^+] = \frac{K_W}{[OH^-]} = \frac{1.00 \times 10^{-14}\ \text{M}^2}{4.29 \times 10^{-2}\ \text{M}} = 2.33 \times 10^{-13}\ \text{M}$$

The pH of the solution is

$$\text{pH} = -\log[H_3O^+] = -\log(2.33 \times 10^{-13}) = 12.63$$

18-12. The H_3O^+ concentration of the $HNO_3(aq)$ solution before adding KOH(aq) is

$$[H_3O^+] = 0.25\ \text{M}$$

and the pH is

$$\text{pH} = -\log[H_3O^+] = -\log(0.25) = 0.60$$

The volume of KOH to reach the equivalence point is determined by the condition

$$\text{moles } H_3O^+ = \text{moles } OH^-$$

$$M_{acid}V_{acid} = M_{base}V_{base}$$

or

$$V_{base} = \frac{M_{acid}V_{acid}}{M_{base}} = \frac{(0.25\ mol\cdot L)(50.0\ mL)}{(0.50\ mol\cdot L)}$$

$$= 25.0\ mL$$

Because $HNO_3(aq)$ is a strong acid and KOH(aq) is a strong base the pH at the equivalence point is 7.0. The pH of the solution after adding 50.0 mL of KOH can be obtained by realizing that after adding 25.0 mL of KOH, the $[H_3O^+]$ has been neutralized and we are adding OH^- to the solution. The number of moles of OH^- in excess of H_3O^+ is

$$\text{moles of } OH^-(aq) \text{ in excess} = MV_{excess} = (0.50\ mol\cdot L^{-1})(0.025\ L)$$

$$= 0.0125\ mol$$

The volume of the solution after adding 50.0 mL of KOH is 50.0 mL + 50.0 mL = 100 mL. The concentration of OH^- is

$$[OH^-] = \frac{0.0125\ mol}{0.100\ L} = 0.125\ M$$

$$[H_3O^+] = \frac{K_w}{[OH^-]} = \frac{1.00 \times 10^{-14}\ M^2}{0.125\ M} = 8.00 \times 10^{-14}\ M$$

$$pH = -\log[H_3O^+] = -\log(8.00 \times 10^{-14}) = 13.10$$

The titration curve looks like

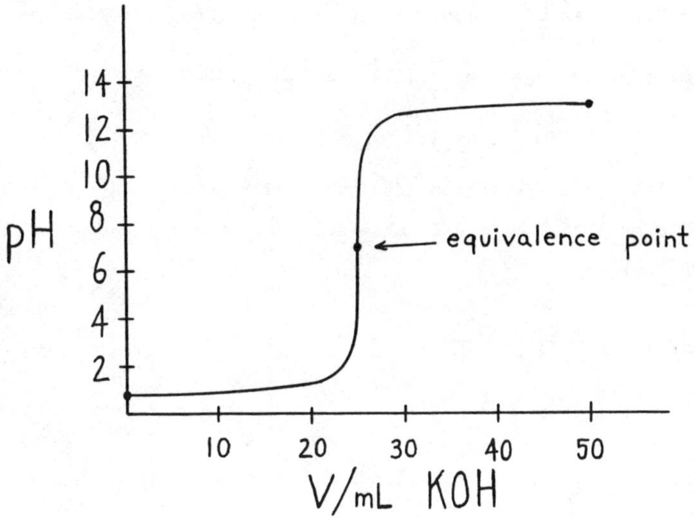

18-14. At the equivalence point,

$$M_aV_a = M_bV_b$$

Therefore,

$$M_b = \frac{M_aV_a}{V_b} = \frac{(0.125\ \text{M})(34.7\ \text{mL})}{(15.0\ \text{mL})}$$

$$= 0.289\ \text{M}$$

18-16. The number of moles in 500 mg of $Al(OH)_3$ is

$$n = (0.500\ \text{g})\left(\frac{1\ \text{mol}}{78.00\ \text{g}}\right) = 0.00641\ \text{mol}$$

At neutralization

$$\text{moles of } OH^-(aq) = \text{moles of } H_3O^+(aq)$$

There are three moles of $OH^-(aq)$ per mole of $Al(OH)_3(aq)$, and so

$$(3)(0.00641\ \text{mol}) = M_aV_a = (0.10\ \text{mol}\cdot\text{L}^{-1})V_a$$

$$V_a = \frac{0.0192\ \text{mol}}{0.10\ \text{mol}\cdot\text{L}^{-1}} = 0.19\ \text{L} = 190\ \text{mL}$$

18-18. The extra drop is added to a neutral solution of NaBr. The number of moles of $OH^-(aq)$ in one drop of 0.250 M NaOH(aq) is

$$\text{moles of } OH^-(aq) = MV = (0.250\ \text{mol}\cdot\text{L}^{-1})(0.05\ \text{mL})\left(\frac{1\ \text{L}}{10^3\ \text{mL}}\right)$$

$$= 1.25 \times 10^{-5}\ \text{mol}$$

The volume of the solution after the addition of NaOH(aq) is 25.0 mL + 25.0 mL + 0.05 mL = 50.1 mL. The concentration of $OH^-(aq)$ is

$$[OH^-] = \frac{1.25 \times 10^{-5}\ \text{mol}}{0.0501\ \text{L}} = 2.5 \times 10^{-4}\ \text{M}$$

The $H_3O^+(aq)$ concentration is

$$[H_3O^+] = \frac{K_w}{[OH^-]} = \frac{1.00 \times 10^{-14}\ \text{M}^2}{2.5 \times 10^{-4}\ \text{M}} = 4.0 \times 10^{-11}\ \text{M}$$

The pH of the solution is

$$\text{pH} = -\log[H_3O^+] = -\log(4.0 \times 10^{-11}) = 10.40$$

18-20. We first calculate the volume of NaOH(aq) added to reach the equivalence point

$$V_b = \frac{M_aV_a}{M_b} = \frac{(0.100\ \text{M})(50.0\ \text{mL})}{(0.100\ \text{M})} = 50.0\ \text{mL}$$

The reaction is

$$HNO_2(aq) + OH^-(aq) = NO_2^-(aq) + H_2O(l)$$

At the equivalence point all the HNO_2 has been converted to NO_2^- and thus

$$\text{moles of } NO_2^- = \text{initial moles of } HNO_2 = MV =$$

$$(0.100\ \text{mol}\cdot\text{L}^{-1})(0.0500\ \text{L}) = 5.00 \times 10^{-3}\ \text{mol}$$

The volume of the solution at the equivalence point is 50.0 mL + 50.0 mL = 100.0 mL. The concentration of $NO_2^-(aq)$ is

$$[NO_2^-] = \frac{5.00 \times 10^{-3}\ \text{mol}}{0.1000\ \text{L}} = 5.00 \times 10^{-2}\ \text{M}$$

The nitrite ion is a weak base because it is the conjugate base of a weak acid. The reaction of $NO_2^-(aq)$ with water is

$$NO_2^-(aq) + H_2O(l) \rightleftharpoons HNO_2(aq) + OH^-(aq)$$

At equilibrium

$$[HNO_2] = [OH^-]$$

$$[NO_2^-] = 5.00 \times 10^{-2}\ M - [OH^-]$$

We shall calculate the value of $[OH^-]$ using the expression for K_b. The value of K_b is

$$K_b = \frac{K_w}{K_a} = \frac{1.00 \times 10^{-14}\ M^2}{4.5 \times 10^{-4}\ M} = 2.22 \times 10^{-11}\ M$$

The expression for K_b is

$$K_b = \frac{[HNO_2][OH^-]}{[NO_2^-]} = \frac{[OH^-]^2}{5.00 \times 10^{-2}\ M - [OH^-]} = 2.22 \times 10^{-11}\ M$$

Neglecting $[OH^-]$ relative to 5.00×10^{-2} M, we have

$$\frac{[OH^-]^2}{5.00 \times 10^{-2}\ M} = 2.22 \times 10^{-11}\ M$$

$$[OH^-]^2 = 1.11 \times 10^{-12}\ M$$

$$[OH^-] = 1.05 \times 10^{-6}\ M$$

$$[H_3O^+] = \frac{K_w}{[OH^-]} = \frac{1.00 \times 10^{-14}\ M^2}{1.05 \times 10^{-6}\ M} = 9.49 \times 10^{-9}\ M$$

The pH of the solution at the equivalence point is

$$pH = -\log[H_3O^+] = -\log(9.49 \times 10^{-9}) = 8.02$$

Referring to Figure 18-2, we see that either thymol blue or phenolphthalein is a suitable indicator.

18-22. At the equivalence point

$$\text{moles of acid} = \text{moles of base} = M_bV_b$$

The number of moles of acid is given by

$$\text{moles of acid} = M_bV_b = (0.250\ \text{mol}\cdot\text{L}^{-1})(0.0253\ \text{L})$$

$$= 0.00633\ \text{mol}$$

We have the correspondence

$$0.772\ \text{g of benzoic acid} \approx 0.00633\ \text{mol of benzoic acid}$$

Dividing by 0.00633, we have

$$122\ \text{g of benzoic acid} \approx \text{one mole of benzoic acid}$$

Thus, the molecular mass of benzoic acid is 122.

18-24. At the equivalence point

$$\text{moles of acid} = \text{moles of base} = M_bV_b$$

$$= (0.100\ \text{mol}\cdot\text{L}^{-1})(0.0624\ \text{L}) = 0.00624\ \text{mol}$$

We have the correspondence

$$0.550\ \text{g of butyric acid} \approx 0.00624\ \text{mol of butyric acid}$$

Dividing both sides by 0.00624, we have

$$88.1\ \text{g of butyric acid} \approx \text{one mole of butyric acid}$$

The molecular mass of butyric acid is 88.1.

18-26. At the equivalence point

$$\text{moles of acid} = \text{moles of base} = M_b V_b$$

$$= (0.200\ \text{mol}\cdot\text{L}^{-1})(0.0555\ \text{L}) = 0.0111\ \text{mol}$$

We have the correspondence

$$2.00\ \text{g of aspirin} \approx 0.0111\ \text{mol of aspirin}$$

Dividing by 0.0111, we have

$$180\ \text{g of aspirin} \approx \text{one mole of aspirin}$$

The molecular mass of aspirin is 180.

18-28. The stoichiometric concentration of the acid and base forms are

$$[\text{acid}]_0 \approx [HC_2H_3O_2] = 0.10\ \text{M} \qquad [\text{base}]_0 \approx [C_2H_3O_2^-] = 0.20\ \text{M}$$

The pK_a of acetic acid is

$$pK_a = -\log K_a = -\log(1.74 \times 10^{-5}) = 4.760$$

From the Henderson-Hasselbalch equation, we have

$$pH = pK_a + \log\frac{[\text{base}]_0}{[\text{acid}]_0} = 4.760 + \log\left(\frac{0.20\ \text{M}}{0.10\ \text{M}}\right)$$

$$= 4.760 + 0.30 = 5.06$$

18-30. The stoichiometric concentrations of the acid and base forms are

$$[\text{acid}]_0 = [HNO_2] = 0.15\ \text{M} \qquad [\text{base}]_0 = [NO_2^-] = 0.25\ \text{M}$$

The pK_a of nitrous acid is

$$pK_a = -\log K_a = -\log(4.5 \times 10^{-4}) = 3.35$$

From the Henderson-Hasselbalch equation, we have

$$pH = pK_a + \log \frac{[base]_o}{[acid]_o} = 3.35 + \log\left(\frac{0.25\ M}{0.15\ M}\right)$$

$$= 3.35 + 0.22 = 3.57$$

18-32. The stoichiometric concentrations of conjugate acid and base are

$$[acid]_o = [NH_4^+] = 0.40\ M$$

$$[base]_o = [NH_3] = 0.20\ M$$

From the Henderson-Hasselbalch equation, we have

$$pH = pK_a + \log \frac{[base]_o}{[acid]_o} = 9.24 + \log\left(\frac{0.20\ M}{0.40\ M}\right)$$

$$= 9.24 - 0.30 = 8.94$$

18-34. The stoichiometric concentrations of the acid and base forms are

$$[acid]_o = [C_5H_5NH^+] = 0.250\ M \qquad [base]_o = [C_5H_5N] = 0.200\ M$$

The pK_a of pyridinium chloride is given

$$pK_a = 5.17$$

From the Henderson-Hasselbalch equation we have

$$pH = pK_a + \log \frac{[base]_o}{[acid]_o} = 5.17 + \log\left(\frac{0.200\ M}{0.250\ M}\right)$$

$$= 5.17 - 0.097 = 5.07$$

18-36. We must first calculate the concentrations of KH_2PO_4 and Na_2HPO_4 in the buffer solution.

$$[KH_2PO_4] = \frac{(10.0\ g)\left(\frac{1\ mol}{136.09\ g}\right)}{1.00\ L} = 0.0735\ M$$

$$[Na_2HPO_4] = \frac{(20.0\ g)\left(\frac{1\ mol}{141.96\ g}\right)}{1.00\ L} = 0.141\ M$$

$$[acid]_o = [H_2PO_4^-]_o = 0.0735\ M$$

$$[base]_o = [HPO_4^{2-}]_o = 0.141\ M$$

From the Henderson-Hasselbalch equation we have

$$pH = pK_a + \log \frac{[base]_o}{[acid]_o} = 7.21 + \log\left(\frac{0.141\ M}{0.0735\ M}\right)$$

$$= 7.21 + 0.28 = 7.49$$

18-38. a) yes c) no (no conjugate base)

b) no (no conjugate base) d) yes

18-40. The number of moles of $OH^-(aq)$ in 10.0 mL of 0.10 M NaOH(aq) is

$$\text{moles of } [OH^-] = MV = (0.10\ mol\cdot L^{-1})(0.0100\ L) = 1.0 \times 10^{-3}\ mol$$

The number of moles of $H_3O^+(aq)$ in the solution before the addition of NaOH is

$$[H_3O^+] = MV = (10^{-pH})V = (10^{-4.76}M)(0.100\ L)$$

$$= (10^{0.24} \times 10^{-5}\ mol\cdot L^{-1})(0.100\ L) = (1.74 \times 10^{-5} mol\cdot L^{-1})(0.100L)$$

$$= 1.74 \times 10^{-6}\ mol$$

The added NaOH neutralizes the HCl in the solution. The excess OH^- is the number of moles of OH^- remaining

$$\text{moles of } OH^- = 1.0 \times 10^{-3}\text{ mol} - 1.74 \times 10^{-6}\text{ mol}$$

$$= 1.0 \times 10^{-3}\text{ mol}$$

The concentration of $OH^-(aq)$ is

$$[OH^-] = \frac{1.0 \times 10^{-3}\text{ mol}}{0.110\text{ L}} = 9.1 \times 10^{-3}\text{ M}$$

$$[H_3O^+] = \frac{K_w}{[OH^-]} = \frac{1.00 \times 10^{-14}\text{ M}^2}{9.1 \times 10^{-3}\text{ M}} = 1.1 \times 10^{-12}\text{ M}$$

The pH of the solution is

$$pH = -\log[H_3O^+] = -\log(1.1 \times 10^{-12}) = 11.96$$

The change in pH is 7.20 pH units.

18-42. If we use equal concentrations of conjugate acid and base, then

$$pH = pK_a + \log\frac{[\text{base}]_o}{[\text{acid}]_o}$$

$$= pK_a$$

To obtain a pH buffered at 5.17, we want $pK_a = 5.17$, or

$$K_a = 10^{-5.17} = 6.7 \times 10^{-6}\text{ M}$$

From Table 17-4, we find that $K_a = 6.7 \times 10^{-6}$ M for pyridium ion, and so a solution of equal concentrations of pyridium ion and pyridine would act as a buffer at pH = 5.17.

18-44. The number of moles of $H_3O^+(aq)$ in 5.00 mL of 0.100 M HCl(aq) is

$$\text{moles of } H_3O^+ = MV = (0.100\ \text{mol}\cdot\text{L}^{-1})(0.00500\ \text{L}) = 5.00 \times 10^{-4}\ \text{mol}$$

The $H_3O^+(aq)$ reacts with $NH_3(aq)$ in the buffer via the reaction

$$NH_3(aq) + H_3O^+(aq) \rightleftharpoons NH_4^+(aq) + H_2O(l)$$

The number of moles of NH_3 in the buffer solution before the addition of HCl(aq) is

$$\text{moles of } NH_3 = MV = (0.100\ \text{mol}\cdot\text{L}^{-1})(0.100\ \text{L}) = 0.0100\ \text{mol}$$

The number of moles of NH_3 in the buffer solution after the addition of HCl(aq) is

$$\text{moles of } NH_3 = \text{moles of } NH_3 \text{ before} - \text{moles of } H_3O^+ \text{ added}$$

$$= 0.0100\ \text{mol} - 5.00 \times 10^{-4}\ \text{mol} = 0.0095\ \text{mol}$$

The concentration of NH_3 after the addition of HCl is

$$[NH_3] = \frac{0.0095\ \text{mol}}{0.105\ \text{L}} = 0.0905\ \text{M}$$

The number of moles of $NH_4^+(aq)$ in the buffer solution before the addition of HCl(aq) is

$$\text{moles of } NH_4^+ = MV = (0.100\ \text{mol}\cdot\text{L}^{-1})(0.100\ \text{L}) = 0.0100\ \text{mol}$$

The number of moles of $NH_4^+(aq)$ in the buffer solution after the addition of HCl(aq) is

$$\text{moles of } NH_4^+ = \text{moles of } NH_4^+ \text{ before} + \text{moles of } H_3O^+ \text{ added}$$

$$= 0.0100\ \text{mol} + 5.00 \times 10^{-4}\ \text{mol} = 0.0105\ \text{mol}$$

The concentration of $NH_4^+(aq)$ after the addition of the HCl is

$$[NH_4^+] = \frac{0.0105\ \text{mol}}{0.105\ \text{L}} = 0.100\ \text{M}$$

The pH of the buffer solution before addition of HCl is

$$pH = pK_a + \log \frac{[\text{base}]_0}{[\text{acid}]_0} = 9.24 + \log\left(\frac{0.100\ M}{0.100\ M}\right) = 9.24$$

The pH of the buffer solution after addition of HCl is

$$pH = 9.24 + \log\left(\frac{0.0905}{0.100}\right) = 9.20$$

The change in pH is 0.04 pH units.

The number of moles of $OH^-(aq)$ in 5.00 mL of 0.100 M NaOH(aq) is

$$\text{moles of } OH^- = MV = (0.100\ \text{mol}\cdot L^{-1})(0.00500\ L) = 5.00 \times 10^{-4}\ \text{mol}$$

The OH^- reacts with $NH_4(aq)$ in the buffer solution via the reaction

$$OH^-(aq) + NH_4^+(aq) \longrightarrow NH_3(aq) + H_2O(l)$$

The number of moles of $NH_3(aq)$ in the buffer solution after the addition of NaOH(aq) is

$$\text{moles of } NH_3 = \text{moles of } NH_3 \text{ before} + \text{moles of } OH^- \text{ added}$$

$$= 0.0100\ \text{mol} + 5.00 \times 10^{-4}\ \text{mol} = 0.0105\ \text{mol}$$

The concentration of $NH_3(aq)$ after the addition of NaOH is

$$[NH_3] = \frac{0.0105\ \text{mol}}{0.105\ L} = 0.100\ M$$

The number of moles of $NH_4^+(aq)$ after the addition of NaOH(aq) is

$$\text{moles of } NH_4^+ = \text{moles of } NH_4^+ \text{ before} - \text{moles of } OH^- \text{ added}$$

$$= 0.0100\ \text{mol} - 5.00 \times 10^{-4}\ \text{mol} = 0.0095\ \text{mol}$$

The concentration of $NH_4^+(aq)$ after the addition of NaOH is

$$[NH_4^+] = \frac{0.0095\ \text{mol}}{0.105\ L} = 0.0905\ M$$

The pH of the solution after the addition of NaOH(aq) is

$$pH = pK_a + \log \frac{[\text{base}]_0}{[\text{acid}]_0} = 9.24 + \log\left(\frac{0.100}{0.0905}\right) = 9.28$$

The change in pH is 0.04 pH units.

18-46. Measure the pH of the original solution. Dilute the solution and measure the pH of the diluted solution. The pH of a buffer would not be affected by dilution.

18-48. a) The two reactions that correspond to K_{a,NH_4^+} and $1/K_{a,HC_2H_3O_2}$ are

$$NH_4^+(aq) + H_2O(l) \rightleftharpoons NH_3(aq) + H_3O^+(aq) \qquad K = K_{a,NH_4^+}$$

$$C_2H_3O_2^-(aq) + H_3O^+(aq) \rightleftharpoons HC_2H_3O_2(aq) + H_2O(l) \qquad K = \frac{1}{K_{a,HC_2H_3O_2}}$$

these are added to give

$$NH_4^+(aq) + C_2H_3O_2^-(aq) \rightleftharpoons NH_3(aq) + HC_2H_3O_2(aq) \qquad K = \frac{K_{a,NH_4^+}}{K_{a,HC_2H_3O_2}}$$

b) By adding the two reactions

$$NH_4^+(aq) + H_2O(l) \rightleftharpoons H_3O^+(aq) + NH_3(aq) \qquad K_{a,NH_4^+}$$

$$HC_2H_3O_2(aq) + H_2O(l) \rightleftharpoons H_3O^+(aq) + C_2H_3O_2^- \qquad K_{a,HC_2H_3O_2}$$

we obtain

$$NH_4^+(aq) + HC_2H_3O_2(aq) + 2H_2O(l) \rightleftharpoons 2H_3O^+(aq) + NH_3(aq) + C_2H_3O_2^-(aq)$$

with

$$K = K_{a,NH_4^+} K_{a,HC_2H_3O_2}$$

$$= \frac{[H_3O^+]^2[NH_3][C_2H_3O_2^-]}{[NH_4^+][HC_2H_3O_2]}$$

But when $NH_4C_2H_3O_2$ is dissolved in water, we have that

$$[NH_4^+] = [C_2H_3O_2^-] \quad \text{and} \quad [NH_3] = [HC_2H_3O_2]$$

Therefore

$$[H_3O^+]^2 = K_{a,NH_4^+}K_{a,HC_2H_3O_2}$$

or

$$[H_3O^+] = (K_{a,NH_4^+}K_{a,HC_2H_3O_2})^{1/2}$$

c) An $NH_4C_2H_3O_2(aq)$ solution acts as a buffer according to the reactions

$$NH_4^+(aq) + OH^-(aq) \rightleftharpoons NH_3(aq) + H_2O(l)$$

$$C_2H_3O_2^-(aq) + H_3O^+(aq) \rightleftharpoons HC_2H_3O_2(aq) + H_2O(l)$$

18-50. a) $H_2CO_3(aq)$

c) $HCO_3^-(aq)$

b) $H_2CO_3(aq)$ and $HCO_3^-(aq)$

d) $HCO_3^-(aq)$ and $CO_3^{2-}(aq)$

18-52. From Figure 18-8, we see that the complete neutralization (second equivalence point) of $H_2CO_3(aq)$ occurs at a pH around 13. From Figure 18-2, we see that appropriate indicators are Poirrier's blue or Clayton yellow.

18-54. Because H_2S is a diprotic acid, it takes two moles of NaOH to neutralize one mole of H_2S. We have that

$$\text{moles of NaOH} = 2 \times \text{moles of } H_2S$$

$$M_bV_b = 2M_aV_a$$

$$V_b = \frac{2M_aV_a}{M_b} = \frac{(2)(0.10\text{ M})(100\text{ mL})}{(0.10\text{ M})}$$

$$V_b = 200\text{ mL}$$

18-56. The moles of NaOH used to neutralize the fumaric acid is

$$\text{moles of NaOH} = MV = (0.300\ \text{mol}\cdot\text{L}^{-1})(0.0690\ \text{L}) = 0.0207\ \text{mol}$$

Because fumaric acid is a diprotic acid, it takes two moles of NaOH to neutralize one mole of fumaric acid. The number of moles of fumaric acid is

$$\text{moles of fumaric acid} = \frac{1}{2}\ \text{moles of NaOH} = 0.01035\ \text{mol}$$

Thus we have the correspondence

$$1.20\ \text{g fumaric acid} \hat{=} 0.01035\ \text{mol fumaric acid}$$

Dividing both sides by 0.01035, we have

$$116\ \text{g fumaric acid} \hat{=} \text{one mole fumaric acid}$$

The molecular mass of fumaric acid is 116.

18-58. Because pK_{a2} is so much greater than pK_{a1}, we can neglect the second dissociation constant of vitamin C. The reaction is

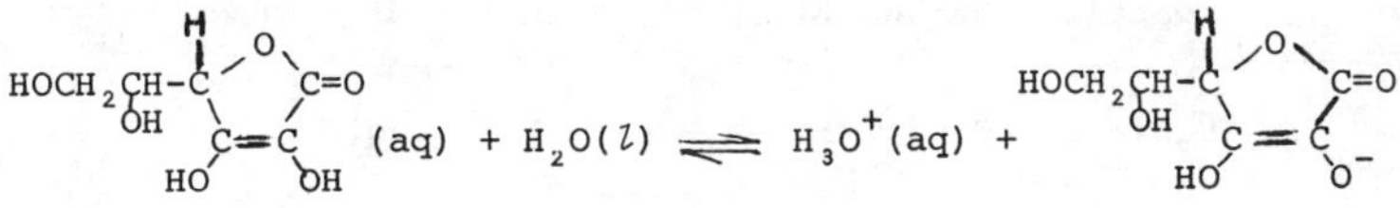

ascorbic acid ascorbate

The K_{a1} expression is

$$K_{a1} = \frac{[H_3O^+][\text{ascorbate}]}{[\text{ascorbic acid}]} = 10^{-4.17} = 10^{0.83} \times 10^{-5} = 6.76 \times 10^{-5}\ \text{M}$$

The concentration of ascorbic acid is

$$[\text{ascorbic acid}] = \frac{(0.500\ \text{g})\left(\frac{1\ \text{mol}}{175.12\ \text{g}}\right)}{1.00\ \text{L}} = 2.86 \times 10^{-3}\ \text{M}$$

The K_{a1} expression then is

$$\frac{[H_3O^+]^2}{2.86 \times 10^{-3}\ M - [H_3O^+]} = 6.76 \times 10^{-5}\ M$$

or

$$[H_3O^+]^2 + 6.76 \times 10^{-5}\ M[H_3O^+] - 1.93 \times 10^{-7}\ M^2 = 0$$

The quadratic formula gives

$$[H_3O^+] = \frac{-6.76 \times 10^{-5}\ M \pm \sqrt{4.57 \times 10^{-9} M^2 - (4)(1)(-1.93 \times 10^{-7} M^2)}}{(2)(1)}$$

$$= \frac{-6.76 \times 10^{-5}\ M \pm 8.81 \times 10^{-4} M}{2}$$

$$= 4.07 \times 10^{-4}\ M$$

The pH of the solution is

$$pH = -\log[H_3O^+] = -\log(4.07 \times 10^{-4}) = 3.39$$

18-60. The reaction for the first dissociation of $H_2S(aq)$ is

$$H_2S(aq) + H_2O(l) \rightleftharpoons H_3O^+(aq) + HS^-(aq)$$

The pH at which $[H_2S] = [HS^-]$ or at which the fraction of H_2S is 0.5 is obtained from

$$K_{a_1} = \frac{[H_3O^+][HS^-]}{[H_2S]} = [H_3O^+] = 9.1 \times 10^{-8}\ M$$

where K_{a_1} is from Table 17-4. The pH is

$$pH = 7.04$$

The pH when the fraction of H_2S is 0.9 is obtained from the requirement

$$[HS^-] = \frac{1}{9}[H_2S]$$

Thus, we have

$$K_{a_1} = \frac{[H_3O^+][HS^-]}{[H_2S]} = \frac{[H_3O^+]\frac{1}{9}[H_2S]}{[H_2S]} = \frac{1}{9}[H_3O^+] = 9.1 \times 10^{-8} \text{ M}$$

or

$$[H_3O^+] = 8.19 \times 10^{-7}$$

The pH is

$$pH = 6.09$$

The pH when the fraction of H_2S is 0.1 is obtained from

$$[HS^-] = 9[H_2S]$$

and

$$K_{a_1} = \frac{[H_3O^+][HS^-]}{[H_2S]} = \frac{9[H_3O^+][H_2S]}{[H_2S]} = 9[H_3O^+] = 9.1 \times 10^{-8}$$

or

$$[H_3O^+] = 1.01 \times 10^{-8} \text{ M}$$

The pH is

$$pH = 8.00$$

The reaction of the second dissociation constant is

$$HS^-(aq) + H_2O(l) \rightleftharpoons S^{2-}(aq) + H_3O^+(aq)$$

$$K_{a2} = \frac{[H_3O^+][S^{2-}]}{[HS^-]} = 1.2 \times 10^{-13} \text{ M}$$

The pH at which $[S^{2-}] = [HS^-]$ or at which the fraction of $[HS^-]$ = 0.5 is obtained from

$$K_{a2} = \frac{[H_3O^+][S^{2-}]}{[HS^-]} = [H_3O^+] = 1.2 \times 10^{-13} \text{ M}$$

or

$$pH = 12.92$$

The pH when the fraction of $[HS^-]$ = 0.1 is given by

$$[S^{2-}] = 9[HS^-]$$

and

$$K_{a2} = \frac{[H_3O^+][S^{2-}]}{[HS^-]} = \frac{9[H_3O^+][HS^-]}{[HS^-]} = 9[H_3O^+] = 1.2 \times 10^{-13} \text{ M}$$

$$[H_3O^+] = 1.33 \times 10^{-14} \text{ M}$$

$$pH = 13.88$$

The pH when the fraction of $[HS^-]$ = 0.9 is obtained from

$$[S^{2-}] = \frac{1}{9}[HS^-]$$

and

$$K_{a2} = \frac{[H_3O^+][S^{2-}]}{[HS^-]} = \frac{[H_3O^+][HS^-]}{9[HS^-]} = \frac{1}{9}[H_3O^+] = 1.2 \times 10^{-13}\text{M}$$

or

$$[H_3O^+] = 1.08 \times 10^{-12} \text{ M}$$

and

$$pH = 11.97$$

Putting all these calculations together gives

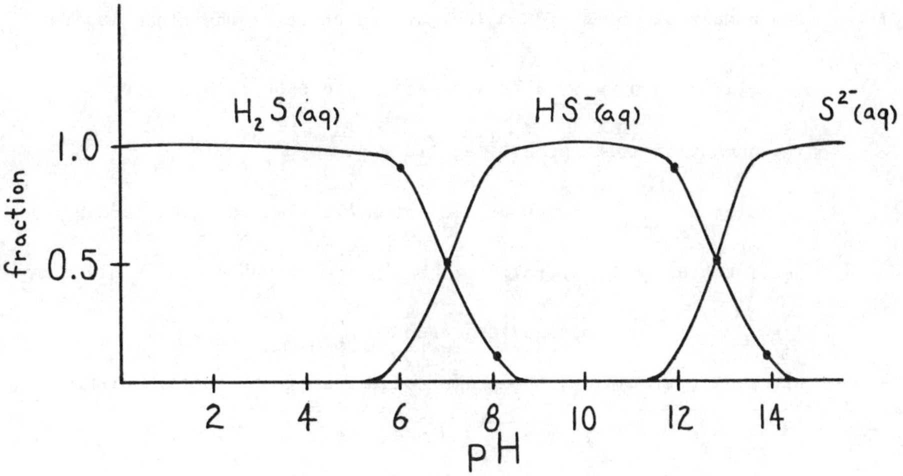

18-62. The number of moles in 2.00 g of citric acid is

$$n = (2.00\ g)\left(\frac{1\ mol}{192.12\ g}\right) = 0.0104\ mol$$

Because citric acid is a triprotic acid, it requires 3 moles of $NaHCO_3$ (a base) to neutralize one mole of citric acid

$$\text{moles of } HCO_3^- = 3 \times \text{moles of citric acid} = 0.0312\ mol$$

The number of grams in 0.0312 mol of $NaHCO_3$ is

$$\text{mass} = (0.0312\ mol)\left(\frac{84.01\ g}{1\ mol}\right) = 2.62\ g$$

18-64. The number of moles of $H_3O^+(aq)$ at the start of the titration is

$$\text{moles of } H_3O^+ = MV = (0.100\ \text{mol}\cdot\text{L}^{-1})(0.0500\ \text{L}) = 0.00500\ \text{mol}$$

The number of moles of $H_3O^+(aq)$ is given by

$$\text{moles of } H_3O^+ = \text{moles of } H_3O^+ \text{ at start} - \text{moles of } OH^- \text{ added}$$

until the HNO_3 is neutralized. The moles of OH^- added is given by

$$\text{moles of } OH^- \text{ added} = M_bV_b$$

After neutralization, the number of moles of OH^- in the solution is

$$\text{moles of } OH^- = \text{moles of } OH^- \text{ added} - \text{moles of } H_3O^+ \text{ at start}$$

volume of NaOH solution added /mL	total volume of resulting solution /mL	moles of $H_3O^+(aq)$ in solution	concentration of $H_3O^+(aq)$ in solution /M	pH of the resulting solution
10.0	60.0	0.00400	0.0667	1.18
20.0	70.0	0.00300	0.0429	1.37
30.0	80.0	0.00200	0.0250	1.60
40.0	90.0	0.00100	0.0111	1.95
45.0	95.0	0.00050	0.0053	2.28
49.0	99.0	0.00010	0.0010	3.00
50.0	100.0	1.0×10^{-8}	1.0×10^{-7}	7.00
51.0	101.0	1.0×10^{-12}	1.0×10^{-11}	11.00
60.0	110.0	1.2×10^{-13}	1.10×10^{-12}	11.96
70.0	120.0	7.2×10^{-14}	6.00×10^{-13}	12.22

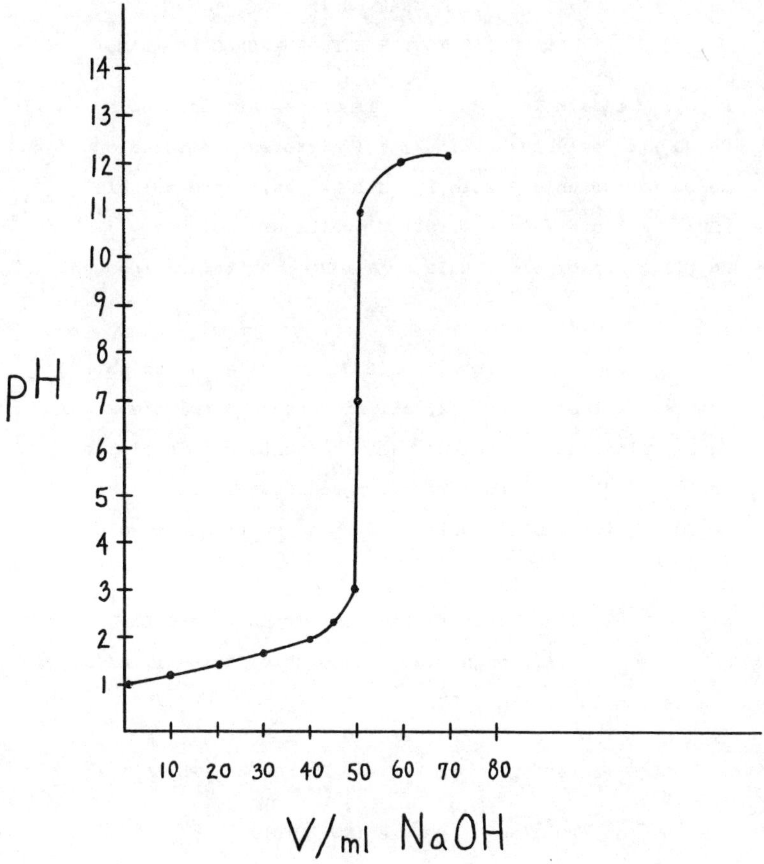

pH
14
13
12
11
10
9
8
7
6
5
4
3
2
1
10 20 30 40 50 60 70 80
V/ml NaOH

CHAPTER 19

SOLUTIONS TO THE EVEN-NUMBERED PROBLEMS

19-2. Fe_2O_3: insoluble - Rule 5, all oxides are insoluble

$Cu(NO_3)_2$: soluble - Rule 2, all nitrates are soluble

Na_2CO_3: soluble - Rule 1, all Na^+ salts are soluble

KOH: soluble - Rule 1, all K^+ salts are soluble

Hg_2Cl_2: insoluble - Rule 3, all Hg_2^{2+} salts are insoluble

19-4. AgCl: insoluble - Rule 3, all Ag^+ salts are insoluble

$AgNO_3$: soluble - Rule 2, all nitrates are soluble

Ag_2S: insoluble - Rule 3, all Ag^+ salts are insoluble

$AgClO_4$: soluble - Rule 2, all perchlorates are soluble

Ag_2CO_3: insoluble - Rule 3, all Ag^+ salts are insoluble

19-6. a) The possible double-replacement products are $LiNO_3$ and $Pb(NO_2)_2$. Although $LiNO_3$ is soluble (Rule 2), $Pb(NO_2)_2$ is insoluble (Rule 3).

$$2LiNO_2(aq)+Pb(NO_3)_2(aq) \longrightarrow 2LiNO_3(aq)+Pb(NO_2)_2(s)$$

$$2NO_2^-(aq)+Pb^{2+}(aq) \longrightarrow Pb(NO_2)_2(s)$$

b) The possible double-replacement products are $CaSO_4$ and $HClO_4$. Although $HClO_4$ is soluble (strong acid) (Rule 2), $CaSO_4$ is insoluble (Rule 6).

$$H_2SO_4(aq)+Ca(ClO_4)_2(aq) \longrightarrow 2HClO_4(aq)+CaSO_4(s)$$

$$SO_4^{2-}(aq)+Ca^{2+}(aq) \longrightarrow CaSO_4(s)$$

c) The possible double-replacement products are $AgClO_4$ and $NaNO_3$. Both $AgClO_4$ and $NaNO_3$ are soluble (Rule 2 and Rule 1)

no reaction

d) The possible double-replacement products are $Hg_2(C_7H_5O_2)_2$ + $NaNO_3$. Although $NaNO_3$ is soluble (Rule 1), $Hg_2(C_7H_5O_2)_2$ is insoluble (Rule 3).

$$Hg_2(NO_3)_2(aq) + 2NaC_7H_5O_2(aq) \longrightarrow 2NaNO_3(aq) + Hg_2(C_7H_5O_2)_2(s)$$

$$Hg_2^{2+}(aq) + 2C_7H_5O_2^-(aq) \longrightarrow Hg_2(C_7H_5O_2)_2(s)$$

e) The possible double-replacement products are NaF and Ag_2SO_4. Although NaF is soluble (Rule 1), Ag_2SO_4 is insoluble (Rule 3).

$$Na_2SO_4(aq) + 2AgF(aq) \longrightarrow 2NaF(aq) + Ag_2SO_4(s)$$

$$SO_4^{2-}(aq) + 2Ag^+(aq) \longrightarrow Ag_2SO_4(s)$$

Note that Rule 3 would predict that AgF is insoluble; however, it is soluble.

19-8. a) $AgNO_3(aq) + NH_4ClO_4(aq) \longrightarrow$ no reaction

b) $Hg_2(NO_3)_2(aq) + 2KCl(aq) \longrightarrow 2KNO_3(aq) + Hg_2Cl_2(s)$ (Rule 3)

c) $Zn(ClO_4)_2(aq) + Na_2S(aq) \longrightarrow 2NaClO_4(aq) + ZnS(s)$ (Rule 5)

d) $CaCl_2(aq) + Na_2CO_3(aq) \longrightarrow 2NaCl(aq) + CaCO_3(s)$ (Rule 5)

e) $Cu(ClO_4)_2(aq) + 2LiOH(aq) \longrightarrow 2LiClO_4(aq) + Cu(OH)_2(s)$ (Rule 5)

19-10. Calcium ion forms an insoluble oxalate, $CaC_2O_4(s)$, which is removed by vomiting. The excess Ca^{2+} is removed by adding $MgSO_4(aq)$ to form $CaSO_4(s)$ which is insoluble in water and in stomach acid. Vomiting of the $CaC_2O_4(s)$ is necessary because the solubility of $CaC_2O_4(s)$ in stomach acid is sufficiently high to permit toxic levels of oxalic acid (a weak acid) to pass through the stomach walls into the bloodstream.

19-12. The reaction is

$$TlCl(s) \rightleftharpoons Tl^{+}(aq) + Cl^{-}(aq)$$

The solubility product constant expression is

$$K_{sp} = [Tl^{+}][Cl^{-}] = 1.7 \times 10^{-4}\ M^2$$

If TlCl is equilibrated with pure water, then at equilibrium we have

$$[Tl^{+}] = [Cl^{-}] = s$$

where s is the solubility of TlCl in pure water.

$$K_{sp} = (s)(s) = s^2 = 1.7 \times 10^{-4}\ M^2$$

$$s = \sqrt{1.7 \times 10^{-4}\ M^2} = 1.3 \times 10^{-2}\ M$$

The solubility in grams per liter is

$$s = (1.3 \times 10^{-2} mol{\cdot}L^{-1})\left(\frac{239.9\ g}{1\ mol}\right) = 3.1\ g{\cdot}L^{-1}$$

19-14. The reaction is

$$Pb(IO_3)_2(s) \rightleftharpoons Pb^{2+}(aq) + 2IO_3^{-}(s)$$

The concentration of $Pb(IO_3)_2$ in pure water is

$$(2.24 \times 10^{-2}\ g{\cdot}L^{-1})\left(\frac{1\ mol}{557.0\ g}\right) = 4.02 \times 10^{-5}\ M$$

From the reaction stoichiometry, we have

$$[Pb^{2+}] = \frac{1}{2}[IO_3^{-}]$$

Using the K_{sp} expression we compute

$$K_{sp} = [Pb^{2+}][IO_3^{-}]^2 = (4.02 \times 10^{-5}\ M)(2 \times 4.02 \times 10^{-5}\ M)^2$$

$$= 2.60 \times 10^{-13}\ M^3$$

19-16. The reaction is

$$PbBr_2(s) \rightleftharpoons Pb^{2+}(aq) + 2Br^-(aq)$$

The K_{sp} expression is

$$K_{sp} = [Pb^{2+}][Br^-]^2 = 4.0 \times 10^{-5}\ M^3$$

When $PbBr_2$ is in equilibrium with pure water,

$$[Br^-] = 2[Pb^{2+}]$$

The solubility of $PbBr_2(s)$ in pure water is equal to $[Pb^{2+}]$ because each $PbBr_2$ that dissolves yields one Pb^{2+}. The solubility of $PbBr_2$ is

$$s = [Pb^{2+}] = \frac{1}{2}[Br^-]$$

so that $[Br^-] = 2s$. The K_{sp} expression is

$$K_{sp} = [Pb^{2+}][Br^-]^2 = (s)(2s)^2 = 4s^3 = 4.0 \times 10^{-5}\ M^3$$

$$s = \left(\frac{4.0 \times 10^{-5}\ M^3}{4}\right)^{1/3} = 2.15 \times 10^{-2}\ M$$

The solubility in grams per liter is

$$s = (2.15 \times 10^{-2}\ mol \cdot L^{-1})\left(\frac{367.0\ g}{1\ mol}\right) = 7.9\ g \cdot L^{-1}$$

19-18. The reaction is

$$LiF(s) \rightleftharpoons Li^+(aq) + F^-(aq)$$

The K_{sp} expression is

$$K_{sp} = [Li^+][F^-]$$

From the reaction stoichiometry, at equilibrium we have that

$$[Li^+] = [F^-] = \text{solubility of LiF}$$

The solubility of LiF is

$$s = \frac{0.27\ g}{0.100\ L} = (2.7\ g\cdot L^{-1})\left(\frac{1\ mol}{25.94\ g}\right) = 0.104\ M$$

Substituting the values of $[Li^+]$ and $[F^-]$ into the K_{sp} expression, we have

$$K_{sp} = (0.10\ M)(0.104\ M) = 1.1 \times 10^{-2}\ M^2$$

19-20. The reaction is

$$Ca(OH)_2(s) \rightleftharpoons Ca^{2+}(aq) + 2OH^-(aq)$$

From the reaction stoichiometry, at equilibrium we have

$$[Ca^{2+}] = \frac{1}{2}[OH^-]$$

The solubility of $Ca(OH)_2(aq)$ is equal to $[Ca^{2+}]$ because each mole of $Ca(OH)_2$ that dissolves yields one mole of Ca^{2+}. We can find the value of $[OH^-]$ from the pH of the solution

$$[H_3O^+] = 10^{-pH} = 10^{-12.45} = 10^{0.55} \times 10^{-13} = 3.55 \times 10^{-13}\ M$$

$$[OH^-] = \frac{K_w}{[H_3O^+]} = \frac{1.00 \times 10^{-14}\ M^2}{3.55 \times 10^{-13}\ M} = 2.82 \times 10^{-2}\ M$$

The solubility of $Ca(OH)_2$ is

$$s = [Ca^{2+}] = \frac{1}{2}[OH^-] = \frac{1}{2}(2.82 \times 10^{-2}\ M) = 1.41 \times 10^{-2}\ M$$

The solubility in grams per liter is

$$s = (1.41 \times 10^{-2}\ mol\cdot L^{-1})\left(\frac{74.10\ g}{1\ mol}\right) = 1.04 g\cdot L^{-1}$$

19-22. The equilibrium reaction that describes the solubility of barium chromate in water is

$$BaCrO_4(s) \rightleftharpoons Ba^{2+}(aq) + CrO_4^{2-}(aq)$$

The solubility product expression is

$$K_{sp} = [Ba^{2+}][CrO_4^{2-}] = 1.2 \times 10^{-10}\ M^2 \quad \text{(Table 19-1)}$$

The only source of $Ba^{2+}(aq)$ is from the $BaCrO_4(s)$ that dissolves. If we let s be the solubility of $BaCrO_4(s)$ in 0.0553 M ammonium chromate, then

$$[Ba^{2+}] = s$$

The $CrO_4^{2-}(aq)$ is due to the 0.0553 M $(NH_4)_2CrO_4(aq)$ and the $BaCrO_4(s)$ that dissolves, thus

$$[CrO_4^{2-}] = 0.0553\ M + s$$

If we substitute the expressions for $[Ba^{2+}]$ and CrO_4^{2-} into the K_{sp} expression, we have

$$K_{sp} = (s)(0.0553\ M + s) = 1.2 \times 10^{-10}\ M^2$$

Because $BaCrO_4$ is only very slightly soluble, we expect the value of s to be small. Therefore we neglect s compared to 0.0553 M and write

$$(s)(0.0553\ M) = 1.2 \times 10^{-10}\ M^2$$

$$s = \frac{1.2 \times 10^{-10}\ M^2}{0.0553\ M} = 2.17 \times 10^{-9}\ M$$

The solubility in grams per liter is

$$s = (2.17 \times 10^{-9}\ mol \cdot L^{-1})\left(\frac{253.3\ g}{1\ mol}\right) = 5.5 \times 10^{-7}\ g \cdot L^{-1}$$

19-24. The reaction that describes the solubility of PbI_2 is

$$PbI_2(s) \rightleftharpoons Pb^{2+}(aq) + 2I^-(aq)$$

The solubility product expression is

$$K_{sp} = [Pb^{2+}][I^-]^2 = 7.1 \times 10^{-9}\ M^3$$

The only source of $I^-(aq)$ is from the $PbI_2(s)$ that dissolves. If we let s be the solubility of PbI_2 in 0.010 M $Pb(ClO_4)_2(aq)$, then

$$[I^-] = 2s$$

The Pb^{2+} is due to the 0.010 M $Pb(ClO_4)_2(aq)$ and the $PbI_2(s)$ that dissolves.

$$[Pb^{2+}] = 0.010\ M + s$$

If we substitute the expressions for $[Pb^{2+}]$ and $[I^-]$ into the K_{sp} expression, we have

$$K_{sp} = (0.010\ M + s)(2s)^2 = 7.1 \times 10^{-9}\ M^3$$

Because PbI_2 is a slightly soluble salt, we expect the value of s to be small. Therefore, we neglect s compared to 0.010 M and write

$$(0.010\ M)(2s)^2 \simeq 7.1 \times 10^{-9}\ M^3$$

Thus

$$s^2 = \frac{7.1 \times 10^{-9}\ M^3}{0.040\ M} = 1.8 \times 10^{-7}\ M^2$$

and

$$s = 4.2 \times 10^{-4}\ M$$

The solubility in grams per liter is

$$s = (4.2 \times 10^{-4}\ mol \cdot L^{-1})\left(\frac{461.0\ g}{1\ mol}\right) = 0.19\ g \cdot L^{-1}$$

19-26. The reaction that describes the solubility of $CaSO_4$ is

$$CaSO_4(s) \rightleftharpoons Ca^{2+}(aq) + SO_4^{2-}(aq)$$

The solubility product expression is

$$K_{sp} = [Ca^{2+}][SO_4^{2-}] = 9.1 \times 10^{-6}\ M^2 \qquad \text{(Table 19-1)}$$

If we let s be the solubility of $CaSO_4$ in the solution, then

$$[Ca^{2+}] = s$$

and in a 0.25 M $Na_2SO_4(aq)$ solution

$$[SO_4^{2-}] = 0.25\ M + s$$

If we substitute the expressions for $[Ca^{2+}]$ and $[SO_4^{2-}]$ in the K_{sp} expression, then we have

$$K_{sp} = s(0.25\ M + s) = 9.1 \times 10^{-6}\ M^2$$

Neglecting s with respect to 0.25 M, we obtain

$$s \approx \frac{9.1 \times 10^{-6}\ M^2}{0.25\ M} = 3.64 \times 10^{-5}\ M$$

The solubility in grams per liter is

$$s = (3.64 \times 10^{-5}\ mol \cdot L^{-1})\left(\frac{136.14\ g}{1\ mol}\right) = \mathbf{5.0} \times 10^{-3}\ g \cdot L^{-1}$$

19-28. The equilibrium constant expression is

$$K = \frac{[Ag(S_2O_3)_2^{3-}][Cl^-]}{[S_2O_3^{2-}]^2} = 5.20 \times 10^3$$

If we let s be the solubility of AgCl(s) in $S_2O_3^{2-}(aq)$, then we have

$$[Cl^-] = s$$

The K_{sp} for AgCl(s) is 1.8×10^{-10} M^2 which is very small compared to the above K value, thus $[Ag^+] << [Ag(S_2O_3)_2^{3-}]$ and therefore

$$[Ag(S_2O_3)^{3-}] \approx [Cl^-] = s$$

Substituting the expressions for $[Cl^-]$ and $[Ag(S_2O_3)^{3-}]$ in the K expression, we have

$$K = \frac{(s)(s)}{[S_2O_3^{2-}]^2} = \frac{s^2}{(0.015\ M)^2} = 5.20 \times 10^3$$

$$s^2 = 1.17\ M^2$$

and

$$s = 1.1\ M$$

19-30. a) Solubility is increased; an increase in the concentration of $OH^-(aq)$ will shift the equilibrium from left to right.

b) Solubility is unaffected; the amount of a solid reactant has no effect on the equilibrium concentrations.

c) Solubility is increased; a decrease in the concentration of $I^-(aq)$ will shift the equilibrium from left to right.

19-32. The concentration of $Ag^+(aq)$ after mixing is

$$[Ag^+]_0 = \frac{(0.20\ M)(0.0500\ L)}{(0.200\ L)} = 0.050\ M$$

The concentration of $SO_4^{2-}(aq)$ after mixing is

$$[SO_4^{2-}]_0 = \frac{(0.10\ M)(0.150\ L)}{(0.200\ L)} = 0.075\ M$$

The value of Q_{sp} is

$$Q_{sp} = [Ag^+]_0^2[SO_4^{2-}]_0 = (0.050\ M)^2(0.075\ M) = 1.9 \times 10^{-4}\ M^3$$

The value of K_{sp} is $1.4 \times 10^{-5}\ M^3$ (Table 19-1) and thus

$$\frac{Q_{sp}}{K_{sp}} = \frac{1.9 \times 10^{-4}\ M^3}{1.4 \times 10^{-5}\ M^3} = 14 > 1$$

Thus $Ag_2SO_4(s)$ will precipitate from the solution.

19-34. The concentration of $Ag^+(aq)$ after mixing is

$$[Ag^+]_0 = \frac{(0.50\ M)(0.0500\ L)}{(0.100\ L)} = 0.25\ M$$

The concentration of $Br^-(aq)$ after mixing is

$$[Br^-]_0 = \frac{(1.00 \times 10^{-4}\ M)(0.0500\ L)}{(0.100\ L)} = 5.0 \times 10^{-5}\ M$$

The value of Q_{sp} is

$$Q_{sp} = [Ag^+]_0[Br^-]_0 = (0.25\ M)(5.0 \times 10^{-5}\ M) = 1.3 \times 10^{-5}\ M^2$$

The value of K_{sp} for AgBr(s) is $5.0 \times 10^{-13}\ M^2$, thus

$$\frac{Q_{sp}}{K_{sp}} = \frac{1.3 \times 10^{-5}\ M^2}{5.0 \times 10^{-13}\ M^2} = 2.6 \times 10^7 > 1$$

Because $Q_{sp}/K_{sp} > 1$, the precipitation of AgBr(s) will occur. Because $[Ag^+]_0 >> [Br^-]_0$, and essentially all of the $Br^-(aq)$ is precipitated as AgBr(s), and the final equilibrium value of $[Ag^+]$ is 0.25 M. Thus we have at equilibrium following the precipitation of AgBr(s)

$$[Ag^+]_0[Br^-] \approx K_{sp} = 5.0 \times 10^{-13}\ M^2$$

and thus the final concentration of $Br^-(aq)$ is

$$[Br^-] \simeq \frac{5.0 \times 10^{-13}\ M^2}{[Ag^+]_0} = \frac{5.0 \times 10^{-13}\ M^2}{(0.25\ M)}$$

$$\simeq 2.0 \times 10^{-12}\ M$$

The moles of AgBr(s) that precipitates is given by

$$\begin{pmatrix}\text{moles of AgBr} \\ \text{precipitated}\end{pmatrix} = \begin{pmatrix}\text{initial moles} \\ \text{of } Br^-\end{pmatrix} - \begin{pmatrix}\text{final moles} \\ \text{of } Br^-\end{pmatrix}$$

$$= (5.0 \times 10^{-5}\ M - 2.0 \times 10^{-12}\ M)(0.100\ L)$$

$$= 5.0 \times 10^{-6}\ \text{mol}$$

The equilibrium concentrations following the precipitation of AgBr(s) are

$$[Ag^+] = [Ag^+]_0 = 0.25\ M \qquad [NO_3^-] = [Ag^+]_0 = 0.25\ M$$

$$[Br^-] = 2.0 \times 10^{-12}\ M \qquad [Na^+] = [Br^-]_0 = 5.0 \times 10^{-5}\ M$$

19-36. The concentration of $Zn^{2+}(aq)$ after mixing is

$$[Zn^{2+}]_0 = \frac{(0.30\ M)(0.0100L)}{0.0200L} = 0.15\ M$$

Neglecting the reaction of $S^{2-}(aq)$ with $H_2O(l)$, we have for the concentration of $S^{2-}(aq)$ after mixing

$$[S^{2-}]_0 = \frac{(2.00 \times 10^{-4}\ M)(0.0100L)}{0.0200L} = 1.0 \times 10^{-4}\ M$$

The value of K_{sp} for ZnS(s) is $1.6 \times 10^{-24}\ M^2$. Because ZnS(aq) is only very slightly soluble and $[Zn^{2+}]_0 \gg [S^{2-}]_0$, essentially all of the $S^{2-}(aq)$ will precipitate from solution as ZnS(s). The number of moles of ZnS(s) that precipitates is essentially equal to the number of moles of $S^{2-}(aq)$ initially present, thus

a) moles of ZnS(s) precipitated = (moles of S^{2-}(aq))$_0$ = (1.0×10^{-4} M)(0.0200 L)

moles of ZnS(s) = 2.0×10^{-6} mol

The number of milligrams of ZnS(s) in 2.0×10^{-6} mol is

$$\text{mass} = (2.0 \times 10^{-6}\,\text{mol})\left(\frac{97.44\text{ g}}{1\text{ mol}}\right)\left(\frac{1000\text{ mg}}{1\text{ g}}\right) = 0.19\text{ mg}$$

b) The concentration of Zn^{2+}(aq) at equilibrium is $[Zn^{2+}]_0$ because only 2.0×10^{-6} mol were used to precipitate ZnS(s)

$$[Zn^{2+}] = 0.15\text{ M}$$

The concentration of S^{2-}(aq) at equilibrium can be found from the K_{sp} expression

$$K_{sp} = [Zn^{2+}][S^{2-}] = (0.15\text{ M})[S^{2-}] = 1.6 \times 10^{-24}\text{ M}^2$$

$$[S^{2-}] = \frac{1.6 \times 10^{-24}\text{ M}^2}{0.15\text{ M}} = 1.1 \times 10^{-23}\text{ M}$$

19-38. The reaction is

$$MgC_2O_4(s) \rightleftharpoons Mg^{2+}(aq) + C_2O_4^{2-}(aq)$$

Recall that $C_2O_4^{2-}$ is the conjugate base of a weak acid ($HC_2O_4^-$).

a) The solubility is increased; the added H_3O^+(aq) reacts with $C_2O_4^{2-}$, thereby decreasing the concentration of $C_2O_4^{2-}$(aq). A decrease in $[C_2O_4^{2-}]$ shifts the equilibrium from left to right.

b) A slight decrease in solubility owing to a shift to the left in the equilibrium

$$C_2O_4^{2-}(aq) + H_2O(l) \rightleftharpoons HC_2O_4^-(aq) + OH^-(aq)$$

which leads to an increase in $[C_2O_4^{2-}]$.

c) The solubility is decreased; an increase in $[Mg^{2+}]$ shifts the equilibrium from right to left.

19-40. FeS; S^{2-} is the conjugate base of the weak acid, $HS^-(aq)$

$ZnCO_3$; CO_3^{2-} is the conjugate base of the weak acid, $HCO_3^-(aq)$

$PbCrO_4$; CrO_4^{2-} is the conjugate base of the weak acid $HCrO_4^-$ (aq), also because of the equilibrium

$$2CrO_4^{2-}(aq) + 2H^+(aq) \rightleftharpoons Cr_2O_7^{2-}(aq) + H_2O(l)$$

$Ag_2C_2O_4$; $C_2O_4^{2-}$ is the conjugate base of the weak acid, $HC_2O_4^-(aq)$

Ag_2O; solubility higher at lower pH because of the reaction

$$Ag_2O(s) + 2H^+(aq) \rightleftharpoons 2Ag^+(aq) + H_2O(l)$$

19-42. The reaction is

(1) $AgC_7H_5O_2(s) \rightleftharpoons Ag^+(aq) + C_7H_5O_2^-(aq)$

The K_{sp} expression is

$$K_{sp} = [Ag^+][C_7H_5O_2^-] = 2.5 \times 10^{-5}\ M^2$$

Let the solubility of $AgC_7H_5O_2$ be s, then

$$s = [Ag^+]$$

and

$$s = [C_7H_5O_2^-] + [HC_7H_5O_2]$$

where the second equation follows from the fact that the equilibrium

(2) $C_7H_5O_2^-(aq) + H_2O(l) \rightleftharpoons HC_7H_5O_2(aq) + OH^-(aq)$

exists in the solution. For reaction (2) we have

$$K = K_b = \frac{K_w}{K_a} = \frac{[HC_7H_5O_2][OH^-]}{[C_7H_5O_2^-]}$$

but at pH = 4.0

$$[OH^-] = \frac{K_w}{[H^+]} = \frac{K_w}{10^{-4.0}\ M}$$

and thus

$$\frac{[HC_7H_5O_2]K_w}{[C_7H_5O_2^-]10^{-4.0}\ M} = \frac{K_w}{6.5 \times 10^{-5}\ M}$$

Therefore

$$[HC_7H_5O_2] = [C_7H_5O_2^-]\frac{10^{-4.0}}{6.5 \times 10^{-5}} = 1.54[C_7H_5O_2^-]$$

thus

$$s = [C_7H_5O_2^-] + 1.54[C_7H_5O_2^-] = 2.54[C_7H_5O_2^-]$$

or

$$[C_7H_5O_2^-] = \frac{s}{2.54}$$

Combination of the above results with the K_{sp} expression yields

$$K_{sp} = 2.5 \times 10^{-5}\ M^2 = [Ag^+][C_7H_5O_2^-]$$

$$= (s)\left(\frac{s}{2.54}\right)$$

and

$$s = \{(2.54)(2.5 \times 10^{-5}\ M^2)\}^{\frac{1}{2}} = 8.0 \times 10^{-3}\ M$$

19-44. The reaction is

$$Cd(OH)_2(s) \rightleftharpoons Cd^{2+}(aq) + 2OH^-(aq)$$

The K_{sp} expression is

$$K_{sp} = [Cd^{2+}][OH^-]^2 = 2.5 \times 10^{-14}\ M^3 \quad \text{(Table 19-1)}$$

Let s be the solubility of $Cd(OH)_2(s)$, then

$$[Cd^{2+}] = s$$

At pH = 9.0

$$[H_3O^+] = 10^{-pH} = 10^{-9.0} = 1.0 \times 10^{-9} \text{ M}$$

and thus

$$[OH^-] = \frac{K_w}{[H_3O^+]} = \frac{1.00 \times 10^{-14} \text{ M}^2}{1.0 \times 10^{-9} \text{ M}} = 1.0 \times 10^{-5} \text{ M}$$

Therefore

$$K_{sp} = (s)(1.0 \times 10^{-5} \text{ M})^2 = 2.5 \times 10^{-14} \text{ M}^3$$

and

$$s = \frac{2.5 \times 10^{-14} \text{ M}^3}{1.0 \times 10^{-10} \text{ M}^2} = 2.5 \times 10^{-4} \text{ M}$$

19-46. a) $\longrightarrow$; an increase in $[H_3O^+]$ shifts the equilibrium from left to right.

b) $\longleftarrow$; OH^- reacts with H_3O^+ and reduces $[H_3O^+]$. A decrease in $[H_3O^+]$ shifts the equilibrium from right to left.

c) no change; same number of solute particles on each side of the reaction.

d) $\longleftarrow$; increasing the pH is the same as decreasing $[H_3O^+]$. A decrease in $[H_3O^+]$ shifts the equilibrium from right to left.

19-48. The solubility product expression for $Cu(OH)_2(s)$ is

$$K_{sp} = [Cu^{2+}][OH^-]^2 = 2.2 \times 10^{-20} \text{ M}^3 \quad \text{(Table 19-1)}$$

and thus the solubility is given by

$$s = [Cu^{2+}] = \frac{2.2 \times 10^{-20} \text{ M}^3}{[OH^-]^2} = \frac{2.2 \times 10^{-20} \text{ M}^3 [H_3O^+]^2}{K_w^2}$$

At pH = 4.0

$$[H_3O^+] = 10^{-pH} = 10^{-4.0} = 1.0 \times 10^{-4} \text{ M}$$

and thus

$$s = \frac{(2.2 \times 10^{-20}\ M^3)(1.0 \times 10^{-4}\ M)^2}{(1.00 \times 10^{-14}\ M^2)^2} = 2.2\ M$$

The solubility product expression for $Zn(OH)_2(s)$ is

$$K_{sp} = [Zn^{2+}][OH^-]^2 = 1.0 \times 10^{-15}\ M^3 \quad \text{(Table 19-1)}$$

and thus the solubility is given by

$$s = [Zn^{2+}] = \frac{1.0 \times 10^{-15}\ M^3}{[OH^-]^2} = \frac{1.0 \times 10^{-15}\ M^3 [H_3O^+]^2}{K_w^2}$$

$$s = \frac{(1.0 \times 10^{-15}\ M^3)(1.0 \times 10^{-4}\ M)^2}{(1.00 \times 10^{-14}\ M^2)^2} = 1.0 \times 10^5\ M$$

Therefore both $Cu(OH)_2(s)$ and $Zn(OH)_2(s)$ are very soluble at pH = 4.0 and a separation cannot be achieved at pH = 4.0 using differences in $M(OH)_2(s)$ solubilities (2.2 M vs. 1×10^5 M).

19-50. The solubility product expression for $Ca(OH)_2(s)$ is

$$K_{sp} = [Ca^{2+}][OH^-]^2 = 5.5 \times 10^{-6}\ M^3 \quad \text{(Table 19-1)}$$

Thus, the solubility as a function of $[H_3O^+]$ is

$$s = [Ca^{2+}] = \frac{5.5 \times 10^{-6}\ M^3}{[OH^-]^2} = \frac{5.5 \times 10^{-6}\ M^3 [H_3O^+]^2}{K_w^2}$$

$$= \frac{5.5 \times 10^{-6}\ M^3 [H_3O^+]^2}{(1.00 \times 10^{-14}\ M^2)^2} = 5.5 \times 10^{22}\ M^{-1} [H_3O^+]^2$$

The solubility product expression for $Mg(OH)_2(s)$ is

$$K_{sp} = [Mg^{2+}][OH^-]^2 = 1.8 \times 10^{-11}\ M^3 \quad \text{(Table 19-1)}$$

Thus, the solubility as a function of $[H_3O^+]$ is

$$s = [Mg^{2+}] = \frac{1.8 \times 10^{-11}\ M^3}{[OH^-]^2} = \frac{1.8 \times 10^{-11}\ M^3 [H_3O^+]^2}{K_w^2}$$

$$= \frac{1.8 \times 10^{-11}\ M^3 [H_3O^+]^2}{(1.00 \times 10^{-14}\ M^2)^2} = 1.8 \times 10^{17}\ M^{-1} [H_3O^+]^2$$

Note from the two expressions for s that $Ca(OH)_2(s)$ is more soluble at a given pH than is $Mg(OH)_2(s)$. The $[H_3O^+]$ at which the solubility of $Mg(OH)_2$ is 1×10^{-6} M is

$$s = 1 \times 10^{-6}\ M = 1.8 \times 10^{17}\ M^{-1} [H_3O^+]^2$$

or

$$[H_3O^+] = 2.4 \times 10^{-12}\ M$$

from which we compute a pH of

$$pH = 11.6$$

The solubility of $Ca(OH)_2(s)$ at pH = 11.6 is

$$s = (5.5 \times 10^{22}\ M^{-1})(10^{-11.6})^2$$

$$s = 0.35\ M$$

Thus an effective separation can be achieved at pH = 11.6 where $Mg(OH)_2(s)$ has a solubility of only 1×10^{-6} M whereas the solubility of $Ca(OH)_2(s)$ is 0.35 M.

19-52. The K_{sp} expression for SnS(s) is

$$K_{sp} = [Sn^{2+}][S^{2-}] = 1.0 \times 10^{-25}\ M^2 \qquad \text{(Table 19-1)}$$

The solubility of SnS(s)

$$s = [Sn^{2+}] = \frac{1.0 \times 10^{-25}\ M^2}{[S^{2-}]}$$

From Equation (19-18) we have

$$[S^{2-}] = \frac{1.1 \times 10^{-21}\ M^3}{[H_3O^+]^2}$$

At pH = 2.0

$$[H_3O^+] = 10^{-2.0} = 1.0 \times 10^{-2}\ M$$

and thus

$$[S^{2-}] = \frac{1.1 \times 10^{-21}\ M^3}{(1.0 \times 10^{-2}\ M)^2} = 1.1 \times 10^{-17}\ M$$

Therefore, the solubility is

$$s = \frac{1.0 \times 10^{-25}\ M^2}{1.1 \times 10^{-17}\ M} = 9.1 \times 10^{-9}\ M$$

19-54. The solubilities of FeS(s) and PbS(s) as a function of $[H_3O^+]$ at $[H_2S]$ = 0.10 M are obtained using Equation (19-18) and the K_{sp} values. Thus for FeS(s) we have

$$s = \frac{K_{sp}[H_3O^+]}{1.1 \times 10^{-21}\ M^3} = \frac{6.3 \times 10^{-18} M^2 [H_3O^+]}{1.1 \times 10^{-21}\ M^3} = (5.7 \times 10^3 M^{-1})[H_3O^+]$$

whereas for PbS(s) we have

$$s = \frac{K_{sp}[H_3O^+]}{1.1 \times 10^{-21}\ M^3} = \frac{8.0 \times 10^{-28} M^2 [H_3O^+]}{1.1 \times 10^{-21}\ M^3} = (7.3 \times 10^{-7} M^{-1})[H_3O^+]$$

Note that at any $[H_3O^+]$ value PbS(s) is much less soluble than FeS(s). The value of $[H_3O^+]$ at which s = 1 x 10^{-6} M for PbS(s) is

$$1 \times 10^{-6}\ M = (7.3 \times 10^{-7}\ M^{-1})[H_3O^+]$$

or

$$[H_3O^+] = \sqrt{1.37\ M^2} = 1.17\ M$$

and the corresponding pH is -log(1.17) = -0.07. The solubility of FeS(s) at $[H_3O^+] =$ 1.17 M is

$$s = (5.7 \times 10^3\ M^{-1})(1.17\ M)^2 = 7.8 \times 10^3\ M$$

Thus if the pH of the solution is adjusted to about -0.1, an effective separation can be achieved. At this pH, the solubility of PbS(s) is 1×10^{-6} M, whereas the solubility of FeS(s) is 8000 M.

19-56. The concentration of $S^{2-}(aq)$ in a saturated solution of $H_2S(aq)$ is given by Equation (19-18) as

$$[S^{2-}] = \frac{1.1 \times 10^{-21}\ M^3}{[H_3O^+]^2} = \frac{1.1 \times 10^{-21}\ M^3}{(0.30\ M)^2} = 1.2 \times 10^{-20}\ M$$

The calculated concentration of $Mn^{2+}(aq)$ is

$$[Mn^{2+}] = \frac{K_{sp}}{[S^{2-}]} = \frac{2.5 \times 10^{-13}\ M^2}{1.2 \times 10^{-20}\ M} = 2.1 \times 10^7\ M$$

Showing that MnS(s) is quite soluble at this pH.

Thus at equilibrium

$$[Mn^{2+}] = 0.01\ M$$

because none of the $Mn^{2+}(aq)$ precipitates. For $[Cd^{2+}]$ we have

$$[Cd^{2+}] = \frac{K_{sp}}{[S^{2-}]} = \frac{8.0 \times 10^{-27}\ M^2}{1.2 \times 10^{-20}\ M} = 6.7 \times 10^{-7}\ M$$

which would be the $[Cd^{2+}]$ at equilibrium. The concentration of $Fe^{2+}(aq)$ is

$$[Fe^{2+}] = \frac{K_{sp}}{[S^{2-}]} = \frac{6.3 \times 10^{-18}\ M^2}{1.2 \times 10^{-20}\ M} = 5.3 \times 10^2\ M$$

Thus at equilibrium

$$[Fe^{2+}] = 0.01\ M$$

because none of the $Fe^{2+}(aq)$ precipitates.

19-58. The precipitation reaction is

$$Zn(ClO_4)_2(aq) + 2KOH(aq) \rightleftharpoons Zn(OH)_2(s) + 2KClO_4(aq)$$

The precipitate dissolves via the reaction

$$Zn(OH)_2(s) + 2OH^-(aq) \rightleftharpoons Zn(OH)_4^{2-}(aq)$$

19-60. The solution is basic and thus the reaction is

$$Pb(OH)_2(s) + OH^-(aq) \rightleftharpoons Pb(OH)_3^-(aq)$$

Letting s = solubility of lead(II)hydroxide = $[Pb(OH)_3^-]$

$$K = \frac{[Pb(OH)_3^-]}{[OH^-]} = \frac{s}{[OH^-]} = 0.08$$

At pH = 13.0, $[H_3O^+] = 1.0 \times 10^{-13}$ M, and

$$[OH^-] = \frac{K_w}{[H_3O^+]} = \frac{1.00 \times 10^{-14}\ M^2}{1.0 \times 10^{-13}\ M} = 1.0 \times 10^{-1}\ M$$

thus the solubility of $Pb(OH)_2$ is given by

$$s = (0.08)[OH^-] = (0.08)(1.0 \times 10^{-1}) = 8 \times 10^{-3}\ M$$

19-62. The solubility of $Zn(OH)_2(s)$ in a basic solution is

$$s = [Zn(OH)_4^{2-}] = K[OH^-]^2 = 0.050\ M^{-1}[OH^-]^2$$

The concentration of $OH^-(aq)$ in a 0.10 M NaOH(aq) solution is

$$[OH^-] = 0.10\ M$$

At equilibrium we have from the reaction stoichiometry

$$[OH^-] = 0.10\ M - 2s$$

thus

$$s = (0.050\ M^{-1})(0.10\ M - 2s)^2$$

Neglecting 2s compared to 0.10 M, we have

$$s = (0.050\ M^{-1})(0.10\ M)^2 = 5.0 \times 10^{-4}\ M$$

19-64. The solubility of $Pb(OH)_2(s)$ in acidic solution is controlled primarily by the reaction

$$Pb(OH)_2(s) \rightleftharpoons Pb^{2+}(aq) + 2OH^-(aq)$$

The solubility of $Pb(OH)_2(s)$ due to the above reaction is

$$s = [Pb^{2+}] = \frac{K_{sp}}{[OH^-]^2} = \frac{1.2 \times 10^{-15}\ M^3}{[OH^-]^2}$$

At pH = 3.0, $[H_3O^+] = 1.0 \times 10^{-3}$ M and

$$[OH^-] = \frac{K_w}{[H_3O^+]} = \frac{1.00 \times 10^{-14}\ M^2}{1.0 \times 10^{-3}\ M} = 1.0 \times 10^{-11}\ M$$

thus

$$s = \frac{1.2 \times 10^{-15}\ M^3}{(1.0 \times 10^{-11}\ M)^2} = 1.2 \times 10^7\ M$$

At pH = 6.0, $[H_3O^+] = 1.0 \times 10^{-6}$ M and

$$[OH^-] = \frac{K_w}{[H_3O^+]} = \frac{1.00 \times 10^{-14}\ M^2}{1.0 \times 10^{-6}\ M} = 1.0 \times 10^{-8}\ M$$

thus

$$s = \frac{1.2 \times 10^{-15}\ M^3}{(1.0 \times 10^{-8}\ M)^2} = 12\ M$$

The solubility of $Pb(OH)_2(s)$ in strongly basic solution is controlled primarily by the reaction

$$Pb(OH)_2(s) + OH^-(aq) \rightleftharpoons Pb(OH)_3^-(aq)$$

The solubility of $Pb(OH)_2(s)$ due to the above reaction is

$$s = [Pb(OH)_3^-] = K[OH^-] = 0.08[OH^-]$$

At pH = 12.0, $[H_3O^+] = 1.0 \times 10^{-12}$ M and

$$[OH^-] = \frac{K_w}{[H_3O^+]} = \frac{1.00 \times 10^{-14}\ M^2}{1.0 \times 10^{-12}\ M} = 1.0 \times 10^{-2}\ M$$

thus

$$s = (0.08)(1.0 \times 10^{-2}\ M) = 8 \times 10^{-4}\ M$$

CHAPTER 20

SOLUTIONS TO THE EVEN-NUMBERED PROBLEMS

20-2. a) S^{2-}. We assign sulfur an oxidation state of -2 because sulfur is in the same family as oxygen (Rule 6).

b) SO_3 We assign oxygen an oxidation state of -2 (Rule 6). The oxidation state, X, of sulfur is given by (Rule 2) $X + 3(-2) = 0$ or $X = +6$. The oxidation state of sulfur in SO_3 is +6.

c) S . We assign sulfur an oxidation state of zero (Rule 1).

d) SO_2. We assign oxygen an oxidation state of -2 (Rule 6). The oxidation state, X, of sulfur is given by $X + 2(-2) = 0$ or $X = +4$. The oxidation state of sulfur in SO_2 is +4.

e) SO_4^{2-}. We assign oxygen an oxidation state of -2 (Rule 6). The oxidation state, X, of sulfur is given by (Rule 2) $X + 4(-2) = -2$ or $X = +6$. The oxidation state of sulfur in SO_4^{2-} is +6.

20-4. a) NaClO. We assign sodium an oxidation state of +1 (Rule 3) and oxygen an oxidation state of -2 (Rule 6). The oxidation state, X, of chlorine is given by (Rule 2) $+1 + X - 2 = 0$ or $X = +1$. The oxidation state of chlorine in NaClO is +1.

b) $KMnO_4$. We assign potassium an oxidation state of +1 (Rule 3) and oxygen an oxidation state of -2 (Rule 6). The oxidation state X of manganese is given by (Rule 2) $+1 + X + 4(-2) = 0$ or $X = +7$. The oxidation state of manganese in $KMnO_4$ is +7.

c) CaH_2 . We assign calcium an oxidation state of +2 (Rule 4). The oxidation state, X, of hydrogen is given by (Rule 2) $+2 + 2X = 0$ or $X = -1$. The oxidation state of hydrogen in CaH_2 is -1.

d) KO_2 . We assign potassium an oxidation state of +1 (Rule 3). The oxidation state, X, of oxygen is given by (Rule 2) $+1 + 2X = 0$ or $X = -\frac{1}{2}$. The oxidation state of oxygen in KO_2 is $-\frac{1}{2}$.

e) Fe_2O_3. We assign oxygen an oxidation state of -2 (Rule 6). The oxidation state, X, of iron is given by (Rule 2) $2X + 3(-2) = 0$ or $X = +3$. The oxidation state of iron in Fe_2O_3 is +3.

20-6. a) Cl_2O. We assign oxygen an oxidation state of -2 (Rule 6). The oxidation state, X, of chlorine is (Rule 2) $2X + 1(-2) = 0$ or $X = +1$.

b) Cl_2O_3. We assign oxygen an oxidation state of -2. The oxidation state, X, of chlorine is (Rule 2) $2X + 3(-2) = 0$ or $X = +3$.

c) ClO_2. We assign oxygen an oxidation state of -2. The oxidation state, X, of chlorine is (Rule 2) $X + 2(-2) = 0$ or $X = +4$.

d) Cl_2O_4. We assign oxygen an oxidation state of -2. The oxidation state, X, of chlorine is (Rule 2) $2X + 4(-2) = 0$ or $X = +4$.

e) Cl_2O_6. We assign oxygen an oxidation state of -2. The oxidation state, X, of chlorine is (Rule 2) $2X + 6(-2) = 0$ or $X = +6$.

20-8. a) P_4O_6. We assign oxygen an oxidation state of -2 (Rule 6). The oxidation state, X, of phosphorus is given by (Rule 2) $4(X) + 6(-2) = 0$ or $X = +3$.

b) P_4O_7. We assign oxygen an oxidation state of -2. The oxidation state, X, of phosphorus is given by $4(X) + 7(-2) = 0$ or $X = +7/2$.

c) P_4O_8. We assign oxygen an oxidation state of -2. The oxidation state, X, of phosphorus is given by $4X + 8(-2) = 0$ or $X = +4$.

d) P_4O_9. We assign oxygen an oxidation state of -2. The oxidation state, X, of phosphorus is given by $4X + 9(-2) = 0$ or $X = +9/2$.

e) P_4O_{10}. We assign oxygen an oxidation state of -2. The oxidation state, X, of phosphorus is given by $4X + 10(-2) = 0$ or $X = +5$.

20-10. a) XeF_2. The Lewis formula for XeF_2 is

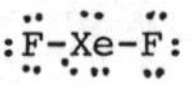

Because fluorine is more electronegative than xenon, we assign the electrons in each covalent bond to fluorine. The oxidation state of xenon is $8 - 6 = +2$.

b) XeF_6. We assign fluorine an oxidation state of -1. The oxidation state, X, of xenon is given by $X + 6(-1) = 0$ or $X = +6$.

c) Xe. We assign xenon an oxidation state of 0 (Rule 1).

d) $XeOF_4$. The Lewis formula for $XeOF_4$ is

Because fluorine is more electronegative than xenon, we assign the electrons in each of the four covalent bonds to fluorine. Because oxygen is more electronegative than xenon, we assign the electrons in the covalent double bond to oxygen. Therefore, the oxidation state of xenon is $8 - 2 = +6$.

e) XeO_2F_2. The Lewis formula of XeO_2F_2 is

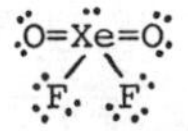

Because fluorine is more electronegative than xenon, we assign the electrons in each of the two covalent bonds to fluorine. Because oxygen is more electronegative than xenon, we assign the electrons in each of the two covalent double bonds to oxygen. Therefore, the oxidation state of xenon is 8 - 2 = +6.

20-12. The oxidation state of sulfur decreases from +6 in Na_2SO_4 to -2 in Na_2S. Thus sulfur is reduced; Na_2SO_4 is the oxidizing agent. The oxidation state of carbon increases from 0 in C to +2 in CO. Thus carbon is oxidized; C is the reducing agent.

20-14. The oxidation state of carbon increases from 0 in C to +4 in $CaCO_3$. Thus carbon is oxidized; and C is the reducing agent. The oxidation state of chlorine decreases from +4 in ClO_2 to +3 in $NaClO_2$. Thus chlorine is reduced, and ClO_2 is the oxidizing agent.

20-16. a) The oxidation state of copper decreases from +2 in Cu^{2+} to 0 in Cu. Thus $Cu^{2+}(aq)$ is reduced and acts as the oxidizing agent. The oxidation state of chromium increases from +2 in Cr^{2+} to +3 in Cr^{3+}. Thus $Cr^{2+}(aq)$ is oxidized and acts as the reducing agent. The half-reactions are

$$Cr^{2+}(aq) \longrightarrow Cr^{3+}(aq) + e^- \quad \text{(oxidation half-reaction)}$$

$$Cu^{2+}(aq) + 2e^- \longrightarrow Cu(s) \quad \text{(reduction half-reaction)}$$

b) The oxidation state of indium increases from +1 in In^+ to +3 in In^{3+}. Thus $In^+(aq)$ is oxidized and acts as a reducing agent. The oxidation state of iron decreases from +3 in Fe^{3+} to +2 in Fe^{2+}. Thus $Fe^{3+}(aq)$ is reduced and acts as an oxidizing agent. The half-reactions are

$$In^{+}(aq) \longrightarrow In^{3+}(aq) + 2e^{-} \quad \text{(oxidation half-reaction)}$$

$$Fe^{3+}(aq) + e^{-} \longrightarrow Fe^{2+}(aq) \quad \text{(reduction half-reaction)}$$

20-18. The hydrogen in $LiAlH_4$ is in an usual oxidation state (-1). The common oxidation state of hydrogen is +1, and so the hydrogen in $LiAlH_4$ is easily oxidized from -1 to +1, thus making $LiAlH_4$ a strong reducing agent.

20-20. a) The oxidation state of cobalt decreases from +2 to 0. The oxidation of sulfur increases +4 to +6. The oxidation half-reaction and reduction half-reaction are

$$SO_3^{2-} \longrightarrow SO_4^{2-} \quad \text{(oxidation)}$$

$$Co(OH)_2 \longrightarrow Co \quad \text{(reduction)}$$

Both half-reactions are balanced with respect to S and Co. To balance the oxygen atoms, we add H_2O to the left side of the oxidation half-reaction and 2 H_2O to the right side of the reduction half-reaction

$$SO_3^{2-} + H_2O \longrightarrow SO_4^{2-} \quad \text{(oxidation)}$$

$$Co(OH)_2 \longrightarrow Co + 2H_2O \quad \text{(reduction)}$$

To balance the hydrogen atoms, we add $2H^+$ to the right side of the oxidation half-reaction and $2H^+$ to the left side of the reduction half-reaction.

$$SO_3^{2-} + H_2O \longrightarrow SO_4^{2-} + 2H^+ \quad \text{(oxidation)}$$

$$Co(OH)_2 + 2H^+ \longrightarrow Co + 2H_2O \text{(reduction)}$$

To balance the charge, we add 2 electrons to the right side of the oxidation half-reaction and 2 electrons to the left side of the reduction half-reaction

$$SO_3^{2-} + H_2O \longrightarrow SO_4^{2-} + 2H^+ + 2e^- \quad \text{(oxidation)}$$

$$Co(OH)_2 + 2H^+ + 2e^- \longrightarrow Co + 2H_2O \quad \text{(reduction)}$$

If we add the two half-reactions and indicate the various phases, then we have the balanced equation

$$Co(OH)_2(s) + SO_3^{2-}(aq) \longrightarrow SO_4^{2-}(aq) + Co(s) + H_2O(l)$$

b) The oxidation state of germanium increases from +2 to +4. The oxidation state of iodine decreases +5 to -1. The oxidation half-reaction and reduction half-reaction are

$$GeO \longrightarrow GeO_2 \qquad \text{(oxidation)}$$

$$IO_3^- \longrightarrow I^- \qquad \text{(reduction)}$$

Both half-reactions are balanced with respect to Ge and I. To balance the oxygen atoms, we add H_2O to the left side of the oxidation half-reaction and $3H_2O$ to the right side of the reduction half-reaction

$$GeO + H_2O \longrightarrow GeO_2 \qquad \text{(oxidation)}$$

$$IO_3^- \longrightarrow I^- + 3H_2O \qquad \text{(reduction)}$$

To balance the hydrogen atoms, we add $2H^+$ to the right side of the oxidation half-reaction and $6H^+$ to the left side of the reduction half-reaction

$$GeO + H_2O \longrightarrow GeO_2 + 2H^+ \qquad \text{(oxidation)}$$

$$IO_3^- + 6H^+ \longrightarrow I^- + 3H_2O \qquad \text{(reduction)}$$

To balance charges, we add 2 electrons to the right side of the oxidation half-reaction and 6 electrons to the left side of the reduction half-reaction

$$GeO + H_2O \longrightarrow GeO_2 + 2H^+ + 2e^- \qquad \text{(oxidation)}$$

$$IO_3^- + 6H^+ + 6e^- \longrightarrow I^- + 3H_2O \qquad \text{(reduction)}$$

If we multiply the oxidation half-reaction by 3, then both reactions involve 6 electrons.

$$3GeO + 3H_2O \longrightarrow 3GeO_2 + 6H^+ + 6e^- \quad \text{(oxidation)}$$

$$IO_3^- + 6H^+ + 6e^- \longrightarrow I^- + 3H_2O \quad \text{(reduction)}$$

We add these two half-reactions to obtain the balanced equation

$$3GeO(s) + IO_3^-(aq) \longrightarrow 3GeO_2(s) + I^-(aq)$$

c) The two half-reactions are

$$PbO \longrightarrow PbO_2 \quad \text{(oxidation)}$$

$$NO_3^- \longrightarrow NO_2^- \quad \text{(reduction)}$$

Adding H_2O to balance the oxygen atoms gives

$$PbO + H_2O \longrightarrow PbO_2 \quad \text{(oxidation)}$$

$$NO_3^- \longrightarrow NO_2^- + H_2O \quad \text{(reduction)}$$

Adding H^+ to balance the hydrogen atoms gives

$$PbO + H_2O \longrightarrow PbO_2 + 2H^+ \quad \text{(oxidation)}$$

$$NO_3^- + 2H^+ \longrightarrow NO_2^- + H_2O \quad \text{(reduction)}$$

Adding electrons to balance the charges gives

$$PbO + H_2O \longrightarrow PbO_2 + 2H^+ + 2e^- \quad \text{(oxidation)}$$

$$NO_3^- + 2H^+ + 2e^- \longrightarrow NO_2^- + H_2O \quad \text{(reduction)}$$

Adding these two half-reactions gives the balanced equation

$$NO_3^-(aq) + PbO(s) \longrightarrow NO_2^-(aq) + PbO_2(s)$$

20-22. a) The two half-reactions are

$$BrO_3^- \longrightarrow BrO_4^- \quad \text{(oxidation)}$$

$$F_2 \longrightarrow 2F^- \quad \text{(reduction)}$$

Add H_2O to balance the oxygen atoms

$$BrO_3^- + H_2O \longrightarrow BrO_4^- \quad \text{(oxidation)}$$

$$F_2 \longrightarrow 2F^- \quad \text{(reduction)}$$

Add H^+ to balance the hydrogen atoms

$BrO_3^- + H_2O \longrightarrow BrO_4^- + 2H^+$ (oxidation)

$F_2 \longrightarrow 2F^-$ (reduction)

Add electrons to balance the charges

$BrO_3^- + H_2O \longrightarrow BrO_4^- + 2H^+ + 2e^-$ (oxidation)

$F_2 + 2e^- \longrightarrow 2F^-$ (reduction)

Add the two half-reactions to obtain the balanced equation (in acid solution)

$$BrO_3^- + F_2 + H_2O \longrightarrow BrO_4^- + 2F^- + 2H^+$$

For basic solution, add $2OH^-$ to both sides of this equation to get

$$BrO_3^- + F_2 + H_2O + 2OH^- \longrightarrow BrO_4^- + 2F^- + \underbrace{2H^+ + 2OH^-}_{2H_2O}$$

or finally

$$BrO_3^-(aq) + F_2(g) + 2OH^-(aq) \longrightarrow BrO_4^-(aq) + 2F^-(aq) + H_2O(l)$$

b) The two half-reactions are

$H_3AsO_3 \longrightarrow H_3AsO_4$ (oxidation)

$I_2 \longrightarrow 2I^-$ (reduction)

Adding H_2O and H^+ to balance the oxygen atoms and hydrogen atoms

$H_3AsO_3 + H_2O \longrightarrow H_3AsO_4 + 2H^+$ (oxidation)

$I_2 \longrightarrow 2I^-$ (reduction)

Balance the charges by adding electrons

$H_3AsO_3 + H_2O \longrightarrow H_3AsO_4 + 2H^+ + 2e^-$ (oxidation)

$I_2 + 2e^- \longrightarrow 2I^-$ (reduction)

Add these two half-reactions to obtain

$$H_3AsO_3(aq) + I_2(aq) + H_2O(l) \longrightarrow H_3AsO_4(aq) + 2I^-(aq) + 2H^+(aq)$$

c) The two half-reactions are

$$C_2O_4^{2-} \longrightarrow CO_2 \qquad \text{(oxidation)}$$

$$MnO_4^- \longrightarrow Mn^{2+} \qquad \text{(reduction)}$$

Balance the carbon atoms to get

$$C_2O_4^{2-} \longrightarrow 2CO_2 \qquad \text{(oxidation)}$$

$$MnO_4^- \longrightarrow Mn^{2+} \qquad \text{(reduction)}$$

Add H_2O and H^+ to balance the oxygen atoms and hydrogen atoms

$$C_2O_4^{2-} \longrightarrow 2CO_2 \qquad \text{(oxidation)}$$

$$MnO_4^- + 8H^+ \longrightarrow Mn^{2+} + 4H_2O \qquad \text{(reduction)}$$

Add electrons to balance the charges

$$C_2O_4^{2-} \longrightarrow 2CO_2 + 2e^- \qquad \text{(oxidation)}$$

$$MnO_4^- + 8H^+ + 5e^- \longrightarrow Mn^{2+} + 4H_2O \qquad \text{(reduction)}$$

Multiply the oxidation half-reaction by 5 and the reduction half-reaction by 2

$$5C_2O_4^{2-} \longrightarrow 10CO_2 + 10e^- \qquad \text{(oxidation)}$$

$$2MnO_4^- + 16H^+ + 10e^- \longrightarrow 2Mn^{2+} + 8H_2O \qquad \text{(reduction)}$$

Add these two half-reactions to obtain the balanced equation

$$2MnO_4^-(aq) + 5C_2O_4^{2-}(aq) + 16H^+(aq) \longrightarrow 2Mn^{2+}(aq) + 10CO_2(aq) + 8H_2O(l)$$

20-24. a) The various steps are

$$2Cl^- \longrightarrow Cl_2 \qquad \text{(oxidation)}$$

$$MnO_2 \longrightarrow Mn^{2+} \qquad \text{(reduction)}$$

$$2Cl^- \longrightarrow Cl_2 \qquad \text{(oxidation)}$$

$$MnO_2 + 4H^+ \longrightarrow Mn^{2+} + 2H_2O \qquad \text{(reduction)}$$

$$2Cl^- \longrightarrow Cl_2 + 2e^- \quad \text{(oxidation)}$$

$$MnO_2 + 4H^+ + 2e^- \longrightarrow Mn^{2+} + 2H_2O \quad \text{(reduction)}$$

$$MnO_2(s) + 2Cl^-(aq) + 4H^+(aq) \longrightarrow Mn^{2+}(aq) + Cl_2(g) + 2H_2O(l)$$

b) The various steps are

$$2I^- \longrightarrow I_2 \quad \text{(oxidation)}$$

$$Cr_2O_7^{2-} \longrightarrow 2Cr^{3+} \quad \text{(reduction)}$$

$$2I^- \longrightarrow I_2 \quad \text{(oxidation)}$$

$$Cr_2O_7^{2-} + 14H^+ \longrightarrow 2Cr^{3+} + 7H_2O \quad \text{(reduction)}$$

$$2I^- \longrightarrow I_2 + 2e^- \quad \text{(oxidation)}$$

$$Cr_2O_7^{2-} + 14H^+ + 6e^- \longrightarrow 2Cr^{3+} + 7H_2O \quad \text{(reduction)}$$

$$6I^- \longrightarrow 3I_2 + 6e^- \quad \text{(oxidation)}$$

$$Cr_2O_7^{2-} + 14H^+ + 6e^- \longrightarrow 2Cr^{3+} + 7H_2O \quad \text{(reduction)}$$

$$Cr_2O_7^{2-}(aq) + 6I^-(aq) + 14H^+(aq) \longrightarrow 2Cr^{3+}(aq) + 3I_2(s) + 7H_2O(l)$$

c) The various steps are

$$CuS \longrightarrow S \quad \text{(oxidation)}$$

$$NO_3^- \longrightarrow NO \quad \text{(reduction)}$$

$$CuS \longrightarrow S + Cu^{2+} \quad \text{(oxidation)}$$

$$NO_3^- + 4H^+ \longrightarrow NO + 2H_2O \quad \text{(reduction)}$$

$$CuS \longrightarrow S + Cu^{2+} + 2e^- \quad \text{(oxidation)}$$

$$NO_3^- + 4H^+ + 3e^- \longrightarrow NO + 2H_2O \quad \text{(reduction)}$$

$$3CuS \longrightarrow 3S + 3Cu^{2+} + 6e^- \quad \text{(oxidation)}$$

$$2NO_3^- + 8H^+ + 6e^- \longrightarrow 2NO + 4H_2O \quad \text{(reduction)}$$

$$3CuS(s) + 2NO_3^-(aq) + 8H^+(aq) \longrightarrow 3Cu^{2+}(aq) + 3S(s) + 2NO(g) + 4H_2O(l)$$

20-26.a) The various steps are

$$CoCl_2 \longrightarrow Co(OH)_3 \quad \text{(oxidation)}$$

$$Na_2O_2 \longrightarrow OH^- \quad \text{(reduction)} \quad \text{in a basic solution}$$

$$CoCl_2 + 3H_2O \longrightarrow Co(OH)_3 + 2Cl^- + 3H^+ \quad \text{(oxidation)}$$

$$Na_2O_2 + 2H^+ \longrightarrow 2OH^- + 2Na^+ \quad \text{(reduction)}$$

$$CoCl_2 + 3H_2O \longrightarrow Co(OH)_3 + 2Cl^- + 3H^+ + e^- \quad \text{(oxidation)}$$

$$Na_2O_2 + 2H^+ + 2e^- \longrightarrow 2OH^- + 2Na^+ \quad \text{(reduction)}$$

$$2CoCl_2 + 6H_2O \longrightarrow 2Co(OH)_3 + 4Cl^- + 6H^+ + 2e^- \quad \text{(oxidation)}$$

$$Na_2O_2 + 2H^+ + 2e^- \longrightarrow 2OH^- + 2Na^+ \quad \text{(reduction)}$$

$$2CoCl_2 + Na_2O_2 + 6H_2O \longrightarrow 2Co(OH)_3 + 4Cl^- + 2Na^+ + 4H^+ + 2OH^-$$

or

$$2CoCl_2 + Na_2O_2 + 6H_2O + 2OH^- \longrightarrow 2Co(OH)_3 + 4Cl^- + 2Na^+ + \underbrace{4H^+ + 2OH^- + 2OH^-}_{4H_2O}$$

$$2CoCl_2(s) + Na_2O_2(aq) + 2H_2O(l) + 2OH^-(aq) \longrightarrow 2Co(OH)_3(s) + 4Cl^-(aq) + 2Na^+(aq)$$

b)

$$N_2H_4 \longrightarrow N_2 \quad \text{(oxidation)}$$

$$Cu(OH)_2 \longrightarrow Cu \quad \text{(reduction)}$$

$$N_2H_4 \longrightarrow N_2 + 4H^+ \quad \text{(oxidation)}$$

$$Cu(OH)_2 + 2H^+ \longrightarrow Cu + 2H_2O \quad \text{(reduction)}$$

$$N_2H_4 \longrightarrow N_2 + 4H^+ + 4e^- \quad \text{(oxidation)}$$

$$Cu(OH)_2 + 2H^+ + 2e^- \longrightarrow Cu + 2H_2O \quad \text{(reduction)}$$

$N_2H_4 \longrightarrow N_2 + 4H^+ + 4e^-$ (oxidation)

$2Cu(OH)_2 + 4H^+ + 4e^- \longrightarrow 2Cu + 4H_2O$ (reduction)

$N_2H_4(aq) + 2Cu(OH)_2(s) \longrightarrow N_2(g) + 2Cu(s) + 4H_2O(l)$

c) $C_2O_4^{2-} \longrightarrow CO_2$ (oxidation)

$MnO_2 \longrightarrow Mn^{2+}$ (reduction)

$C_2O_4^{2-} \longrightarrow 2CO_2$ (oxidation)

$MnO_2 + 4H^+ \longrightarrow Mn^{2+} + 2H_2O$ (reduction)

$C_2O_4^{2-} \longrightarrow 2CO_2 + 2e^-$ (oxidation)

$MnO_2 + 4H^+ + 2e^- \longrightarrow Mn^{2+} + 2H_2O$ (reduction)

$C_2O_4^{2-}(aq) + MnO_2(s) + 4H^+(aq) \longrightarrow Mn^{2+}(aq) + 2CO_2(g) + 2H_2O(l)$

20-28. a) The various steps are

$ZnS \longrightarrow S$ (oxidation)

$NO_3^- \longrightarrow NO$ (reduction)

$ZnS \longrightarrow S + Zn^{2+}$ (oxidation)

$NO_3^- + 4H^+ \longrightarrow NO + 2H_2O$ (reduction)

$ZnS \longrightarrow S + Zn^{2+} + 2e^-$ (oxidation)

$NO_3^- + 4H^+ + 3e^- \longrightarrow NO + 2H_2O$ (reduction)

$3ZnS \longrightarrow 3S + 3Zn^{2+} + 6e^-$ (oxidation)

$2NO_3^- + 8H^+ + 6e^- \longrightarrow 2NO + 4H_2O$ (reduction)

$3ZnS(s) + 2NO_3^-(aq) + 8H^+(aq) \longrightarrow 3Zn^{2+}(aq) + 3S(s) + 2NO(g) + 4H_2O(l)$

electron donor	ZnS
electron acceptor	NO_3^-
oxidizing agent	NO_3^-
reducing agent	ZnS
species oxidized	sulfur in ZnS
species reduced	nitrogen in NO_3^-

b) The various steps are

$HNO_2 \longrightarrow NO_3^-$ (oxidation)

$MnO_4^- \longrightarrow Mn^{2+}$ (reduction)

$HNO_2 + H_2O \longrightarrow NO_3^- + 3H^+$ (oxidation)

$MnO_4^- + 8H^+ \longrightarrow Mn^{2+} + 4H_2O$ (reduction)

$HNO_2 + H_2O \longrightarrow NO_3^- + 3H^+ + 2e^-$ (oxidation)

$MnO_4^- + 8H^+ + 5e^- \longrightarrow Mn^{2+} + 4H_2O$ (reduction)

$5HNO_2 + 5H_2O \longrightarrow 5NO_3^- + 15H^+ + 10e^-$ (oxidation)

$2MnO_4^- + 16H^+ + 10e^- \longrightarrow 2Mn^{2+} + 8H_2O$ (reduction)

$$2MnO_4^-(aq)+5HNO_2(aq)+H^+(aq) \longrightarrow 5NO_3^-(aq)+2Mn^{2+}(aq)+3H_2O(l)$$

electron donor	HNO_2
electron acceptor	MnO_4^-
oxidizing agent	MnO_4^-
reducing agent	HNO_2
species oxidized	nitrogen in HNO_2
species reduced	manganese in MnO_4^-

20-30. $Cd \longrightarrow Cd(OH)_2$ (oxidation)

$NiO_2H \longrightarrow Ni(OH)_2$ (reduction)

$Cd + 2H_2O \longrightarrow Cd(OH)_2 + 2H^+$ (oxidation)

$NiO_2H + H^+ \longrightarrow Ni(OH)_2$ (reduction)

$$Cd + 2H_2O \longrightarrow Cd(OH)_2 + 2H^+ + 2e^- \quad \text{(oxidation)}$$

$$NiO_2H + H^+ + e^- \longrightarrow Ni(OH)_2 \quad \text{(reduction)}$$

$$Cd + 2H_2O \longrightarrow Cd(OH)_2 + 2H^+ + 2e^- \quad \text{(oxidation)}$$

$$2NiO_2H + 2H^+ + 2e^- \longrightarrow 2Ni(OH)_2 \quad \text{(reduction)}$$

$$2NiO_2H(s) + Cd(s) + 2H_2O(l) \longrightarrow Cd(OH)_2(s) + 2Ni(OH)_2(s)$$

20-32.

$$\left.\begin{array}{l} C_2H_5OH \longrightarrow 2CO_2 \\ C_2H_5OH \longrightarrow 2CHO_2^- \\ C_2H_5OH \longrightarrow 2HCI_3 \end{array}\right\} \text{(oxidation)}$$

$$I_3^- \longrightarrow 3I^- \quad \text{(reduction)}$$

$$\left.\begin{array}{l} C_2H_5OH + 3H_2O \longrightarrow 2CO_2 + 12H^+ \\ C_2H_5OH + 3H_2O \longrightarrow 2CHO_2^- + 10H^+ \\ C_2H_5OH + 2I_3^- \longrightarrow 2HCI_3 + H_2O + 2H^+ \end{array}\right\} \text{(oxidation)}$$

$$I_3^- \longrightarrow 3I^- \quad \text{(reduction)}$$

$$\left.\begin{array}{l} C_2H_5OH + 3H_2O \longrightarrow 2CO_2 + 12H^+ + 12e^- \\ C_2H_5OH + 3H_2O \longrightarrow 2CHO_2^- + 10H^+ + 8e^- \\ C_2H_5OH + 2I_3^- \longrightarrow 2HCI_3 + H_2O + 2H^+ + 4e^- \end{array}\right\} \text{(oxidation)}$$

$$\left.\begin{array}{l} I_3^- + 2e^- \longrightarrow 3I^- \\ 12I_3^- + 24e^- \longrightarrow 36I^- \end{array}\right. \quad \text{(reduction)}$$

$$3C_2H_5OH + 6H_2O + 14I_3^- \longrightarrow 2CO_2 + 2CHO_2^- + 2HCI_3 + 36I^- + 24H^+ + H_2O$$

$$3C_2H_5OH(aq) + 5H_2O(l) + 14I_3^-(aq) \longrightarrow 2CO_2(g) + 2CHO_2^-(aq) + 2HCI_3(aq) + 36I^-(aq) + 24H^+(aq)$$

(answer not unique)

20-34. a) Add H_2O and H^+ to balance the oxygen atoms and hydrogen atoms

$$H_2BO_3^-(aq) + 8H^+(aq) \longrightarrow BH_4^-(aq) + 3H_2O(l)$$

Add electrons to balance the charge

$$H_2BO_3^-(aq) + 8H^+(aq) + 8e^- \longrightarrow BH_4^-(aq) + 3H_2O(l)$$

b) First balance the chlorine atoms

$2ClO_3^- \longrightarrow Cl_2$

Add H_2O and H^+ to balance the oxygen atoms and hydrogen atoms

$2ClO_3^- + 12H^+ \longrightarrow Cl_2 + 6H_2O$

Add electrons to balance the charge

$2ClO_3^-(aq) + 12H^+(aq) + 10e^- \longrightarrow Cl_2(g) + 6H_2O(l)$

c) First balance the half-reaction with respect to chlorine

$Cl_2 \longrightarrow 2HClO$

Add H_2O to balance the oxygen atoms and H^+ to balance the hydrogen atoms

$Cl_2 + 2H_2O \longrightarrow 2HClO + 2H^+$

Add two electrons to balance the charge

$Cl_2(g) + 2H_2O(l) \longrightarrow 2HClO(aq) + 2H^+(aq) + 2e^-$

20-36. These half-reactions are balanced in a manner similar to those in Problem 20-34. The steps are

a) $OsO_4 + 8H^+ \longrightarrow Os + 4H_2O$

$OsO_4(s) + 8H^+(aq) + 8e^- \longrightarrow Os(s) + 4H_2O(l)$

b) $S + 3H_2O \longrightarrow SO_3^{2-} + 6H^+$

$S + 3H_2O \longrightarrow SO_3^{2-} + 6H^+ + 4e^-$

$$S + 3H_2O + 6OH^- \longrightarrow SO_3^{2-} + \underbrace{6H^+ + 6OH^-}_{6H_2O} + 4e^-$$

$S(s) + 6OH^-(aq) \longrightarrow SO_3^{2-}(aq) + 3H_2O(l) + 4e^-$

c) $Sn + 2H_2O \longrightarrow HSnO_2^- + 3H^+$

$Sn + 2H_2O \longrightarrow HSnO_2^- + 3H^+ + 2e^-$

$$Sn + 2H_2O + 3OH^- \longrightarrow HSnO_2^- + \underbrace{3H^+ + 3OH^-}_{3H_2O} + 2e^-$$

$Sn(s) + 3OH^-(aq) \longrightarrow HSnO_2^-(aq) + H_2O(l) + 2e^-$

20-38. The steps are

a) $Au(CN)_2^- \longrightarrow Au + 2CN^-$

$Au(CN)_2^-(aq) + e^- \longrightarrow Au(s) + 2CN^-(aq)$

b) $PbO_2 + SO_4^{2-} + 4H^+ \longrightarrow PbSO_4 + 2H_2O$

$PbO_2(s) + SO_4^{2-}(aq) + 4H^+(aq) + 2e^- \longrightarrow PbSO_4(s) + 2H_2O(l)$

c) $MnO_4^- + 4H^+ \longrightarrow MnO_2 + 2H_2O$

$MnO_4^-(aq) + 4H^+(aq) + 3e^- \longrightarrow MnO_2(s) + 2H_2O(l)$

d) $Cr(OH)_3 + H_2O \longrightarrow CrO_4^{2-} + 5H^+$

$Cr(OH)_3 + H_2O \longrightarrow CrO_4^{2-} + 5H^+ + 3e^-$

$$Cr(OH)_3 + H_2O + 5OH^- \longrightarrow CrO_4^{2-} + \underbrace{5H^+ + 5OH^-}_{5H_2O} + 3e^-$$

$Cr(OH)_3(s) + 5OH^-(aq) \longrightarrow CrO_4^{2-}(aq) + 4H_2O(l) + 3e^-$

20-40. The steps are

$C \longrightarrow CO_2$ (oxidation)

$NO_3^- \longrightarrow N_2$ (reduction)

$S \longrightarrow S^{2-}$ (reduction)

$C + 2H_2O \longrightarrow CO_2 + 4H^+$ (oxidation)

$2NO_3^- + 12H^+ \longrightarrow N_2 + 6H_2O$ (reduction)

$S \longrightarrow S^{2-}$ (reduction)

$C + 2H_2O \longrightarrow CO_2 + 4H^+ + 4e^-$ (oxidation)

$2NO_3^- + 12H^+ + 10e^- \longrightarrow N_2 + 6H_2O$ (reduction)

$S + 2e^- \longrightarrow S^{2-}$ (reduction)

$3C + 6H_2O \longrightarrow 3CO_2 + 12H^+ + 12e^-$ (oxidation)

$2NO_3^- + 12H^+ + 10e^- \longrightarrow N_2 + 6H_2O$ (reduction)

$S + 2e^- \longrightarrow S^{2-}$ (reduction)

$$3C + 2NO_3^- + S \longrightarrow 3CO_2 + N_2 + S^{2-}$$

or

$$3C(s) + 2KNO_3(s) + S(s) \longrightarrow 3CO_2(g) + N_2(g) + K_2S(s)$$

The reducing agent is carbon, the oxidizing agents are KNO_3 and S.

20-42. $KO_2 \longrightarrow O_2$ (oxidation)

$KO_2 \longrightarrow 2OH^-$ (reduction)

$KO_2 \longrightarrow O_2 + K^+$ (oxidation)

$KO_2 + 2H_2O \longrightarrow 4OH^- + K^+$ (reduction)

$KO_2 \longrightarrow O_2 + K^+ + e^-$ (oxidation)

$KO_2 + 2H_2O + 3e^- \longrightarrow 4OH^- + K^+$ (reduction)

$3KO_2 \longrightarrow 3O_2 + 3K^+ + 3e^-$ (oxidation)

$KO_2 + 2H_2O + 3e^- \longrightarrow 4OH^- + K^+$ (reduction)

$4KO_2(s) + 2H_2O(l) \longrightarrow 4KOH(s) + 3O_2(g)$

and

$KOH(s) + CO_2(g) \longrightarrow KHCO_3(s)$

20-44. $2Fe \longrightarrow Fe_2O_3$ (oxidation)

$2CrO_4^{2-} \longrightarrow Cr_2O_3$ (reduction)

$2Fe + 3H_2O \longrightarrow Fe_2O_3 + 6H^+$ (oxidation)

$2CrO_4^{2-} + 10H^+ \longrightarrow Cr_2O_3 + 5H_2O$ (reduction)

$2Fe + 3H_2O \longrightarrow Fe_2O_3 + 6H^+ + 6e^-$ (oxidation)

$2CrO_4^{2-} + 10H^+ + 6e^- \longrightarrow Cr_2O_3 + 5H_2O$ (reduction)

$$2Fe + 2CrO_4^{2-} + 4H^+ \longrightarrow Fe_2O_3 + Cr_2O_3 + 2H_2O$$

$$2Fe + 2CrO_4^{2-} + \underbrace{4H^+ + 4OH^-}_{4H_2O} \longrightarrow Fe_2O_3 + Cr_2O_3 + 2H_2O + 4OH^-$$

$$2Fe(s) + 2CrO_4^{2-}(aq) + 2H_2O(l) \longrightarrow Fe_2O_3(s) + Cr_2O_3(s) + 4OH^-(aq)$$

20-46. $Al \longrightarrow Al^{3+}$ (oxidation)

$H_2O \longrightarrow H_2$ (reduction)

$Al \longrightarrow Al^{3+}$ (oxidation)

$2H_2O \longrightarrow H_2 + 2OH^-$ (reduction)

$Al \longrightarrow Al^{3+} + 3e^-$ (oxidation)

$2H_2O + 2e^- \longrightarrow H_2 + 2OH^-$ (reduction)

$2Al \longrightarrow 2Al^{3+} + 6e^-$ (oxidation)

$6H_2O + 6e^- \longrightarrow 3H_2 + 6OH^-$ (reduction)

$$2Al(s) + 6H_2O(l) \longrightarrow 2Al^{3+}(aq) + 3H_2(g) + 6OH^-(aq)$$

$Al(OH)_3$ is an insoluble hydroxide.

$$2Al(s) + 6H_2O(l) \longrightarrow 2Al(OH)_3(s) + 3H_2(g)$$

However, in alkaline solution, $Al(OH)_3$ is soluble due to the formation of $Al(OH)_4^-$ (Section 19-8)

$$2Al(s) + 6H_2O(l) + 2OH^-(aq) \longrightarrow 2Al(OH)_4^-(aq) + 3H_2(g)$$

20-48. The number of moles of $KMnO_4$ is

$$\text{moles of } KMnO_4 = MV = (0.0512\ mol\cdot L^{-1})(0.0395\ L)$$

$$= 2.022\times 10^{-3}\ mol$$

The number of moles of $FeCl_2$ that reacts with 2.02 mol of $KMnO_4$ is

$$\text{moles of } FeCl_2 = (2.022\times 10^{-3}\ mol\ KMnO_4)\left(\frac{5\ mol\ FeCl_2}{1\ mol\ KMnO_4}\right)$$

$$= 1.011\times 10^{-2}\ mol$$

The number of moles of iron in 1.01 x 10^{-2} mol of $FeCl_2$ is

$$\text{moles of Fe} = (1.011 \times 10^{-2} \text{ mol } FeCl_2)\left(\frac{1 \text{ mol Fe}}{1 \text{ mol } FeCl_2}\right)$$

$$= 1.011 \times 10^{-2} \text{ mol}$$

$$\text{mass of Fe} = (1.011 \times 10^{-2} \text{ mol})\left(\frac{55.85 \text{ g}}{1 \text{ mol}}\right)$$

$$= 0.565 \text{ g}$$

$$\% \text{ Fe} = \frac{0.565 \text{ g Fe}}{5.00 \text{ g ore}} \times 100 = 11.3\%$$

20-50. The two half-reactions are

$C \longrightarrow CO_3^{2-}$ (oxidation)

$ClO_2 \longrightarrow ClO_2^-$ (reduction)

Add H_2O and H^+ to balance the oxygen atoms and hydrogen atoms

$C + 3H_2O \longrightarrow CO_3^{2-} + 6H^+$ (oxidation)

$ClO_2 \longrightarrow ClO_2^-$ (reduction)

Balance the charges by adding electrons

$C + 3H_2O \longrightarrow CO_3^{2-} + 6H^+ + 4e^-$ (oxidation)

$ClO_2 + e^- \longrightarrow ClO_2^-$ (reduction)

Multiply the reduction half-reaction by 4

$C + 3H_2O \longrightarrow CO_3^{2-} + 6H^+ + 4e^-$

$4ClO_2 + 4e^- \longrightarrow 4ClO_2^-$

Add these two half-reactions to get a balanced equation (in acid solution)

$$C + 4ClO_2 + 3H_2O \longrightarrow CO_3^{2-} + 4ClO_2^- + 6H^+$$

Add OH^-(aq) to eliminate the H^+(aq)

$$C + 4ClO_2 + 3H_2O + 6OH^- \longrightarrow CO_3^{2-} + 4ClO_2^- + \underbrace{6H^+ + 6OH^-}_{6H_2O}$$

to get the final balanced equation in basic solution

$$4NaOH(aq) + Ca(OH)_2(aq) + C(s) + 4ClO_2(g) \longrightarrow$$
$$4NaClO_2(aq) + CaCO_3(s) + 3H_2O(l)$$

The number of moles of $NaClO_2$ in 1.00 metric ton is

$$\text{moles of } NaClO_2 = (1.00 \text{ metric ton})\left(\frac{10^3 \text{ kg}}{1 \text{ metric ton}}\right)\left(\frac{10^3 \text{ g}}{1 \text{ kg}}\right)\left(\frac{1 \text{ mol}}{90.44 \text{ g}}\right)$$

$$= 1.106 \times 10^4 \text{ mol}$$

$$\text{moles of } ClO_2 = (1.106 \times 10^4 \text{ mol } NaClO_2)\left(\frac{4 \text{ mol } ClO_2}{4 \text{ mol } NaClO_2}\right)$$

$$= 1.106 \times 10^4 \text{ mol}$$

$$\text{mass of } ClO_2 = (1.106 \times 10^4 \text{ mol})\left(\frac{67.45 \text{ g}}{1 \text{ mol}}\right)$$

$$= 7.46 \times 10^5 \text{ g} = 7.46 \times 10^2 \text{ kg}$$

$$= 746 \text{ kg}$$

20-52. $\text{moles of } Ce^{4+} = MV = (0.105 \text{ mol}\cdot L^{-1})(0.0295 \text{ L}) = 3.10 \times 10^{-3} \text{ mol}$

$$\text{moles of } Fe^{2+} = (3.10 \times 10^{-3} \text{ mol } Ce^{4+})\left(\frac{1 \text{ mol } Fe^{2+}}{1 \text{ mol } Ce^{4+}}\right)$$

$$= 3.10 \times 10^{-3} \text{ mol}$$

$$[Fe^{2+}] = \frac{\text{moles of } Fe^{2+}}{\text{volume of solution}} = \frac{3.10 \times 10^{-3} \text{ mol}}{0.0250 \text{ L}} = 0.124 \text{ M}$$

$$\text{mass of } Fe^{2+} = (3.10 \times 10^{-3} \text{ mol})\left(\frac{55.85 \text{ g}}{1 \text{ mol}}\right) = 0.173 \text{ g}$$

$$= 173 \text{ mg}$$

CHAPTER 21

SOLUTIONS TO THE EVEN-NUMBERED PROBLEMS

21-2.

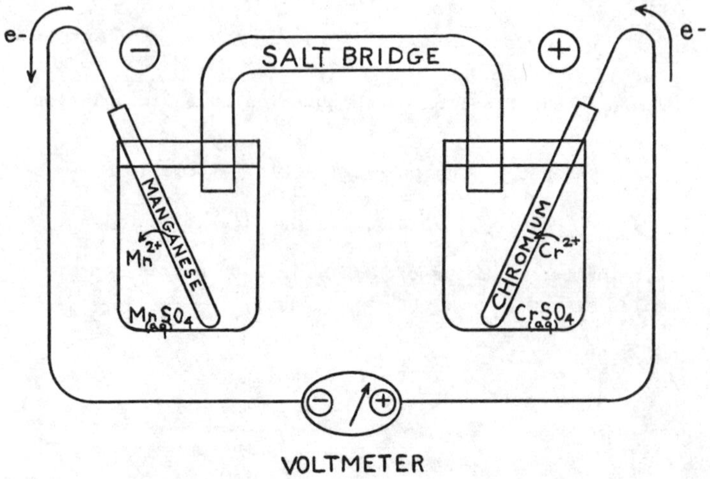

The reaction at the negative electrode is

$$Mn(s) \longrightarrow Mn^{2+}(aq) + 2e^-$$

The reaction at the positive electrode is

$$Cr^{2+}(aq) + 2e^- \longrightarrow Cr(s)$$

The cell diagram is

$$Mn(s)|MnSO_4(aq)||CrSO_4(aq)|Cr(s)$$

21-4.

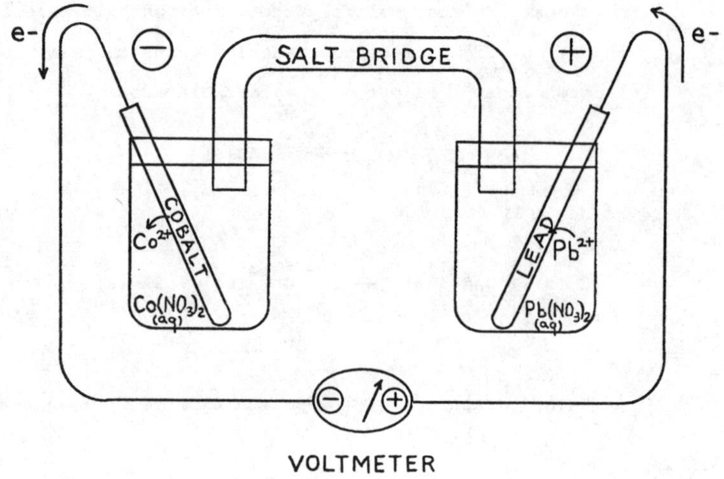

The reaction at the negative electrode is

$$Co(s) \longrightarrow Co^{2+}(aq) + 2e^-$$

The reaction at the positive electrode is

$$Pb^{2+}(aq) + 2e^- \longrightarrow Pb(s)$$

The cell diagram is

$$Co(s)|Co(NO_3)_2(aq)||Pb(NO_3)_2(aq)|Pb(s)$$

21-6. Oxidation takes place at the left electrode. The reaction at the left electrode is

$$Cu(s) \longrightarrow Cu^{2+}(aq) + 2e^-$$

Reduction takes place at the right electrode. The reaction at the right electrode is

$$Ag^+(aq) + e^- \longrightarrow Ag(s)$$

The net cell reaction is the sum of the two electrode half-reactions. We must multiply the reduction half-reaction by 2 so that the right electrode reaction gains the same number of electrons that are lost at the left electrode

$$2Ag^+(aq) + 2e^- \longrightarrow 2Ag(s)$$

The net cell reaction

$$Cu(s) + 2Ag^+(aq) \longrightarrow Cu^{2+}(aq) + 2Ag(s)$$

or

$$Cu(s) + 2AgClO_4(aq) \longrightarrow Cu(ClO_4)_2(aq) + 2Ag(s)$$

21-8. The half-reaction that occurs at the left electrode is

$$Sn(s) \longrightarrow Sn^{2+}(aq) + 2e^- \qquad \text{(oxidation)}$$

The half-reaction that occurs at the right electrode is

$$Ag^+(aq) + e^- \longrightarrow Ag(s) \qquad \text{(reduction)}$$

The net cell reaction is

$$Sn(s) + 2Ag^+(aq) \longrightarrow Sn^{2+}(aq) + 2Ag(s)$$

A sketch of the cell is as follows

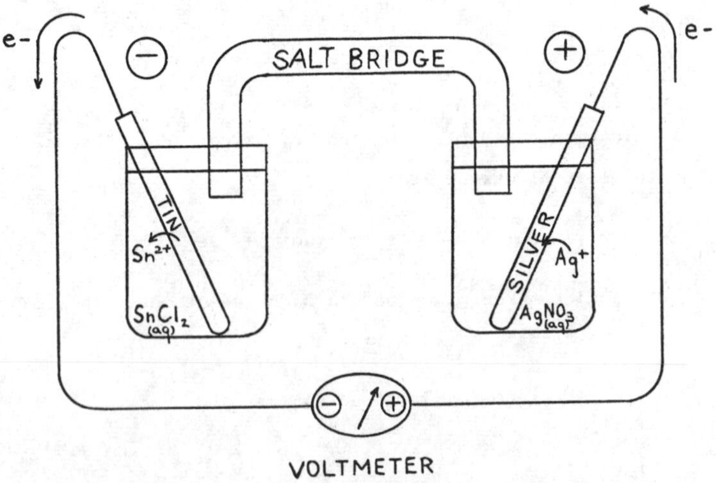

21-10. The reaction at the left electrode is

$$Pb(s) + SO_4^{2-}(aq) \longrightarrow PbSO_4(s) + 2e^- \quad \text{(oxidation)}$$

The $SO_4^{2-}(aq)$ come from the $K_2SO_4(aq)$ solution in contact with $Pb(s)$ and $PbSO_4(s)$. The reaction at the right electrode is

$$Hg_2SO_4(s) + 2e^- \longrightarrow 2Hg(l) + SO_4^{2-}(aq) \quad \text{(reduction)}$$

The net reaction is

$$Pb(s) + Hg_2SO_4(s) \longrightarrow 2Hg(l) + PbSO_4(s)$$

21-12. The oxidation reaction is

$$H_2(g) \longrightarrow 2H^+(aq) + 2e^- \quad \text{(left electrode)}$$

The reduction reaction is

$$AgCl(s) \longrightarrow Ag(s) + Cl^-(aq) \quad \text{(right electrode)}$$

The oxidation of $H_2(g)$ takes place on a platinum electrode. The cell diagram is

$$Pt(s)|H_2(g)|HCl(aq)|AgCl(s)|Ag(s)$$

21-14. a) An increase in the concentration of $Ag^+(aq)$ drives the reaction from left to right, thus the cell voltage increases.

b) An increase in $PbSO_4(s)$ has no effect on the driving force of the reaction and thus does not affect the cell voltage.

c) An increase in the concentration of $SO_4^{2-}(aq)$ increases the driving force of the cell reaction from left to right and thus increases the cell voltage.

21-16. a) An increase in $[HClO]$ drives the reaction from left to right. The cell voltage increases.

b) The size of the electrode has no effect on the cell voltage.

c) An increase in the pH corresponds to a decrease in $[H^+]$. A decrease in $[H^+]$ drives the reaction from right to left. The cell voltage decreases.

d) Dissolving KCl in the cell solution containing $Cl^-(aq)$ increases $[Cl^-]$. An increase in $[Cl^-]$ drives the reaction from right to left. The cell voltage decreases.

21-18. a) An increase in $[Ag^+]$ drives the reaction from left to right. The cell voltage increases.

b) An increase in $[Fe^{3+}]$ drives the reaction from right to left. The cell voltage decreases.

c) A two-fold decrease in both $[Fe^{3+}]$ and $[Fe^{2+}]$ does not change Q and thus has no effect on the cell voltage.

d) A decrease in the amount of Ag(s) has no effect on the cell voltage.

e) A decrease in $[Fe^{2+}]$ drives the reaction from right to left. The cell voltage decreases.

f) The added NaCl precipitates $Ag^+(aq)$ as AgCl(s) thereby decreasing $[Ag^+]$. A decrease in $[Ag^+]$ drives the reaction from right to left. The cell voltage decreases.

21-20. a) The oxidation half-reaction is

$$Cu(s) \longrightarrow Cu^{2+}(aq) + 2e^-$$

The reduction half-reaction is

$$Mg^{2+}(aq) + 2e^- \longrightarrow Mg(s)$$

The number of electrons transferred is two or n = 2.

b) The oxidation half-reaction is

$$2Na(s) \longrightarrow 2Na^+(aq) + 2e^-$$

The reduction half-reaction is

$$2H_2O(l) \longrightarrow H_2(g) + 2OH^-(aq) + 2e^-$$

The number of electrons transferred is two or n = 2.

21-22. The relation between E^0 and K is

$$E^0 = \left[\frac{0.0592\ V}{n}\right] \log K$$

Solving for log K, we have

$$\log K = \frac{nE^0}{0.0592\ V}$$

We know that n = 2 for this reaction and thus

$$\log K = \frac{(2)(0.22\ V)}{(0.0592\ V)} = 7.43$$

or

$$K = 10^{7.43} = 2.7 \times 10^7$$

21-24. The Nernst equation

$$E = E^0 - \left(\frac{0.0592\ V}{n}\right) \log Q$$

Thus

$$E = E^0 - \left(\frac{0.0592\ V}{n}\right) \log \frac{[Co^{2+}]}{[Sn^{2+}]}$$

For the cell reaction n = 2, thus

$$0.168\ V = E^0 - \left(\frac{0.0592\ V}{2}\right) \log \left(\frac{0.020\ M}{0.18\ M}\right)$$

$$0.168\ V = E^0 + 0.028\ V$$

The value of E^0 is

$$E^0 = 0.168\ V - 0.028\ V = 0.140\ V$$

The relationship between E^0 and K is

$$E^0 = \left(\frac{0.0592\ V}{n}\right) \log K$$

Thus

$$\log K = \frac{(2)(0.140\ V)}{0.0592\ V} = 4.73$$

$$K = 10^{4.73} = 5.4 \times 10^4$$

21-26. Application of the Nernst equation to the cell reaction (n = 2) yields

$$E = E^0 - \left(\frac{0.0592\ V}{2}\right) \log \frac{[Zn^{2+}]}{[Hg_2^{2+}]}$$

Thus

$$1.54\ V = E^0 - \left(\frac{0.0592\ V}{2}\right) \log \left(\frac{0.50\ M}{0.30\ M}\right)$$

$$1.54\ V = E^0 - 0.007\ V$$

and

$$E^0 = 1.54\ V + 0.007\ V = 1.55\ V$$

The relationship between E^0 and K is

$$\log K = \frac{nE^0}{0.0592\ V}$$

Thus

$$\log K = \frac{(2)(1.55\ V)}{0.0592\ V} = 52.36$$

$$K = 10^{52.36} = 2.3 \times 10^{52}$$

21-28. When the voltage of the cell is negative, oxidation occurs spontaneously at the right electrode

$$Cd(s) \longrightarrow Cd^{2+}(aq) + 2e^-$$

and reduction occurs at the left electrode

$$2H^+(aq) + 2e^- \longrightarrow H_2(g)$$

The reaction that occurs spontaneously is thus

$$Cd(s) + 2H^+(aq) \longrightarrow Cd^{2+}(aq) + H_2(g)$$

21-30. The voltage measured between the Zn(s) and Cu(s) rods will be zero because the cell is short circuited. The reducing agent, Zn(s) must be physically separated from the oxidizing agent $CuSO_4$(aq). In this case, the Zn(s) rod is in the Cu^{2+}(aq) solution.

21-32. We see that copper is oxidized

$$Cu^+(aq) \longrightarrow Cu^{2+}(aq) + e^- \qquad \text{(left electrode)}$$

and the nitrate ion is reduced

$$NO_3^-(aq) + 2H^+ + e^- \longrightarrow NO_2(g) + H_2O(l) \quad \text{(right electrode)}$$

The E^0_{cell} value for the complete cell is given by

$$E^0_{cell} = E^0_{right} - E^0_{left}$$

or

$$E^0_{cell} = E^0_{NO_3^-/NO_2} - E^0_{Cu^{2+}/Cu^+}$$

From Table 21-1 we obtain

$$E^0_{Cu^{2+}/Cu^+} = 0.15 \text{ V}$$

Thus

$$0.65 \text{ V} = E^0_{NO_3^-/NO_2} - 0.15 \text{ V}$$

$$E^0_{NO_3^-/NO_2} = 0.65 \text{ V} + 0.15 \text{ V} = 0.80 \text{ V}$$

21-34. The oxidation half-reaction is

$$2H_2O(l) \longrightarrow H_2O_2(aq) + 2e^- + 2H^+(aq)$$

The reduction half-reaction is

$$S_2O_8^{2-}(aq) + 2e^- \longrightarrow 2SO_4^{2-}(aq)$$

The value of E^0_{cell} is given by

$$E^0_{cell} = E^0_{right}(\text{reduction}) - E^0_{left}(\text{oxidation})$$

thus

$$E^0_{cell} = E^0_{S_2O_8^{2-}/SO_4^{2-}} - E^0_{H_2O_2/H_2O}$$

From Table 21-1, we obtain

$$E^0_{cell} = 2.01\text{ V} - (1.78\text{ V}) = 0.23\text{ V}$$

The reaction

$$S_2O_8^{2-}(aq) + 2H_2O(l) \longrightarrow H_2O_2(aq) + 2SO_4^{2-}(aq) + 2H^+(aq)$$

has a positive value for E^0_{cell} and thus is spontaneous when $Q = 1$. An aqueous solution of potassium persulfate is not stable over a long period of time because $S_2O_8^{2-}(aq)$ is capable of oxidizing water.

21-36. The value of E^0_{cell} is given by

$$E^0_{cell} = E^0_{right}(\text{reduction}) - E^0_{left}(\text{oxidation})$$

thus

$$E^0_{cell} = E^0_{Ag^+/Ag} - E^0_{Cu^{2+}/Cu}$$

From Table 21-1, we obtain

$$E^0_{cell} = 0.80\text{ V} - (0.34\text{ V}) = 0.46\text{ V}$$

From the Nernst equation, we have

$$\log K = \frac{nE^0}{0.0592\text{ V}}$$

For the reaction we have n = 2, thus

$$\log K = \frac{(2)(0.46\ V)}{0.0592\ V} = 15.54$$

$$K = 10^{15.54} = 3.5 \times 10^{15}$$

21-38. The value of E^0_{cell} is given by

$$E^0_{cell} = E^0_{right}(\text{reduction}) - E^0_{left}(\text{oxidation})$$

$$E^0_{cell} = E^0_{Pb^{2+}/Pb} - E^0_{Cd^{2+}/Cd}$$

From Table 21-1, we obtain

$$E^0_{cell} = -0.13\ V - (-0.40\ V) = 0.27\ V$$

From the Nernst equation

$$E = E^0 - \left(\frac{0.0592\ V}{n}\right) \log Q$$

we obtain (n = 2)

$$E = E^0 - \left(\frac{0.0592\ V}{2}\right) \log\left(\frac{[Cd^{2+}]}{[Pb^{2+}]}\right)$$

Thus

$$E = 0.27\ V - \left(\frac{0.0592\ V}{2}\right) \log\left(\frac{0.010\ M}{0.10\ M}\right)$$

$$E = 0.27\ V + 0.03\ V = 0.30\ V$$

21-40. The oxidation half-reaction in acidic solution is

$$BH_4^-(aq) + 3H_2O \longrightarrow H_2BO_3^-(aq) + 8H^+ + 8e^-$$

Since the reaction is carried out in basic solution, we convert the above half-reaction to basic solution

$$BH_4^-(aq) + 8OH^-(aq) \longrightarrow H_2BO_3^-(aq) + 5H_2O(l) + 8e^-$$

The reduction half-reaction is

$$O_2(g) + e^- \longrightarrow O_2^-(aq)$$

From Table 21-1, we obtain

$$E^0_{O_2/O_2^-} = -0.56\ V$$

We can obtain $E^0_{H_2BO_3^-/BH_4^-}$ from E^0_{cell}

$$E^0_{cell} = E^0_{right} - E^0_{left}$$

$$E^0_{cell} = E^0_{O_2/O_2^-} - E^0_{H_2BO_3^-/BH_4^-}$$

Thus

$$0.68\ V = -0.56\ V - E^0_{H_2BO_3^-/BH_4^-}$$

$$E^0_{H_2BO_3^-/BH_4^-} = -0.56\ V - 0.68\ V = -1.24\ V$$

21-42. The reaction at the negative electrode is

$$Cd(Hg) \longrightarrow Cd^{2+}(aq) + 2e^-$$

The reaction at the positive electrode is

$$Hg_2SO_4(s) + 2e^- \longrightarrow 2Hg(l) + SO_4^{2-}(aq)$$

The net reaction is

$$Cd(Hg) + Hg_2SO_4(s) \longrightarrow 2Hg(l) + CdSO_4(aq)$$

The voltage of the cell is given by the Nernst equation

$$E = E^0 - \left(\frac{0.0592\ V}{n}\right) \log Q$$

$$E = E^0 - \left(\frac{0.0592\ V}{n}\right) \log [Cd^{2+}][SO_4^{2-}]$$

Because the $CdSO_4(aq)$ solution is saturated, the concentrations of $Cd^{2+}(aq)$ and $SO_4^{2-}(aq)$ do not change, $[Cd^{2+}][SO_4^{2-}] = K_{sp}$, during the reaction. Thus E, the cell voltage, does not change.

21-44. a) The reaction at the left (negative) electrode is

$$Zn(s) \longrightarrow ZnO_2^{2-}(s)$$

$$Zn(s) + 2H_2O(l) \longrightarrow ZnO_2^{2-}(s) + 4H^+(aq) + 2e^-$$

Converting to basic solution

$$Zn(s) + 4KOH(aq) \longrightarrow K_2ZnO_2(s) + 2H_2O(l) + 2K^+(aq) + 2e^-$$

b) The reaction at the right (positive) electrode is

$$Ag_2O_2(s) \longrightarrow 2Ag(s)$$

$$Ag_2O_2(s) + 4H^+(aq) + 4e^- \longrightarrow 2Ag(s) + 2H_2O(l)$$

Converting to basic solution

$$Ag_2O_2(s) + 4K^+(aq) + 2H_2O(l) + 4e^- \longrightarrow 2Ag(s) + 4KOH(aq)$$

c) Multiplying the oxidation half-reaction by 2 yields

$$2Zn(s) + 8KOH(aq) \longrightarrow 2K_2ZnO_2(s) + 4H_2O(l) + 4K^+(aq) + 4e^-$$

Adding the two half-reactions yields the complete cell reaction

$$2Zn(s) + Ag_2O_2(s) + 4KOH(aq) \longrightarrow 2Ag(s) + 2K_2ZnO_2(s) + 2H_2O(l)$$

21-46. The reaction at the negative electrode is

$$Mg(s) + 2Cl^-(soln) \longrightarrow MgCl_2(s) + 2e^-$$

The reaction at the positive electrode is

$$2AgCl(s) + 2e^- \longrightarrow 2Ag(s) + 2Cl^-(soln)$$

The cell reaction is

$$Mg(s) + 2AgCl(s) \rightleftharpoons MgCl_2(s) + 2Ag(s)$$

The cell diagram is

$$\ominus Mg(s)|MgCl_2(s)|KSCN(NH_3)|AgCl(s)|Ag(s)\oplus$$

The Cl^-(soln) arises from the slight solubility of the $MgCl_2$ and AgCl in the solution. The $KSCN(NH_3)$ is used as the electrolytic solution to reduce the internal resistance of the cell. Water is not used as the solvent because it is a solid at $-50^{\circ}C$.

21-48. The total charge is given by

$$\underset{\text{(coulombs)}}{\text{charge}} = \underset{\text{(amperes)}}{\text{current}} \times \underset{\text{(seconds)}}{\text{time}}$$

$$= (0.150 \text{ A})(20.0 \text{ min})\left(\frac{60 \text{ s}}{1 \text{ min}}\right)$$

$$= 180 \text{ C}$$

The number of moles of electrons that corresponds to 180 C is

$$\text{moles of electrons} = (180 \text{ C})\left(\frac{1}{96,500 \text{ C}\cdot\text{mol}^{-1}}\right)$$

$$= 0.001865 \text{ mol}$$

The reaction is

$$Ag^+(aq) + e^- \longrightarrow Ag(s)$$

The number of moles of silver that are deposited is

$$\text{moles of silver} = (0.001865 \text{ mol electrons})\left(\frac{1 \text{ mol Ag}}{1 \text{ mol e}^-}\right)$$

$$= 0.001865 \text{ mol}$$

The mass of the silver deposited is

$$\text{mass of Ag(s)} = (0.001865\ \text{mol})\left(\frac{107.9\ \text{g}}{1\ \text{mol}}\right)$$

$$= 0.201\ \text{g}$$

21-50. The total charge is

$$\text{charge} = \text{current} \times \text{time}$$

$$= (5.0\ \text{C}\cdot\text{s}^{-1})(1.0\ \text{hr})\left(\frac{60\ \text{min}}{1\ \text{hr}}\right)\left(\frac{60\ \text{s}}{1\ \text{min}}\right)$$

$$= 1.8 \times 10^4\ \text{C}$$

The number of moles of electrons that correspond to 1.8×10^4 C is

$$\text{moles of electrons} = (1.8 \times 10^4\ \text{C})\left(\frac{1}{96{,}500\ \text{C}\cdot\text{mol}^{-1}}\right)$$

$$= 0.187\ \text{mol}$$

The reaction is

$$Be^{2+} + 2e^- \longrightarrow Be(s)$$

The number of moles of beryllium that are deposited is

$$\text{moles of Be(s)} = (0.187\ \text{mol electrons})\left(\frac{1\ \text{mol Be}}{2\ \text{mol}\ e^-}\right)$$

$$= 0.0935\ \text{mol}$$

The mass of beryllium deposited is

$$\text{mass of Be(s)} = (0.0935\ \text{mol})\left(\frac{9.012\ \text{g}}{1\ \text{mol}}\right)$$

$$= 0.84\ \text{g}$$

21-52. The number of moles of zinc deposited is

$$\text{moles of Zn} = (65\ \text{g})\left(\frac{1\ \text{mol}}{65.38\ \text{g}}\right) = 0.994\ \text{mol}$$

The number of moles of electrons required to deposit 0.994 mol of zinc is

$$\text{moles of electrons} = (0.994\ \text{mol Zn})\left(\frac{2\ \text{mol e}^-}{1\ \text{mol Zn}}\right)$$

$$= 1.99\ \text{mol}$$

The charge that corresponds to 1.99 moles of electrons is

$$\text{charge} = (1.99\ \text{mol})(96{,}500\ \text{C}\cdot\text{mol}^{-1})$$

$$= 1.92 \times 10^5\ \text{C}$$

The total charge used in the 30-day period is

$$\frac{1.92 \times 10^5\ \text{C}}{0.11} = 1.74 \times 10^6\ \text{C}$$

The average current in amperes is

$$\text{current} = \frac{\text{charge}}{\text{time}} = \frac{1.74 \times 10^6\ \text{C}}{(30\ \text{day})\left(\frac{24\ \text{hr}}{1\ \text{day}}\right)\left(\frac{60\ \text{min}}{1\ \text{hr}}\right)\left(\frac{60\ \text{s}}{1\ \text{min}}\right)}$$

$$= 0.671\ \text{A}$$

The number of ampere-hours is the current in amperes times the number of hours the current flows, thus

$$\text{ampere-hours} = (0.671\ \text{A})(30\ \text{day})\left(\frac{24\ \text{hr}}{1\ \text{day}}\right) = 483\ \text{A}\cdot\text{hr}$$

21-54. The total charge is

$$\text{charge} = (0.600\ \text{C}\cdot\text{s}^{-1})(1\ \text{hr})\left(\frac{60\ \text{min}}{1\ \text{hr}}\right)\left(\frac{60\ \text{s}}{1\ \text{min}}\right)$$

$$= 2.16 \times 10^3\ \text{C}$$

The number of moles of electrons that corresponds to 2.16×10^3 C is

$$\text{moles of electrons} = \frac{2160 \text{ C}}{96{,}500 \text{ C}\cdot\text{mol}^{-1}} = 0.02238 \text{ mol}$$

The number of moles of metal deposited is (n = 1)

$$\text{moles of metal} = \text{moles of electrons} = 0.02238 \text{ mol}$$

We have the correspondance

$$2.38 \text{ g} \approx 0.02238 \text{ mol}$$

Dividing both sides by 0.02238, we have

$$106 \text{ g} \approx 1 \text{ mol}$$

The atomic mass of the metal is 106. The metal with an atomic mass of 106 is palladium, Pd.

21-56. The reaction for the electrochemical production of aluminum is

$$2Al_2O_3(soln) + 3C(s) \longrightarrow 4Al(l) + 3CO_2(g)$$

The number of moles of aluminum produced each year is

$$\text{moles of Al} = (5 \times 10^6 \text{tons})\left(\frac{2000 \text{ lbs}}{1 \text{ ton}}\right)\left(\frac{1 \text{ kg}}{2.205 \text{ lb}}\right)\left(\frac{10^3 \text{ g}}{1 \text{ kg}}\right)\left(\frac{1 \text{ mol}}{26.98 \text{ g}}\right)$$

$$= 1.68 \times 10^{11} \text{ mol Al}$$

The mass of Al_2O_3 required to produce 5×10^6 tons of Al is

$$\text{mass of } Al_2O_3 = (1.68 \times 10^{11} \text{mol Al})\left(\frac{2 \text{ mol } Al_2O_3}{4 \text{ mol Al}}\right)\left(\frac{101.96 \text{ g } Al_2O_3}{1 \text{ mol } Al_2O_3}\right)$$

$$= 8.56 \times 10^{12} \text{ g}$$

The mass of bauxite is

$$(0.55)(\text{mass of bauxite}) = \text{mass of } Al_2O_3 = 8.56 \times 10^{12} \text{ g}$$

$$\text{mass of bauxite} = 1.56 \times 10^{13} \text{ g}$$

$$\text{mass of bauxite} = (1.56 \times 10^{13}\text{g})\left(\frac{1 \text{ kg}}{10^3 \text{ g}}\right)\left(\frac{2.205 \text{ lb}}{1 \text{ kg}}\right)\left(\frac{1 \text{ ton}}{2000 \text{ lbs}}\right)$$

$$= 1.72 \times 10^7 \text{ tons}$$

$$= 17 \text{ million tons}$$

21-58. We want the net cell reaction to be

$$AgCl(s) \rightleftharpoons Ag^+(aq) + Cl^-(aq)$$

The oxidation half-reaction is

$$Ag(s) \longrightarrow Ag^+(aq) + e^-$$

The reduction half-reaction is

$$AgCl(s) + e^- \longrightarrow Ag(s) + Cl^-(aq)$$

We want to set up the cell so that the oxidation half-reaction

is separated from the reduction half-reaction, thus

$$Ag(s)|AgNO_3(aq)||NaCl(aq)|AgCl(s)|Ag(s)$$

21-60. The oxidation half-reaction is

$$H_2(g) \longrightarrow 2H^+(aq) + 2e^-$$

The reduction half-reaction is

$$Hg_2Cl_2(s) + 2e^- \longrightarrow 2Hg(l) + 2Cl^-(aq)$$

We showed in Problem 21-59 that the cell voltage of a cell using these two half-reactions is related to the concentration of $H^+(aq)$

$$E = E_0 - \frac{0.0592\ V}{n}[H^+]^2$$

or

$$E = 0.2415\ V - (0.0592\ V)[H^+]$$

For the reaction

$$HC_2H_3O_2(aq) \rightleftharpoons H^+(aq) + C_2H_3O_2^-(aq)$$

we have

$$K_a = \frac{[H^+][C_2H_3O_2^-]}{[HC_2H_3O_2]}$$

Recall that in Section 18-4 on the Henderson-Hasselbalch equation, it was shown that in a solution containing an acid and its conjugate base

$$[HC_2H_3O_2]_{eq} = [HC_2H_3O_2]_0$$

$$[C_2H_3O_2^-]_{eq} = [C_2H_3O_2^-]_0$$

CHAPTER 22

SOLUTIONS TO THE EVEN-NUMBERED PROBLEMS

22-2. We calculate $\Delta\overline{S}_{fus}$ using the relationship

$$\Delta\overline{S}_{fus} = \frac{\Delta\overline{H}_{fus}}{T_m}$$

Using the values of T_m and $\Delta\overline{H}_{fus}$ given, we have

$$\text{HF} \quad \Delta\overline{S}_{fus} = \frac{4.577 \times 10^3 \text{ J}\cdot\text{mol}^{-1}}{190.04 \text{ K}} = 24.08 \text{ J}\cdot\text{K}^{-1}\cdot\text{mol}^{-1}$$

$$\text{HCl} \quad \Delta\overline{S}_{fus} = \frac{1.991 \times 10^3 \text{ J}\cdot\text{mol}^{-1}}{158.9 \text{ K}} = 12.53 \text{ J}\cdot\text{K}^{-1}\cdot\text{mol}^{-1}$$

$$\text{HBr} \quad \Delta\overline{S}_{fus} = \frac{2.406 \times 10^3 \text{ J}\cdot\text{mol}^{-1}}{186.19 \text{ K}} = 12.92 \text{ J}\cdot\text{K}^{-1}\cdot\text{mol}^{-1}$$

$$\text{HI} \quad \Delta\overline{S}_{fus} = \frac{2.871 \times 10^3 \text{ J}\cdot\text{mol}^{-1}}{222.24 \text{ K}} = 12.92 \text{ J}\cdot\text{K}^{-1}\cdot\text{mol}^{-1}$$

We calculate $\Delta\overline{S}_{vap}$ using the relationship

$$\Delta\overline{S}_{vap} = \frac{\Delta\overline{H}_{vap}}{T_b}$$

Using the values of T_b and $\Delta\overline{H}_{vap}$ given, we have

$$\text{HF} \quad \Delta\overline{S}_{vap} = \frac{25.18 \times 10^3 \text{ J}\cdot\text{mol}^{-1}}{292.69 \text{ K}} = 86.03 \text{ J}\cdot\text{K}^{-1}\cdot\text{mol}^{-1}$$

$$\text{HCl} \quad \Delta\overline{S}_{vap} = \frac{17.53 \times 10^3 \text{ J}\cdot\text{mol}^{-1}}{188.3 \text{ K}} = 93.10 \text{ J}\cdot\text{K}^{-1}\cdot\text{mol}^{-1}$$

$$\text{HBr} \quad \Delta\overline{S}_{vap} = \frac{19.27 \times 10^3 \text{ J}\cdot\text{mol}^{-1}}{206.2 \text{ K}} = 93.45 \text{ J}\cdot\text{K}^{-1}\cdot\text{mol}^{-1}$$

$$\text{HI} \quad \Delta\overline{S}_{vap} = \frac{21.16 \times 10^3 \text{ J}\cdot\text{mol}^{-1}}{237.77 \text{ K}} = 88.99 \text{ J}\cdot\text{K}^{-1}\cdot\text{mol}^{-1}$$

22-4. We calculate $\Delta\overline{S}_{fus}$ using the relationship

$$\Delta\overline{S}_{fus} = \frac{\Delta\overline{H}_{fus}}{T_m}$$

Using the values of T_m and $\Delta\overline{H}_{fus}$ given, we have

$$CH_3OH \quad \Delta\overline{S}_{fus} = \frac{3.177 \times 10^3 \text{ J}\cdot\text{mol}^{-1}}{175.4 \text{ K}} = 18.11 \text{ J}\cdot\text{K}^{-1}\cdot\text{mol}^{-1}$$

$$C_2H_5OH \quad \Delta\overline{S}_{fus} = \frac{5.021 \times 10^3 \text{ J}\cdot\text{mol}^{-1}}{158.7 \text{ K}} = 31.64 \text{ J}\cdot\text{K}^{-1}\cdot\text{mol}^{-1}$$

$$C_3H_7OH \quad \Delta\overline{S}_{fus} = \frac{5.195 \times 10^3 \text{ J}\cdot\text{mol}^{-1}}{147.1 \text{ K}} = 35.32 \text{ J}\cdot\text{K}^{-1}\cdot\text{mol}^{-1}$$

We calculate $\Delta\overline{S}_{vap}$ using the relationship

$$\Delta\overline{S}_{vap} = \frac{\Delta\overline{H}_{vap}}{T_b}$$

Using the values of T_b and $\Delta\overline{H}_{fus}$ given, we have

$$CH_3OH \quad \Delta\overline{S}_{vap} = \frac{37.57 \times 10^3 \text{ J}\cdot\text{mol}^{-1}}{338.11 \text{ K}} = 111.1 \text{ J}\cdot\text{K}^{-1}\cdot\text{mol}^{-1}$$

$$C_2H_5OH \quad \Delta\overline{S}_{fus} = \frac{40.48 \times 10 \text{ J}\cdot\text{mol}^{-}}{351.7 \text{ K}} = 115.1 \text{ J}\cdot\text{K}^{-}\cdot\text{mol}^{-}$$

$$C_3H_7OH \quad \Delta\overline{S}_{fus} = \frac{43.60 \times 10^3 \text{ J}\cdot\text{mol}^{-1}}{370.6 \text{ K}} = 117.7 \text{ J}\cdot\text{K}^{-1}\cdot\text{mol}^{-1}$$

22-6. We calculate $\Delta\overline{S}_{vap}$ using the relationship

$$\Delta\overline{S}_{vap} = \frac{\Delta\overline{H}_{vap}}{T_b}$$

Using the values of T_b and $\Delta\overline{H}_{vap}$ given, we have

$$\text{Li} \quad \Delta\overline{S}_{vap} = \frac{134.7 \times 10^3 \text{ J}\cdot\text{mol}^{-1}}{1615 \text{ K}} = 83.41 \text{ J}\cdot\text{K}^{-1}\cdot\text{mol}^{-1}$$

$$\text{Na} \quad \Delta\overline{S}_{vap} = \frac{89.6 \times 10^3 \text{ J}\cdot\text{mol}^{-1}}{1156 \text{ K}} = 77.5 \text{ J}\cdot\text{K}^{-1}\cdot\text{mol}^{-1}$$

K $\Delta\overline{S}_{vap} = \dfrac{77.1 \times 10^3\ J\cdot mol^{-1}}{1033\ J} = 74.6\ J\cdot K^{-1}\cdot mol^{-1}$

Rb $\Delta\overline{S}_{vap} = \dfrac{69 \times 10^3\ J\cdot mol^{-1}}{956\ K} = 72\ J\cdot K^{-1}\cdot mol^{-1}$

Cs $\Delta\overline{S}_{vap} = \dfrac{66 \times 10^3\ J\cdot mol^{-1}}{942\ K} = 70\ J\cdot K^{-1}\cdot mol^{-1}$

22-8.

$$CH_4 < NH_3 < H_2O$$

Hydrogen bonding is not present in $CH_4(l)$, whereas the extent of hydrogen bonding in $H_2O(l)$ is greater than in $NH_3(l)$ because hydrogen bonds to oxygen are stronger than hydrogen bonds to nitrogen and there are more hydrogen bonds in $H_2O(l)$ than in $NH_3(l)$.

22-10. a) The mass of PCl_5 is greater than the mass of PCl_3; the compound PCl_5 contains a greater number of atoms than PCl_3. Thus we predict that

$$\overline{S}^0(PCl_3) < \overline{S}^0(PCl_5)$$

b) Ethylene oxide has less freedom of movement than ethanol. Thus we predict that

$$\overline{S}^0(\text{ethylene oxide}) < \overline{S}^0(\text{ethanol})$$

c) Pyrrolidine has less freedom of movement than butyl amine because of its ring structure. Thus we predict that

$$\overline{S}^0(\text{pyrrolidine}) < \overline{S}^0(\text{butyl amine})$$

22-12. The molecular masses are approximately

CH_3Cl (50.5) CH_4 (16.0) CH_3OH (32)

The structure and numbers of atoms are about the same, thus mass is the dominant factor and we predict

$$\overline{S}^0(CH_4) < \overline{S}^0(CH_3OH) < \overline{S}^0(CH_3Cl)$$

22-14. The compound $Fe_3O_4(s)$ contains more atoms and has a greater mass than $Fe_2O_3(s)$. Thus we would expect that

$$\overline{S}^0(Fe_2O_3) < \overline{S}^0(Fe_3O_4)$$

22-16. Gaseous molecules have much greater freedom of movement than liquid molecules (higher translational and rotational disorder), and thus $H_2O(g)$ at 1 atm and 100°C has a greater entropy than $H_2O(l)$ at 1 atm and 100°C.

22-18. a) The volume of water vapor decreases when the pressure increases at constant temperature. The water molecules have a greater freedom of movement in a larger volume. The entropy will decrease.

b) The bromine molecules have a greater freedom of movement in the gaseous state. The entropy will increase.

c) The thermal disorder of iodine is greater at a higher temperature. The entropy will increase.

d) The thermal disorder of iron is greater at a higher temperature. The entropy will decrease.

22-20. a) We have one mole of gaseous product and no gaseous reactants ($\Delta n = +1$).

b) We have no gaseous products and two moles of gaseous reactants ($\Delta n = -2$).

c) We have no gaseous product and one mole of gaseous reactant ($\Delta n = -1$).

d) We have two moles of gaseous products and four moles of gaseous reactants ($\Delta n = -2$).

The value of ΔS^0_{rxn} increases as the net change in the number of moles of gas increases, thus we predict

$$\Delta S^0_{rxn}(b) \approx \Delta S^0_{rxn}(d) < \Delta S^0_{rxn}(c) < \Delta S^0_{rxn} \quad (a)$$

22-22. The value of ΔS^0_{rxn} is given by

$$\Delta S^0_{rxn} = \bar{S}^0(\text{products}) - \bar{S}^0(\text{reactants})$$

a) $\Delta S^0_{rxn} = \bar{S}^0[N_2(g)] + 4\bar{S}^0[H_2O(g)] - 2\bar{S}^0[H_2O\ (l)] - \bar{S}^0[N_2H_4(l)]$

$= (1\text{ mol})(191.5\text{ J}\cdot\text{K}^{-1}\cdot\text{mol}^{-1}) + (4\text{ mol})(188.7\text{ J}\cdot\text{K}^{-1}\cdot\text{mol}^{-1})$

$-(2\text{ mol})(110.0\text{ J}\cdot\text{K}^{-1}\cdot\text{mol}^{-1}) - (1\text{ mol})(121\text{ J}\cdot\text{K}^{-1}\cdot\text{mol}^{-1})$

$= 605\text{ J}\cdot\text{K}^{-1}$

b) $\Delta S^0_{rxn} = 2\bar{S}^0[NO(g)] - \bar{S}^0[N_2(g)] - \bar{S}^0[O_2(g)]$

$= (2\text{ mol})(210.6\text{ J}\cdot\text{K}^{-1}\cdot\text{mol}^{-1}) - (1\text{ mol})(191.5\text{ J}\cdot\text{K}^{-1}\cdot\text{mol}^{-1})$

$-(1\text{ mol})(205.0\text{ J}\cdot\text{K}^{-1}\cdot\text{mol}^{-1})$

$= 24.7\text{ J}\cdot\text{K}^{-1}$

c) $\Delta S^0_{rxn} = 2\bar{S}^0[CH_3OH(l)] - 2\bar{S}^0[CH_4(g)] - \bar{S}^0[O_2(g)]$

$= (2\text{ mol})(126.9\text{ J}\cdot\text{K}^{-1}\cdot\text{mol}^{-1}) - (2\text{ mol})(186.2\text{ J}\cdot\text{K}^{-1}\cdot\text{mol}^{-1})$

$-(1\text{ mol})(205.0\text{ J}\cdot\text{K}^{-1}\cdot\text{mol}^{-1})$

$= -323.6\text{ J}\cdot\text{K}^{-1}$

d) $\Delta S^0_{rxn} = \bar{S}^0[C_2H_6(g)] - \bar{S}^0[C_2H_4(g)] - \bar{S}^0[H_2(g)]$

$= (1\text{ mol})(229.5\text{ J}\cdot\text{K}^{-1}\cdot\text{mol}^{-1}) - (1\text{ mol})(219.6\text{ J}\cdot\text{K}^{-1}\cdot\text{mol}^{-1})$

$-(1\text{ mol})(130.6\text{ J}\cdot\text{K}^{-1}\cdot\text{mol}^{-1})$

$= -120.7\text{ J}\cdot\text{K}^{-1}$

22-24. The value of ΔS^0_{rxn} is given by

$$\Delta S^0_{rxn} = \overline{S}^0(\text{products}) - \overline{S}^0(\text{reactants})$$

a) $\Delta S^0_{rxn} = \overline{S}^0[I_2(g)] - \overline{S}^0[I_2(s)]$

$= (1\ \text{mol})(260.6\ \text{J}\cdot\text{K}^{-1}\cdot\text{mol}^{-1}) - (1\ \text{mol})(116.5\ \text{J}\cdot\text{K}^{-1}\cdot\text{mol}^{-1})$

$= 144.1\ \text{J}\cdot\text{K}^{-1}$

b) $\Delta S^0_{rxn} = \overline{S}^0[BaO(s)] + \overline{S}^0[CO_2(g)] - \overline{S}^0[BaCO_3(s)]$

$= (1\ \text{mol})(70.2\ \text{J}\cdot\text{K}^{-1}\cdot\text{mol}^{-1}) + (1\ \text{mol})(213.6\ \text{J}\cdot\text{K}^{-1}\cdot\text{mol}^{-1})$

$- (1\ \text{mol})(112.2\ \text{J}\cdot\text{K}^{-1}\cdot\text{mol}^{-1})$

$= 171.6\ \text{J}\cdot\text{K}^{-1}$

c) $\Delta S^0_{rxn} = \overline{S}^0[CH_3Cl(g)] + \overline{S}^0[HCl(g)] - \overline{S}^0[CH_4(g)] - \overline{S}^0[Cl_2(g)]$

$= (1\ \text{mol})(234.8\ \text{J}\cdot\text{K}^{-1}\cdot\text{mol}^{-1}) + (1\ \text{mol})(186.8\ \text{J}\cdot\text{K}^{-1}\cdot\text{mol}^{-1})$

$- (1\ \text{mol})(186.2\ \text{J}\cdot\text{K}^{-1}\cdot\text{mol}^{-1}) - (1\ \text{mol})(222.9\ \text{J}\cdot\text{K}^{-1}\cdot\text{mol}^{-1})$

$= 12.5\ \text{J}\cdot\text{K}^{-1}$

d) $\Delta S^0_{rxn} = \overline{S}^0[NaCl(s)] + \overline{S}^0[Br_2(l)] - \overline{S}^0[NaBr(s)] - \overline{S}^0[Cl_2(g)]$

$= (1\ \text{mol})(72.13\ \text{J}\cdot\text{K}^{-1}\cdot\text{mol}^{-1}) + (1\ \text{mol})(152.2\ \text{J}\cdot\text{K}^{-1}\cdot\text{mol}^{-1})$

$- (1\ \text{mol})(86.82\ \text{J}\cdot\text{K}^{-1}\cdot\text{mol}^{-1}) - (1\ \text{mol})(222.9\ \text{J}\cdot\text{K}^{-1}\cdot\text{mol}^{-1})$

$= -85.4\ \text{J}\cdot\text{K}^{-1}$

22-26. The reaction is spontaneous. Because it is spontaneous the sign of ΔG_{rxn} is negative. We learned in Chapter 6 that energy is required for sublimation. Thus the sign of ΔH_{rxn}, which is equal to ΔH_{sub}, is positive. The value of ΔS_{sub}(solid $\rightarrow$ gas) is also positive, thus $T\Delta S_{rxn}$ is positive. The reaction is entropy driven.

22-28. $\Delta S^0_{rxn} = \bar{S}^0(\text{products}) - \bar{S}^0(\text{reactants})$

$$= \bar{S}^0[C_2H_5OH(l)] - \bar{S}^0[C_2H_4(g)] - \bar{S}^0[H_2O(l)]$$

$$= (1\ \text{mol})(160.8\ J\cdot K^{-1}\cdot mol^{-1}) - (1\ \text{mol})(219.6\ J\cdot K^{-1}\cdot mol^{-1})$$

$$-(1\ \text{mol})(69.9\ J\cdot K^{-1}\cdot mol^{-1})$$

$$= -128.7\ J\cdot K^{-1}$$

We calculate ΔG^0_{rxn} using

$$\Delta G^0_{rxn} = \Delta H^0_{rxn} - T\Delta S^0_{rxn}$$

$$\Delta G^0_{rxn} = -35.4\ J - (298\ K)(-128.7\ J\cdot K^{-1})\left(\frac{1\ kJ}{1000\ J}\right)$$

$$= 3.0\ kJ$$

At 1 atm and 25°C, the reaction is spontaneous in the direction

$$C_2H_5OH(l) \longrightarrow C_2H_4(g) + H_2O(l)$$

22-30. $\Delta S^0_{rxn} = \bar{S}^0(\text{products}) - \bar{S}^0(\text{reactants})$

$$\Delta S^0_{rxn} = 2\bar{S}^0[CO(g)] - \bar{S}^0[C(s,\text{graphite})] - \bar{S}^0[CO_2(g)]$$

$$= (2\ \text{mol})(197.8\ J\cdot K^{-1}\cdot mol^{-1}) - (1\ \text{mol})(5.74\ J\cdot K^{-1}\cdot mol^{-1})$$

$$-(1\ \text{mol})(213.6\ J\cdot K^{-1}\cdot mol^{-1})$$

$$= 176.3\ J\cdot K^{-1}$$

We calculate ΔG^0_{rxn} using

$$\Delta G^0_{rxn} = \Delta H^0_{rxn} - T\Delta S^0_{rxn}$$

$$= 172\ kJ - (298\ K)(176.3\ J\cdot K^{-1})\left(\frac{1\ kJ}{1000\ J}\right)$$

$$= 119\ kJ$$

At 1 atm and 25°C, the reaction is spontaneous in the direction

$$2CO(g) \longrightarrow C(s) + CO_2(g)$$

22-32. A catalyst has no effect on the reaction stoichiometry and thus has no effect on ΔG^0_{rxn}, which is determined once the reaction is given.

22-34. The value of ΔG_{rxn} is given by

$$\Delta G_{rxn} = -nFE$$

In the reaction one mole of iron is oxidized from an oxidation state of zero to an oxidation state of +2. Thus the reaction requires two moles of electrons or n = 2

$$\Delta G_{rxn} = -(2\text{ mol})(96{,}500\text{ C}\cdot\text{mol}^{-1})(2.03\text{ V})$$

$$= -392{,}000\text{ J} = -392\text{ kJ}$$

22-36. The value of ΔG^0_{rxn} is given by

$$\Delta G^0_{rxn} = -nFE^0$$

In the reaction six moles of iron are oxidized from an oxidation state of +2 to an oxidation state of +3. Thus the reaction requires six moles of electrons or n = 6.

$$\Delta G^0_{rxn} = -(6\text{ mol})(96{,}500\text{ C}\cdot\text{mol}^{-1})(0.56\text{ V})$$

$$= -320{,}000\text{ J} = -320\text{ kJ}$$

22-38. a) The oxidation half-reaction is

$$Zn(s) \longrightarrow Zn^{2+}(aq) + 2e^-$$

The reduction half-reaction is

$$Cu^{2+}(aq) + 2e^- \longrightarrow Cu(s)$$

The value of the standard cell voltage is

$$E^0 = E^0_{Cu^{2+}/Cu} - E^0_{Zn^{2+}/Zn}$$

$$= 0.34\ V - (-0.76\ V) = 1.10\ V$$

The value of ΔG^0_{rxn} is given by

$$\Delta G^0_{rxn} = -nFE^0$$

Two electrons are transferred in this reaction; thus

$$= -(2\ mol)(96{,}500\ C\cdot mol^{-1})(1.10\ V)$$

$$= -212{,}000\ J = -212\ kJ$$

b) The oxidation half-reaction is

$$Ag(s) \longrightarrow Ag^+(aq) + e^-$$

The reduction half-reaction is

$$Fe^{3+}(aq) + e^- \longrightarrow Fe^{2+}(aq)$$

The value of the standard cell voltage is

$$E^0 = E^0_{Fe^{3+}/Fe^{2+}} - E^0_{Ag^+/Ag}$$

$$= 0.77\ V - (0.80\ V) = -0.03\ V$$

The value of ΔG^0_{rxn} is

$$\Delta G^0_{rxn} = -nFE^0$$

$$= -(1\ mol)(96{,}500\ C\cdot mol^{-1})(-0.03\ V)$$

$$= 3000\ J = 3\ kJ$$

22-40. The cell reaction is

$$Co(s) + 2Ag^+(aq) \longrightarrow Co^{2+}(aq) + 2Ag(s)$$

The value of ΔG^0_{rxn} is given by

$$\Delta G^0_{rxn} = nFE^0$$

$$= -(2\ mol)(96{,}500\ C\cdot mol^{-1})(1.08\ V) = -208\ kJ$$

The value of ΔG_{rxn} is given by

$$\Delta G_{rxn} = \Delta G^0_{rxn} + 2.30RT\log Q$$

Thus

$$\Delta G_{rxn} = -208 \times 10^3 \text{ J} + (2.30)(8.31 \text{ J}\cdot\text{K}^{-1})(298 \text{ K})\log\left\{\frac{[Co^{2+}]}{[Ag^+]^2}\right\}$$

$$= -208 \times 10^3 \text{ J} + (5.696 \times 10^3 \text{ J})\log\left\{\frac{0.0155 \text{ M}}{(1.5 \text{ M})^2}\right\}$$

$$= -208 \times 10^3 \text{ J} - 12.3 \times 10^3 \text{ J} = -220 \times 10^3 \text{ J} = -220 \text{ kJ}$$

The cell voltage is calculated from

$$\Delta G_{rxn} = -nFE$$

Thus

$$E = \frac{-220 \times 10^3 \text{ J}}{2 \text{ mol} \times 96{,}500 \text{ C}\cdot\text{mol}^{-1}} = 1.14 \text{ V}$$

22-42. The value of ΔG^0_{rxn} and the equilibrium constant are related by

$$\Delta G^0_{rxn} = -2.30RT\log K$$

a) The sign of the log of K is positive, thus ΔG^0_{rxn} is negative.

b) The sign of the log of K is negative, thus ΔG^0_{rxn} is positive.

22-44. The value of ΔG^0_{rxn} is given by

$$\Delta G^0_{rxn} = -2.30RT\log K$$

$$= -(2.30)(8.31 \text{ J}\cdot\text{K}^{-1})(800 \text{ K})\log(4.63 \times 10^{-3})$$

$$= +3.57 \times 10^4 \text{ J} = +35.7 \text{ kJ}$$

The reaction is spontaneous right to left when CO_2, Cl_2, and $COCl_2$ are at standard conditions. The value of ΔG_{rxn} at any other conditions is given by

$$\Delta G_{rxn} = 2.30RT\log \frac{Q}{K}$$

Thus

$$G_{rxn} = 2.30\ RT\ \log \left\{ \frac{\frac{[Cl_2][CO]}{[COCl_2]}}{K} \right\}$$

$$= 2.30(8.31\ J\cdot K^{-1})(800\ K)\log \left[\frac{\frac{(1.0\ M)(1.0\ M)}{(0.025\ M)}}{4.63 \times 10^{-3}\ M} \right]$$

$$= 6.02 \times 10^{4}\ J = 60.2\ kJ$$

The reaction is spontaneous from right to left.

22-46. The value of ΔG^{0}_{rxn} is given by

$$\Delta G^{0}_{rxn} = -2.30RT\log K$$

$$= -(2.30)(8.31\ J\cdot K^{-1})(298\ K)\log(3.0 \times 10^{-8})$$

$$= 4.28 \times 10^{4}\ J = 42.8\ kJ$$

Hypochlorous acid will not dissociate spontaneously when $[ClO^-] = [H^+] = [HClO] = 1.0\ M$. The value of ΔG_{rxn} at any other conditions is given by

$$\Delta G_{rxn} = 2.30RT\log \frac{Q}{K}$$

Thus

$$\Delta G_{rxn} = 2.30RT\log \left\{ \frac{\frac{[ClO^-][H^+]}{[HClO]}}{K} \right\}$$

$$= (2.30)(8.31\ J\cdot K^{-1})(298\ K)\log \left[\frac{\frac{(1.0 \times 10^{-6}M)(1.0 \times 10^{-6}M)}{(0.10\ M)}}{3.0 \times 10^{-8}\ M} \right]$$

$$= -1.98 \times 10^{4}\ J = -19.8\ kJ$$

Hypochlorous acid will dissociate spontaneously at these conditions.

22-48. The value of ΔG^0_{rxn} is given by

$$\Delta G^0_{rxn} = -2.30RT\log K$$

Thus

$$\Delta G^0_{rxn} = -(2.30)(8.31\ \ J\cdot K^{-1})(298\ K)\log(1.75\ \times\ 10^{-5})$$

$$= +2.71\ \times\ 10^4\ J\ =\ 27.1\ kJ$$

The reaction is spontaneous from right to left when $NH_3(aq)$, $NH_4^+(aq)$, and $OH^-(aq)$ are at standard conditions. The value of ΔG_{rxn} at any other conditions is given by

$$\Delta G_{rxn} = 2.30RT\log \frac{Q}{K}$$

Thus

$$\Delta G_{rxn} = 2.30RT\log \left\{ \frac{\frac{[NH_4^+][OH^-]}{[NH_3]}}{K} \right\}$$

$$= (2.30)(8.31\ \ J\cdot K^{-1})(298\ K)\log \left[\frac{\frac{(1.0x10^{-6})(1.0x10^{-6})}{(0.050)}}{1.75\ \times\ 10^{-5}} \right]$$

$$= -3.38\ \times\ 10^4\ J\ =\ -33.8\ kJ$$

Ammonia will react with water at these conditions.

22-50. The value of ΔG^0_{rxn} is given by

$$\Delta G^0_{rxn} = -2.30RT\log K$$

Thus

$$\Delta G^0_{rxn} = -(2.30)(8.31\ \ J\cdot K^{-1})(298\ K)\log(2.8\ \times\ 10^{-9})$$

$$= 4.87\ \times\ 10^4\ J\ =\ 48.7\ kJ$$

when $[Ca^{2+}] = [CO_3^{2-}] = 1.0$ M (standard conditions). When a solution is prepared in which $[Ca^{2+}] = [CO_3^{2-}] = 1.0$ M, insoluble $CaCO_3(s)$ will precipitate from the solution.

22-52. The value of ΔG^0_{rxn} is given by

$$\Delta G^0_{rxn} = -2.30RT\log K$$

Thus

$$\Delta G^0_{rxn} = -(2.30)(8.31 \ J\cdot K^{-1})(298 \ K)\log(2.0 \times 10^7)$$

$$= -4.16 \times 10^4 \ J = -41.6 \ kJ$$

The reaction is spontaneous left to right when $Co^{3+}(aq)$, $NH_3(aq)$, and $Co(NH_3)_6^{3+}(aq)$ are at standard conditions. The value of ΔG_{rxn} at any other condition is given by

$$\Delta G_{rxn} = 2.30 \ RT \log \frac{Q}{K}$$

Thus

$$\Delta G_{rxn} = 2.30RT\log \left\{ \frac{\dfrac{[Co(NH_3)_6^{3+}]}{[Co^{3+}][NH_3]^6}}{K} \right\}$$

$$= (2.30)(8.31 \ J\cdot K^{-1})(298 \ K)\log\left[\frac{\dfrac{(0.0010)}{(0.025)(0.50)^6}}{2.0 \times 10^7}\right]$$

$$= -3.93 \times 10^4 \ J = -39.3 \ kJ$$

The reaction is spontaneous left to right.

22-54. No, because if $K = 0$, then ΔG^0_{rxn} is negative infinite and thus an infinite amount of work could be obtained from the reaction, which is impossible.

22-56. The value of ΔG^0_{rxn} is calculated using the data in Table 22-1 and the relationship

$$\Delta G^0_{rxn} = \Delta\overline{G}^0_f(\text{products}) - \Delta\overline{G}^0_f(\text{reactants})$$

The value of the equilibrium constant, K, is given by

$$\Delta G^0_{rxn} = -2.30\ RT \log K$$

Thus

a) $$\Delta G^0_{rxn} = \Delta\overline{G}^0_f[N_2(g)] + 4\Delta\overline{G}^0_f[H_2O(g)] - 2\Delta\overline{G}^0_f[H_2O_2(l)] - \Delta\overline{G}^0_f[N_2H_4(l)]$$

$$= 0 + (4\ \text{mol})(-228.6\ \text{kJ}\cdot\text{mol}^{-1}) - (2\ \text{mol})(-120.4\ \text{kJ}\cdot\text{mol}^{-1})$$

$$-(1\ \text{mol})(149.0\ \text{kJ}\cdot\text{mol}^{-1})$$

$$= -822.6\ \text{kJ}$$

The value of log K is

$$\log K = -\frac{\Delta G^0_{rxn}}{2.30\ RT} = -\frac{(-822.6 \times 10^3\ \text{J})}{(2.30)(8.31\ \text{J}\cdot\text{K}^{-1})(298\ \text{K})}$$

$$= 144.43$$

$$K = 2.7 \times 10^{144}$$

b) $$\Delta G^0_{rxn} = 2\Delta\overline{G}^0_f[NO(g)] - \Delta\overline{G}^0_f[N_2(g)] - \Delta\overline{G}^0_f[O_2(g)]$$

$$= (2\ \text{mol})(86.69\ \text{kJ}\cdot\text{mol}^{-1}) - 0 - 0$$

$$= 173.38\ \text{kJ}$$

Thus

$$\log K = -\frac{\Delta G^0_{rxn}}{2.30\ RT} = -\frac{(173.38 \times 10^3\ \text{J})}{(2.30(8.31\ \text{J}\cdot\text{K}^{-1})(298\ \text{K})}$$

$$= -30.41$$

$$K = 3.62 \times 10^{-31}$$

c) $\Delta G^0_{rxn} = 2\Delta\overline{G}^0_f[CH_3OH(l)] - 2\Delta\overline{G}^0_f[CH_4(g)] - \Delta\overline{G}^0_f[O_2(g)]$

$= (2\text{ mol})(-166.3\text{ kJ}\cdot\text{mol}^{-1})-(2\text{ mol})(-50.75\text{ kJ}\cdot\text{mol}^{-1}) - 0$

$= -231.1\text{ kJ}$

Thus

$$\log K = \frac{\Delta G^0_{rxn}}{2.30\ RT} = -\frac{(-231.1 \times 10^3\text{ J})}{(2.30)(8.31\ \text{J}\cdot\text{K}^{-1})(298\text{ K})}$$

$= 40.75$

$K = 3.76 \times 10^{40}$

22-58. The value of ΔG^0_{rxn} is given by

$$\Delta G^0_{rxn} = \overline{G}^0_f(\text{products}) - \overline{G}^0_f(\text{reactants})$$

Thus

$$\Delta G^0_{rxn} = 3\Delta\overline{G}^0_f[Fe(s)] + 2\Delta\overline{G}^0_f[CO_2(g)] - \Delta\overline{G}^0_f[Fe_3O_4(s)]$$
$$-2\Delta\overline{G}^0_f[C(s,\text{graphite})]$$

$$= (3\text{ mol})(0)+(2\text{ mol})(-394.4\text{ kJ}\cdot\text{mol}^{-1})$$
$$-(1\text{ mol})(-1015\text{ kJ}\cdot\text{mol}^{-1})-(2\text{ mol})(0)$$

$$= 226\text{ kJ}$$

The value of ΔH^0_{rxn} is given by

$$\Delta H^0_{rxn} = \overline{H}^0_f(\text{products}) - \overline{H}^0_f(\text{reactants})$$

Thus

$$\Delta H^0_{rxn} = 3\Delta\overline{H}^0_f[Fe(s)] + 2\Delta\overline{H}^0_f[CO_2(g)] - \Delta\overline{H}^0_f[Fe_3O_4(s)]$$
$$-4\Delta\overline{H}^0_f[C(s,\text{graphite})]$$

$$= (3\text{ mol})(0)+(2\text{ mol})(-393.5\text{ kJ}\cdot\text{mol}^{-1})$$
$$-(1\text{ mol})(-1118\text{ kJ}\cdot\text{mol}^{-1})-(2\text{ mol})(0)$$

$$= 331\text{ kJ}$$

The value of log K is given by

$$\log K = -\frac{\Delta G^0_{rxn}}{2.30\ RT} = -\frac{(226 \times 10^3\ J)}{(2.30)(8.31\ J\cdot K^{-1})(298\ K)}$$

$$= -39.679$$

Thus

$$K = 2.09 \times 10^{-40}$$

22-60. The value of ΔG^0_{rxn} is given by

$$\Delta G^0_{rxn} = \Delta\bar{G}^0_f(\text{products}) - \Delta\bar{G}^0_f(\text{reactants})$$

Thus

$$\Delta G^0_{rxn} = 2\Delta\bar{G}^0_f[HI(g)] - \Delta\bar{G}^0_f[H_2(g)] - \Delta\bar{G}^0_f[I_2(g)]$$

$$= (2\ mol)(1.7\ kJ\cdot mol^{-1})-0-(1\ mol)(+19.36\ kJ\cdot mol^{-1})$$
$$= -16.0\ kJ$$

The value of ΔH^0_{rxn} is given by

$$\Delta H^0_{rxn} = \Delta\bar{H}^0_f(\text{products}) - \Delta\bar{H}^0_f(\text{reactants})$$

Thus

$$\Delta H^0_{rxn} = 2\Delta\bar{H}^0_f[HI(g)] - \Delta\bar{H}^0_f[H_2(g)] - \Delta\bar{H}^0_f[I_2(g)]$$

$$= (2\ mol)(26.1\ kJ\cdot mol^{-1})-0-(1\ mol)(62.4\ kJ\cdot mol^{-1})$$
$$= -10.2\ kJ$$

Therefore

$$\log K = \frac{-\Delta G^0_{rxn}}{2.30\ RT} = -\frac{(-16.0 \times 10^3\ J)}{(2.30)(8.31\ J\cdot K^{-1})(298\ K)}$$

$$= 2.809$$

$$K = 6.44 \times 10^2$$

The value of ΔS^0_{rxn} is given by

$$\Delta S^0_{rxn} = \overline{S}^0(\text{products}) - \overline{S}^0(\text{reactants})$$

$$= 2\overline{S}^0[HI(g)] - \overline{S}^0[H_2(g)] - \overline{S}^0[I_2(g)]$$

$$= (2\text{ mol})(206.4\text{ J}\cdot\text{K}^{-1}\cdot\text{mol}^{-1}) - (1\text{ mol})(130.6\text{ J}\cdot\text{K}^{-1}\cdot\text{mol}^{-1})$$
$$-(1\text{ mol})(260.6\text{ J}\cdot\text{K}^{-1}\cdot\text{mol}^{-1})$$
$$= 21.6\text{ J}\cdot\text{K}^{-1}$$

The value of ΔG^0_{rxn} at 100^0C is given by

$$\Delta G^0_{rxn} = \Delta H^0_{rxn} - T\Delta S^0_{rxn}$$

$$= -10.2\text{ kJ} - (373\text{ K})(0.0216\text{ kJ}\cdot\text{K}^{-1})$$

$$= -18.3\text{ kJ}$$

The value of the equilibrium constant at 100^0C is

$$\log K = \frac{\Delta G^0_{rxn}}{2.30\ RT} = -\frac{(-18.3 \times 10^3\text{ J})}{(2.30)(8.31\text{ J}\cdot\text{mol}^{-1}\cdot\text{K}^{-1})(373\text{ K})}$$

$$= 2.567$$

Thus

$$K = 3.69 \times 10^2$$

22-62. The value of ΔG^0_{rxn} is given by

$$\Delta G^0_{rxn} = \Delta\overline{G}^0_f[CCl_4(l)] + 4\Delta\overline{G}^0_f[HCl(g)] - \Delta\overline{G}^0_f[CH_4(g)] - 4\Delta\overline{G}^0_f[Cl_2(g)]$$

Thus

$$-395.7\text{ kJ} = (1\text{ mol})\Delta\overline{G}^0_f[CCl_4(l)] + (4\text{ mol})(-95.30\text{ kJ}\cdot\text{mol}^{-1})$$
$$-(1\text{ mol})(-50.75\text{ kJ}\cdot\text{mol}^{-1}) - (4\text{ mol})(0)$$

Therefore

$$-395.7 \text{ kJ} = (1 \text{ mol})\Delta\bar{G}^0_f[CCl_4(l)] - 330.45 \text{ kJ}$$

$$\Delta\bar{G}^0_f[CCl_4(l)] = -65.3 \text{ kJ}\cdot\text{mol}^{-1}$$

which agrees with the value of $\Delta G^0_f[CCl_4(l)]$ given in Table 22-1. The value of ΔH^0_{rxn} is given by

$$\Delta H^0_{rxn} = \Delta\bar{H}^0_f[CCl_4(l)] + 4\Delta\bar{H}^0_f[HCl(g)] - \Delta\bar{H}^0_f[CH_4(g)] - 4\Delta\bar{H}^0_f[Cl_2(g)]$$

Thus

$$-429.8 \text{ kJ} = (1 \text{ mol})\Delta\bar{H}^0_f[CCl_4(l)] + (4 \text{ mol})(-92.31 \text{ kJ}\cdot\text{mol}^{-1}) - (1 \text{ mol})(-74.86 \text{ kJ}\cdot\text{mol}^{-1}) - 0$$

$$= (1 \text{ mol})\Delta\bar{H}^0_f[CCl_4(l)] - 294.38 \text{ kJ}$$

Therefore

$$\Delta\bar{H}^0_f[CCl_4(l)] = -135.4 \text{ kJ}\cdot\text{mol}^{-1}$$

which agrees with the value given in Table 22-1.

22-64. $\Delta G_{rxn} = -2.30RT\log K$

Thus at 25°C

$$\Delta G^0_{rxn} = -(2.30)(8.31 \text{ J}\cdot\text{K}^{-1})(298 \text{ K})\log K$$

$$= -(5.70 \text{ kJ})\log K$$

The difference in the values of ΔG^0_{rxn} when there is a tenfold increase in K is

$$\Delta G^0_{rxn_2} - \Delta G^0_{rxn_1} = -(5.70 \text{ kJ})\log K_2 + (5.70 \text{ kJ})\log K_1$$

$$= (5.70 \text{ kJ})\log \frac{K_1}{K_2}$$

where we have used the property of logarithms

$$\log a - \log b = \log a/b$$

We are given that $K_2 = 10 K_1$, thus

$$\Delta G^0_{rxn_2} - \Delta G^0_{rxn_1} = (5.70 \text{ kJ})\log \frac{K_1}{10 K} = (5.70 \text{ kJ})\log\left(\frac{1}{10}\right)$$

A tenfold increase in K corresponds to a decrease of -5.70 kJ in ΔG^0_{rxn}.

22-66. The value of ΔG^0_{rxn} is given by

$$\Delta G^0_{rxn} = 6\Delta\overline{G}^0_f[CO_2(g)] + 6\Delta\overline{G}_f[H_2O(l)] - \Delta\overline{G}^0_f[C_6H_{12}O_6(s)]$$
$$-6\Delta\overline{G}^0_f[O_2(g)]$$

Thus

$$\Delta G^0_{rxn} = (6 \text{ mol})(-394.4 \text{ kJ}\cdot\text{mol}^{-1}) + (6 \text{ mol})(-237.2 \text{ kJ}\cdot\text{mol}^{-1})$$
$$-(1 \text{ mol})(-916 \text{ kJ}\cdot\text{mol}^{-1}) - 0$$
$$= -2874 \text{ kJ}$$

The maximum work that can be obtained is equal to ΔG_{rxn}. Thus for the complete combustion of glucose under standard conditions, the maximum work that can be obtained is 2874 kJ.

22-68. The reaction for the combustion of methane is

$$CH_4(g) + 2O_2(g) \longrightarrow CO_2(g) + 2H_2O(l)$$

The value of ΔG^0_{rxn} is given by

$$\Delta G^0_{rxn} = \Delta\overline{G}^0_f(\text{products}) - \Delta\overline{G}^0_f(\text{reactants})$$

Thus

$$\Delta G^0_{rxn} = \Delta\overline{G}^0_f[CO_2(g)] + 2\Delta\overline{G}^0_f[H_2O(l)] - \Delta\overline{G}^0_f[CH_4(g)] - 2\Delta\overline{G}^0_f[O_2(g)]$$
$$= (1 \text{ mol})(-394.4 \text{ kJ}\cdot\text{mol}^{-1}) + (2 \text{ mol})(-237.2 \text{ kJ}\cdot\text{mol}^{-1})$$
$$-(1 \text{ mol})(-50.75 \text{ kJ}\cdot\text{mol}^{-1}) - 0$$
$$= -818.1 \text{ kJ}$$

The maximum amount of work that can be obtained when the reaction is carried out under standard conditions is 818.1 kJ.

22-70. The reaction that takes place in a methane-oxygen fuel cell is

$$CH_4(g) + 2O_2(g) \longrightarrow CO_2(g) + 2H_2O(l)$$

The value of ΔG^0_{rxn} is given by

$$\Delta G^0_{rxn} = \Delta\bar{G}^0_f[CO_2(g)] + 2\Delta\bar{G}^0_f[H_2O(l)] - \Delta\bar{G}^0_f[CH_4(g)] - 2\Delta\bar{G}^0_f[O_2(g)]$$

Thus

$$\Delta G^0_{rxn} = (1\ \text{mol})(-394.4\ \text{kJ}\cdot\text{mol}^{-1}) + (2\ \text{mol})(-237.2\ \text{kJ}\cdot\text{mol}^{-1})$$
$$-(1\ \text{mol})(-50.75\ \text{kJ}\cdot\text{mol}^{-1}) - 0 = -818.1\ \text{kJ}$$

The value of the standard voltage of the cell is given by

$$E^0 = \frac{-\Delta G^0_{rxn}}{nF}$$

The oxidation state of carbon increases from -4 to +4, thus eight electrons are involved, and

$$E^0 = \frac{-(-818.1 \times 10^3\ \text{J})}{(8\ \text{mol})(96{,}500\ \text{C}\cdot\text{mol}^{-1})}$$

$$= 1.06\ \text{V}$$

22-72. The reaction

$$AgBr(s) \rightleftharpoons Ag^+(aq) + Br^-(aq) \qquad K = K_{sp}$$

The value of ΔG^0_{rxn} is given by

$$\Delta G^0_{rxn} = \Delta\bar{G}^0_f[Ag^+(aq)] + \Delta\bar{G}^0_f[Br^-(aq)] - \Delta\bar{G}^0_f[AgBr(s)]$$

Thus

$$\Delta G^0_{rxn} = 77.1\ \text{kJ} + (-102.8\ \text{kJ}) - (-96.8\ \text{kJ}) = 71.1\ \text{kJ}$$

The solubility product, K_{sp}, can be calculated using the relation

$$\log K_{sp} = \frac{-\Delta G^0_{rxn}}{2.30\ RT} = -\frac{(71.1 \times 10^3\ J}{(2.30)(8.31\ J\cdot K^{-1})(298\ K)}$$

$$= -12.483$$

Thus

$$K_{sp} = 3.29 \times 10^{-13}\ M^2$$

22-74. The value of ΔG^0_{rxn} is given by

$$\Delta G^0_{rxn} = -2.30\ RT \log K$$

At 25°C

$$\Delta G^0_{rxn} = -(2.30)(8.31\ J\cdot K^{-1})(298\ K)\log(0.140)$$

$$= 4860\ J = 4.86\ kJ$$

At 100°C

$$\Delta G^0_{rxn} = -(2.30)(8.31\ J\cdot K^{-1})(373\ K)\log(0.365)$$

$$= 3120\ J = 3.12\ kJ$$

We calculate ΔH^0_{rxn} using the equation

$$\log \frac{K_2}{K_1} = \frac{\Delta H^0_{rxn}}{2.30\ R}\left[\frac{T_2 - T_1}{T_1 T_2}\right]$$

$$\log\left(\frac{0.365}{0.140}\right) = \frac{\Delta H^0_{rxn}}{(2.30)(8.31\ J\cdot K^{-1})}\left[\frac{373\ K - 298\ K}{(298\ K)(373\ K)}\right]$$

$$0.4162 = (3.53 \times 10^{-5}\ J^{-1})\Delta H^0_{rxn}$$

$$\Delta H^0_{rxn} = \frac{0.4162}{3.53 \times 10^{-5}\ J^{-1}} = 1.18 \times 10^4\ J$$

$$= 11.8\ kJ$$

The value of ΔS^0 can be found using the equation

$$\Delta G^0_{rxn} = \Delta H^0_{rxn} - T\Delta S^0_{rxn}$$

At 25°C

$$4.86\ kJ = 11.8\ kJ - (298\ K)\Delta S^0_{rxn}$$

Thus

$$\Delta S^0_{rxn} = \frac{6.94\ kJ}{298\ K} = 0.0233\ kJ\cdot K^{-1}$$

$$= 23.3\ J\cdot K^{-1}$$

22-76. A plot of the data in the form 1/T versus log K_p is a straight line

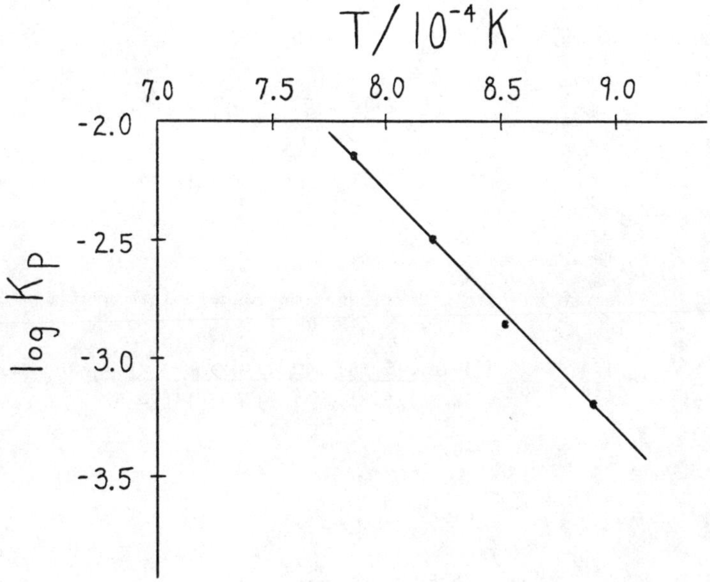

Thus we can use any set of data and the van't Hoff equation to calculate ΔH^0_{rxn}

$$\log\left(\frac{6.88 \times 10^{-3}\ \text{atm}}{3.26 \times 10^{-3}\ \text{atm}}\right) = \frac{\Delta H^0_{rxn}(1273\ \text{K} - 1223\ \text{K})}{(2.30)(8.31\ \text{J}\cdot\text{K}^{-1})(1273\ \text{K})(1223\ \text{K})}$$

$$0.32437 = (1.680 \times 10^{-6}\ \text{J}^{-1})\Delta H^0_{rxn}$$

$$\Delta H^0_{rxn} = 1.93 \times 10^5\ \text{J} = 193\ \text{kJ}$$

22-78. A plot of the data in the form 1/T versus log K_p is a straight line. Thus we can use any set of data and the van't Hoff equation to calculate ΔH^0_{rxn}.

$$\log\left(\frac{3.46\ \text{atm}^{-1}}{43.1\ \text{atm}^{-1}}\right) = \frac{\Delta H^0_{rxn}(1000\ \text{K} - 900\ \text{K})}{(2.30)(8.31\ \text{J}\cdot\text{K}^{-1})(1000\ \text{K})(900\ \text{K})}$$

$$-1.095 = (5.8134 \times 10^{-6}\ \text{J}^{-1})\Delta H^0_{rxn}$$

$$\Delta H^0_{rxn} = -1.88 \times 10^5\ \text{J} = -188\ \text{kJ}$$

22-80. The value of ΔH^0_{rxn} is given by

$$\Delta H^0_{rxn} = 2\Delta \overline{H}^0_f[\text{HI}(g)] - \Delta \overline{H}^0_f[\text{H}_2(g)] - \Delta \overline{H}^0_f[\text{I}_2(g)]$$

$$= (2\ \text{mol})(26.1\ \text{kJ}\cdot\text{mol}^{-1}) - 0 - (1\ \text{mol})(62.4\ \text{kJ}\cdot\text{mol}^{-1})$$

$$= -10.2\ \text{kJ}$$

We use the van't Hoff equation to calculate K at 500°C

$$\log\left(\frac{K}{58.0}\right) = \frac{(-10.2 \times 10^3\ \text{J})(773\ \text{K} - 673\ \text{K})}{(2.30)(8.31\ \text{J}\cdot\text{K}^{-}\ )(773\ \text{K})(673\ \text{K})}$$

$$= -0.10258$$

$$\frac{K}{58.0} = 0.7896$$

$$K = 45.8$$

22-82. For an endothermic reaction, $\Delta H^0_{rxn} > 0$. Thus $\log K_2/K_1 > 0$ when $T_2 > T_1$. Because $\log K_2/K_1$ is positive, $K_2/K_1 > 1$. In other words, $K_2 > K_1$ or the equilibrium constant at a higher temperature is greater than that at a lower temperature.

22-84. Using the data in Problem 22-75, we have

$1/T\ \ 10^{-4}\ K^{-1}$	$\log K_p$
5.000	-3.3893
4.762	-3.1637
4.546	-2.9586
4.348	-2.7721
4.167	-2.6003

A plot of $\log K_p$ versus $1/T$ is a straight line. The equation for the straight line (see Problem 22-83) is

$$\log K = \left(\frac{-\Delta H^0_{rxn}}{2.30\ R}\right)\left(\frac{1}{T}\right) + b$$

The slope of the line is $\dfrac{-\Delta H^0_{rxn}}{2.30\ R}$. The slope can be calculated using

$$\text{slope} = \frac{Y_2 - Y_1}{X_2 - X_1} = \frac{\log K_2 - \log K_1}{\left(\frac{1}{T_2} - \frac{1}{T_1}\right)}$$

$$= \frac{-3.1637 - (-3.3893)}{(4.762 - 5.000)(10^{-4}\ K^{-1})} = -9.4790 \times 10^3\ K$$

$$\text{slope} = \frac{-\Delta H^0_{rxn}}{2.30\ R} = -9.4790 \times 10^3\ K$$

$$\Delta H^0_{rxn} = -(-9.4790 \times 10^3\ K)(2.30)(8.31\ J \cdot K^{-1})$$

$$= +181 \times 10^5\ J = +181\ kJ$$

CHAPTER 23

SOLUTIONS TO THE EVEN-NUMBERED PROBLEMS

23-2. a) There are 28 - 2 = 26 electrons in Ni^{2+}. The electron configuration of $Ni^{2+}(g)$ is $1s^2 2s^2 2p^6 3s^2 3p^6 3d^8$.

b) There are 44 - 3 = 41 electrons in Ru^{3+}. The electron configuration of $Ru^{3+}(g)$ is $1s^2 2s^2 2p^6 3s^2 3p^6 3d^{10} 4s^2 4p^6 4d^5$.

c) There are 73 - 4 = 69 electrons in Ta^{4+}. The electron configuration of $Ta^{4+}(g)$ is $1s^2 2s^2 2p^6 3s^2 3p^6 3d^{10} 4s^2 4p^6 4d^{10} 4f^{14} 5s^2 5p^6 5d^1$.

d) There are 79 - 1 = 78 electrons in Au^+. The electron configuration of $Au^+(g)$ is $1s^2 2s^2 2p^6 3s^2 3p^6 3d^{10} 4s^2 4p^6 4d^{10} 4f^{14} 5s^2 5p^6 5d^{10}$.

e) There are 48 - 2 = 46 electrons in Cd^{2+}. The electron configuration of $Cd^{2+}(g)$ is $1s^2 2s^2 2p^6 3s^2 3p^6 3d^{10} 4s^2 4p^6 4d^{10}$.

23-4. a) Silver is below copper in the periodic table and Ag^+ has one more electron than Ag^{2+}. Thus Ag^+ has ten d electrons.

b) Vanadium(III) has one less electron than V^{2+} and so V^{3+} has two d electrons.

c) Zirconium is below titanium in the periodic table and Zr^{3+} has one less electron than Zr^{2+}. Thus Zr^{3+} has one d electron.

d) Ruthenium is below iron in the periodic table and Ru^{4+} has two less electrons than Ru^{2+}. Thus Ru^{4+} has four d electrons.

e) Mercury is below zinc in the Periodic Table and so Hg^{2+} has ten d electrons.

23-6. a) The M(II) d^4 ions are Cr(II), Mo(II) and W(II).

b) For an M(III) ion to be a d^8 ion, the corresponding M(II) ion must be a d^9 ion. The M(III) d^8 ions are Cu(III), Ag(III) and Au(III).

c) For an M(IV) ion to be a d^0 ion, the corresponding M(II) ion must be a d^2 ion. The M(IV) d^0 ions are Ti(IV), Zr(IV) and Hf(IV).

23-8. a) The charge on the SO_4^{2-} ligand is -2. Denoting the charge on Hg as X, we have X + 2(-2) = 2- or X = +2. The oxidation state of mercury is +2.

b) The charge on the Cl^- ligand is -1. Denoting the charge on Cu as X, we have X + 4(-1) = 2- or X = +2. The oxidation state of copper is +2.

c) The charge on the OH^- ligand is -1. Denoting the charge on Al as X, we have X + 1(-1) = 2+ or X = +3. The oxidation state of aluminum is +3.

d) The charge on the Cl^- ligand is -1. Denoting the charge on Fe as X, we have X + 4(-1) = -1 or X = +3. The oxidation state of iron is +3.

e) The charge on the CN^- ligand is -1. Denoting the charge on Ni as X, we have X + 4(-1) = -2 or X = +2. The oxidation state of nickel is +2.

23-10. a) The charge on the NH_3 ligand is zero; the charge on the Br^- ligand is -1. Denoting the charge on Ir as X, we have X + 3(0) + 2(-1) = 1+ or X = +3. The oxidation state of iridium is +3.

b) The charge on the Cl^- ligand is -1. Denoting the charge on Rh as X, we have X + 6(-1) = 3- or X = +3. The oxidation state of rhodium is +3.

c) The charge on the Cl^- ligand is -1. Denoting the charge on Ir as X, we have X + 6(-1) = 3- or X = +3. The oxidation state of iridium is +3.

d) The charge on the NO_2^- ligand is -1. Denoting the charge on V as X, we have X + 6(-1) = 3- or X = +3. The oxidation state of vanadium is +3.

e) The charge on the NH_3 ligand is zero; the charge on the CO ligand is zero. Denoting the charge on Co as X, we have X + 3(0) + 3(0) = 3+ or X = +3. The oxidation state of cobalt is +3.

23-12. The reaction for the dissolving the compound in water is

$$K_2[Ni(CN)_4](s) \xrightarrow[H_2O(l)]{} 2K^+(aq) + [Ni(CN)_4]^{2-}(aq)$$

Thus there is a total of three moles of ions in solution when one mole of $K_2[Ni(CN)_4]$ is dissolved in water.

23-14. a) One mole of $[Co(NH_3)_6]^{3+}(aq)$ and three moles of $Br^-(aq)$.

b) One mole of $[Pt(NH_3)_3Cl_3]^+(aq)$ and one mole of $Cl^-(aq)$.

c) One mole of $[Cr(H_2O)_6]^{3+}(aq)$ and three moles of $Br^-(aq)$.

d) Four moles of $K^+(aq)$ and one mole of $[Cr(CN)_6]^{4-}(aq)$.

23-16. The chloride ions that exist in solution as $Cl^-(aq)$ but not the chloride ions that are complexed with the platinum atoms are precipitated by $Ag^+(aq)$ as AgCl(s).

$PtCl_2 \cdot 4NH_3$ Both chloride ions exist as $Cl^-(aq)$ because they are precipitated by $Ag^+(aq)$. The chemical formula of the complex salt must be $[Pt(NH_3)_4]Cl_2$.

$PtCl_2 \cdot 3NH_3$ One of the two chloride ions is complexed to the platinum atom because it is not precipitated by $Ag^+(aq)$. The chemical formula of the complex salt must be $[Pt(NH_3)_3Cl]Cl$.

$PtCl_2 \cdot 2NH_3$ Both of the chloride ions must be complexed to the platinum atom because neither is precipitated by $Ag^+(aq)$. The chemical formula of the complex must be $[Pt(NH_3)_2Cl_2]$.

23-18. a) Denoting the oxidation state of iron as X, we have $X + 6(-1) = -3$ or $X = +3$. The complex ion is called hexacyanoferrate(III).

b) Denoting the oxidation state of nickel as X, we have $X + 4(0) = 0$ or $X = 0$. The complex ion is called tetracarbonylnickel(0).

c) Denoting the oxidation state of iron as X, we have $X + 5(0) + (-1) = +2$ or $X = +3$. The complex ion is called pentaaquathiocyanatoiron(III).

d) Denoting the oxidation state of cobalt as X, we have $X + 2(0) + 4(0) = +2$ or $X = +2$. The complex ion is called diamminetetraaquacobalt(II).

23-20. a) Denoting the oxidation state of gold as X, we have $X + 4(-1) = 1-$ or $X = +3$. The compound is called sodium tetracyanoaurate(III).

b) Denoting the oxidation state of chromium as X, we have $X + 6(0) = +3$ or $X = +3$. The compound is called hexaaquachromium(III) chloride.

c) Denoting the oxidation state of manganese as X, we have $X + 5(0) = +1$ or $X = +1$. The compound is called pentacarbonylmanganese(I) chloride.

d) Denoting the oxidation state of copper as X, we have $X + 6(0) = +2$ or $X = +2$. The compound is called hexaamminecopper(II) chloride.

23-22. a) Denoting the oxidation state of ruthenium as X, we have $X + 5(0) + (-1) = +2$ or $X = +3$. The compound is called pentaamminechlororuthenium(III) chloride.

b) Denoting the oxidation state of cobalt as X, we have $X + 6(-1) = -3$ or $X = +3$. The compound is called sodium hexacyanocobaltate(III).

c) Denoting the oxidation state of rhenium as X, we have $X + 6(-1) = -2$ or $X = +4$. The compound is called sodium hexafluororhenate(IV).

d) Denoting the oxidation state of iron as X, we have $X + 5(0) + (-1) = +2$ or $X = +3$. The compound is called pentaaquathiocyanatoiron(III) nitrate.

23-24. a) The anion has one Br^- ligand with a -1 charge, one Cl^- ligand with a -1 charge, two CN^- ligands with a charge of 2(-1) = -2 charge, and one nickel atom in an oxidation state +2. The net charge on the complex anion is (-1) + (-1) + (-2) + (+2) = -2. The formula for the complex anion is $[NiBrCl(CN)_2]^{2-}$.

b) The anion has four SCN^- ligands with a charge of 4(-1) = -4 and a cobalt atom in an oxidation state +2. The net charge on the complex anion is (-4) + (+2) = -2. The formula for the complex anion is $[Co(SCN)_4]^{2-}$.

c) The anion has six Cl^- ligands with a charge of 6(-1) = -6 and one vanadium atom in an oxidation state +3. The net charge on the complex anion is (-6) + (+3) = -3. The formula for the complex anion is $[VCl_6]^{3-}$.

d) The cation has five NH_3 ligands with zero charge, one Cl^- ligand with a charge of -1, and one chromium atom in an oxidation state +3. The net charge on the complex cation is (-1) + (+3) = +2. The formula for the complex cation is $[Cr(NH_3)_5Cl]^{2+}$.

23-26. a) The complex cation has six H_2O ligands with zero charge and one cobalt atom in an oxidation state +2. The net charge on the complex cation is +2. The formula for the complex cation is $[Co(H_2O)_6]^{2+}$.

b) The complex cation has four NH_3 ligands with zero charge and one copper atom in an oxidation state +2. The net charge on the complex cation is +2 The formula for the complex cation is $[Cu(NH_3)_4]^{2+}$.

c) The complex has six CO ligands with zero charge and one chromium atom in an oxidation state 0. The net charge on the complex is 0. The formula for the complex is $[Cr(CO)_6]$.

d) The complex has two Cl^- ligands with a charge of 2(-1) = -2, two O^{2-} ligands with a charge of 2(-2) = -4, and one tungsten atom in an oxidation state +6. The net charge on the complex is (-2) + (-4) + (+6) = 0. The formula for the complex is $[WCl_2O_2]$.

23-28. a) The charge on the anion is 4(-1) + (+2) = -2. The formula for the compound is $(NH_4)_2[CoCl_4]$.

b) The charge on the anion is 6(-1) + (+2) = -4. The formula for the compound is $K_4[Fe(CN)_6]$.

c) The charge on the anion is 4(-1)+2(-2)+(+6) = -2. The formula for the compound is $Na_2[Os(OH)_4O_2]$.

d) The charge on the cation is 5(0) + (-1) + (+3) = +2. The formula for the compound is $[Fe(H_2O)_5SCN]Cl_2$.

e) The charge on the anion is 6(-1) + (+3) = -3. The formula for the compound is $K_3[IrCl_6]$.

23-30. a) The oxidation state, X, of cobalt is given by X + 2(-1) + 2(0) = +1 or X = +3. The name of the complex ion is trans-dichlorobis (ethylenediamine) cobalt(III)

b) The oxidation state, X, of ruthenium is given by X + 1(0) + 4(0) = +3 or X = +3. The name of the complex ion is tetraammineethylenediamineruthenium(III).

c) The oxidation state, X, of molybdenum is given by X + 3(-2) = -3 or X = +3. The name of the complex ion is tris(oxalato)molybdate(III).

d) The oxidation state, X, of iron is given by X + 2(0) + 2(0) = +3 or X = +3. The name of the complex ion is diaquabis(ethylenediamine)iron(III).

23-32. a) The complex cation has one OH^- ligand, one Cl^- ligand, two $H_2NCH_2CH_2NH_2$ ligands, and one cobalt atom in an oxidation state +3. The net charge on the complex cation is (-1)+(-1)+2(0)+(+3) = +1. The formula for the complex cation is $[CoCl(OH)(H_2NCH_2CH_2NH_2)_2]^+$ or $[CoCl(OH)(en)_2]^+$.

b) The complex cation has two $H_2NCH_2CH_2NH_2$ ligands, one $C_2O_4^{2-}$ ligand and one cadmium atom in an oxidation state +2. The net charge on the complex is 2(0)+(-2)+(+2) = 0. The formula for the complex is $[Cd(H_2NCH_2CH_2NH_2)_2(C_2O_4)]$ or $[Cd(en)_2(ox)]$.

c) The complex anion has two NO_2^- ligands, two $C_2O_4^{2-}$ ligands, and one platinum atom in an oxidation state +4. The net charge on the complex anion is 2(-1)+2(-2)+(+4) = -2. The formula for the complex anion is $[Pt(NO_2)_2(C_2O_4)_2]^{2-}$ or $[Pt(NO_2)_2(ox)_2]^{2-}$.

d) The complex cation has two $H_2NCH_2CH_2NH_2$ ligands, one $C_2O_4^{2-}$ ligand, and one vanadium atom in an oxidation state +3. The net charge on the complex cation is 2(0)+(-2)+(+3) = +1. The formula for the complex cation is $[V(H_2NCH_2CH_2NH_2)_2(C_2O_4)]^+$ or $[V(en)_2(ox)]^+$.

23-34. The structure of the complex is square planar. The possible arrangements around the central platinum atom are

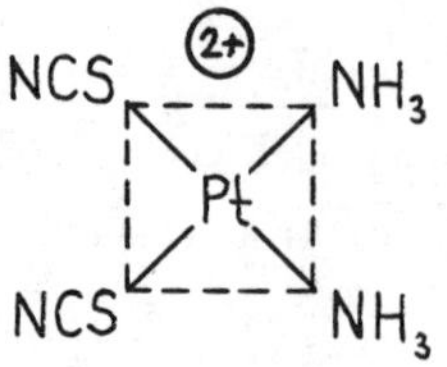

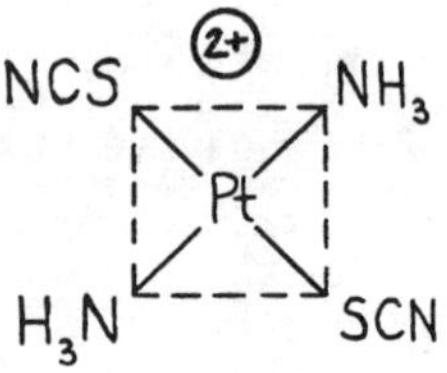

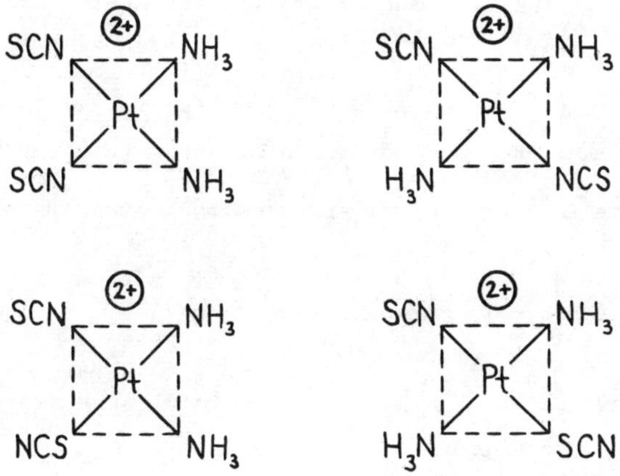

23-36. a) The structure of the complex is octahedral. The possible arrangements around the central paladium ion are

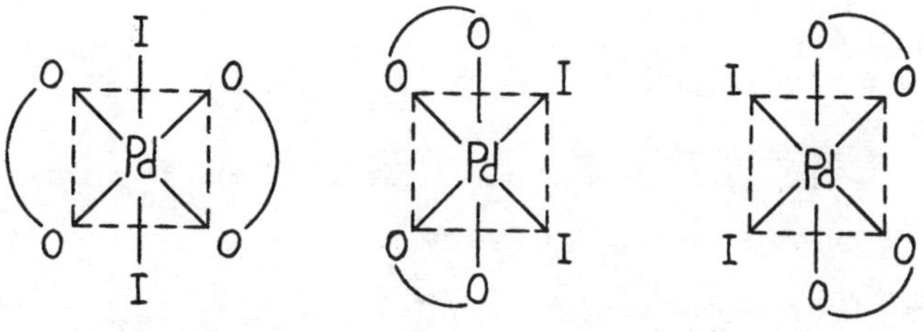

b) The structure of the complex is octahedral. The possible arrangements around the central platinum ion are

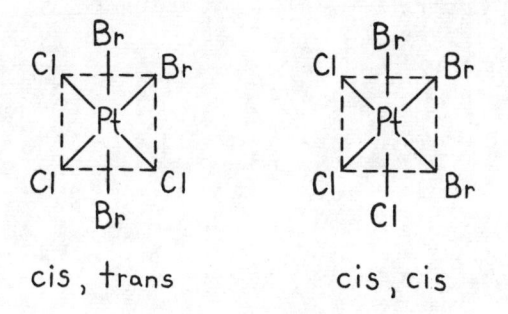

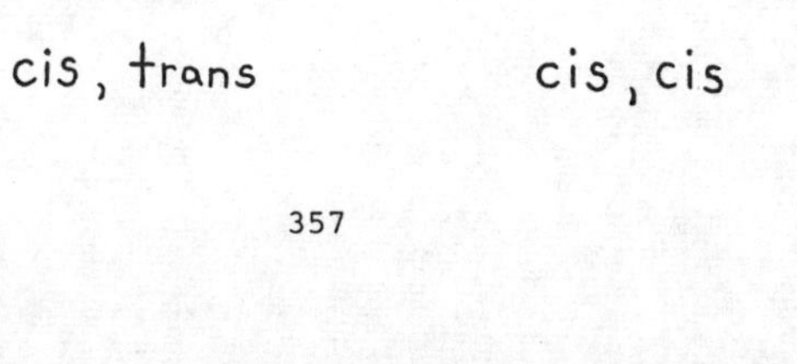

cis, trans cis, cis

23-38. a) The complex is octahedral. There are two isomers, cis and trans.

b) The complex is octahedral. There is only one form because all the vertices on an octahedral are equivalent.

c) There is only one form with bonding through the nitrogen atom of NCS^-.

d) There are two isomers, cis and trans.

23-40. a) Cobalt(II) is a d^7 ion. The d-orbital electron configuration of a high-spin Co^{2+} complex is

$$\begin{array}{ccc} \uparrow & \uparrow & \\ \overline{x^2-y^2 \quad z^2} & & e_g^2 \\[6pt] \uparrow\downarrow \quad \uparrow\downarrow \quad \uparrow & & t_{2g}^5 \\ \overline{xy \quad xz \quad yz} & & \end{array}$$

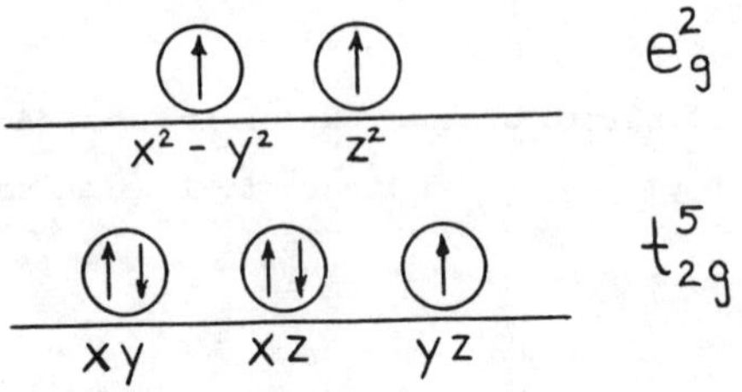

or $t_{2g}^5 e_g^2$.

b) Manganese(II) is a d^5 ion. The d-orbital electron configuration of a high-spin Mn^{2+} complex is

$$\begin{array}{ccc} \uparrow & \uparrow & \\ \overline{x^2-y^2 \quad z^2} & & e_g^2 \\[6pt] \uparrow \quad \uparrow \quad \uparrow & & t_{2g}^3 \\ \overline{xy \quad xz \quad yz} & & \end{array}$$

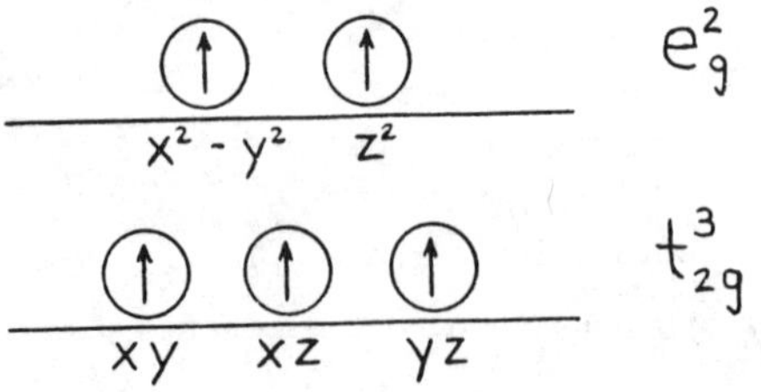

or $t_{2g}^3 e_g^2$.

c) Iron(III) is a d^5 ion. The d-orbital electron configuration of a low-spin Fe^{3+} complex is

$$\begin{array}{ccc} \bigcirc & \bigcirc & \\ x^2-y^2 & z^2 & \end{array} \quad e_g^0$$

$$\begin{array}{ccc} (\uparrow\downarrow) & (\uparrow\downarrow) & (\uparrow) \\ xy & xz & yz \end{array} \quad t_{2g}^5$$

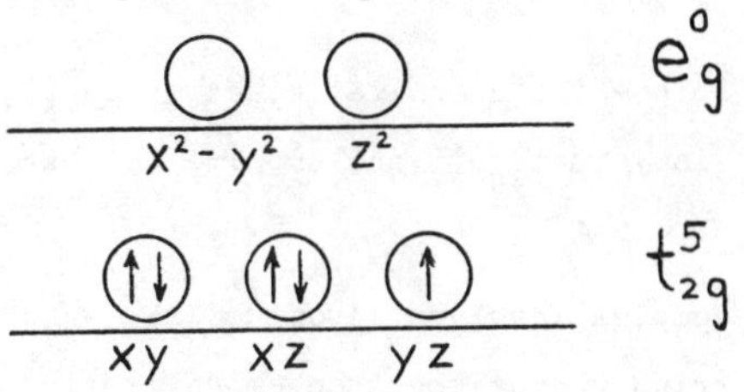

or $t_{2g}^5 e_g^0$.

d) Titanium(IV) is a d^0 ion. The d-orbital electron configuration of Ti^{4+} is

$$\begin{array}{ccc} \bigcirc & \bigcirc & \\ x^2-y^2 & z^2 & \end{array} \quad e_g^0$$

$$\begin{array}{ccc} \bigcirc & \bigcirc & \bigcirc \\ xy & xz & yz \end{array} \quad t_{2g}^0$$

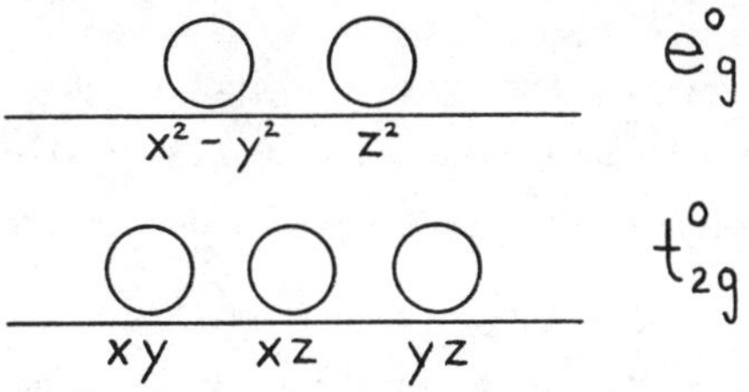

or $t_{2g}^0 e_g^0$.

e) Nickel(II) is a d^8 ion. The d-orbital electron configuration of Ni^{2+} is

$$\begin{array}{ccc} (\uparrow) & (\uparrow) & \\ x^2-y^2 & z^2 & \end{array} \quad e_g^2$$

$$\begin{array}{ccc} (\uparrow\downarrow) & (\uparrow\downarrow) & (\uparrow\downarrow) \\ xy & xz & yz \end{array} \quad t_{2g}^6$$

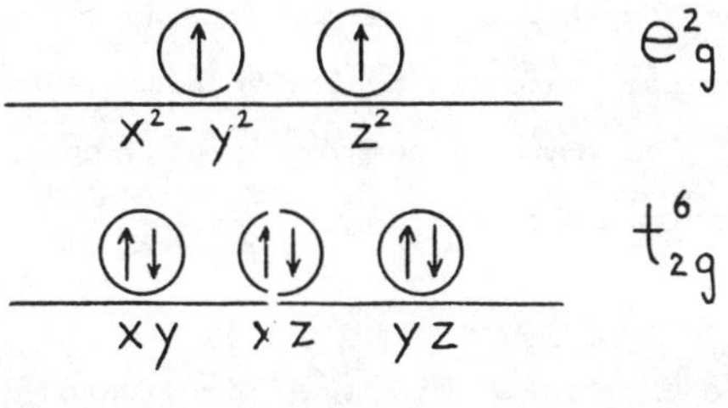

or $t_{2g}^6 e_g^2$.

23-42. a) Low spin - the d-orbital electron configuration of Mn(III), a d^4 ion, is $t_{2g}^4 e_g^0$ (two unpaired electrons).

b) Low spin - the d-orbital electron configuration of Rh(III), a d^6 ion, is $t_{2g}^6 e_g^0$ (no unpaired electrons).

c) High spin - the d-orbital electron configuration of Co(II), a d^7 ion, is $t_{2g}^5 e_g^2$ (three unpaired electrons).

d) High spin - the d-orbital electron configuration of Ir(II), a d^7 ion, is $t_{2g}^5e_g^2$ (three unpaired electrons).

e) Low spin - the d-orbital electron configuration of Ru(III), a d^5 ion, is $t_{2g}^5e_g^0$ (one unpaired electron).

23-44. a) The complex involves chromium(II), which is a d^4 ion. Referring to the spectrochemical series, we see that NO_2^- produces a relatively large Δ_0 value. Therefore, we predict that the complex is low-spin, with a $t_{2g}^4e_g^0$ d-electron configuration.

b) The complex involves cobalt(III), which is a d^6 ion. Because F^- produces a relatively small Δ_0 value, we predict that the complex is high-spin, with a d-electron configuration $t_{2g}^4e_g^2$.

c) The complex involves rhodium(III), which is a d^6 ion. Because CN^- produces a relatively large Δ_0 value, we predict that the complex is low-spin, with a d-electron configuration $t_{2g}^6e_g^0$.

d) The complex involves manganese(II), which is a d^5 ion. Because there are no low-spin tetrahedral complexes, we predict that the complex is high-spin, with a d-electron configuration $e_g^2t_2^3$.

23-46. a) The CN^- ligand is a low-spin ligand. The d-electron configuration of low-spin iron(III), a d^5 ion, is $t_{2g}^5e_g^0$. Thus, we predict that $[Fe(CN)_6]^{3-}$ is an inert complex.

b) The d-electron configuration of chromium(III), a d^3 ion, is $t_{2g}^3e_g^0$. We predict that $[CrF_6]^{3-}$ is an inert complex.

c) The d-electron configuration of vanadium(III), a d^2 ion, is $t_{2g}^2e_g^0$. We predict that $[V(NH_3)_6]^{3+}$ is a labile complex.

d) The d-electron configuration of copper(II), a d^9 ion, is $t_{2g}^6e_g^3$. We predict that $[Cu(en)_3]^{2+}$ is a labile complex.

23-48. a) The d-electron configuration of manganese(IV), a d^3 ion, is $t_{2g}^3 e_g^0$. We predict that $[MnCl_6]^{2-}$ is an inert complex.

b) The d-electron configuration of titanium(III), a d^1 ion, is $t_{2g}^1 e_g^0$. We predict that $[Ti(H_2O)_6]^{3+}$ is a labile complex.

c) The d-electron configuration of high-spin cobalt(II), a d^7 ion, is $t_{2g}^5 e_g^2$. We predict that $[Co(H_2O)_6]^{2+}$ is a labile complex.

d) The d-electron configuration of zirconium(II), a d^2 ion, is $t_{2g}^2 e_g^0$. We predict that $[Zr(H_2O)_6]^{2+}$ is a labile complex.

23-50. a) Vanadium(III) is a d^2 ion. The d-electron configuration is

$$\begin{array}{ccc} \bigcirc & \bigcirc & \\ x^2-y^2 & z^2 & \\ \uparrow & \uparrow & \bigcirc \\ xy & xz & yz \end{array}$$

We predict that there are two unpaired electrons in $[V(H_2O)_6]^{3+}$.

b) Cobalt(II) is a d^7 ion. The d-electron configuration is

$$\begin{array}{ccc} \uparrow & \uparrow & \uparrow \\ xy & xz & yz \\ \uparrow\downarrow & \uparrow\downarrow & \\ x^2-y^2 & z^2 & \end{array}$$

We predict that there are three unpaired electrons in $[CoF_4]^{2-}$.

c) Chromium(II) is a d^4 ion. The CO ligand is a high spin ligand, and the d-electron configuration is

$$\begin{array}{ccc} \bigcirc & \bigcirc & \\ x^2-y^2 & z^2 & \\ \uparrow\downarrow & \uparrow & \uparrow \\ xy & xz & yz \end{array}$$

We predict that there are two unpaired electrons in $[Cr(CO)_6]^{2+}$.

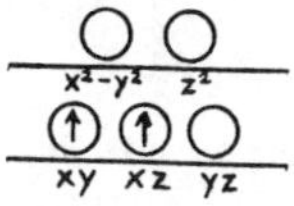

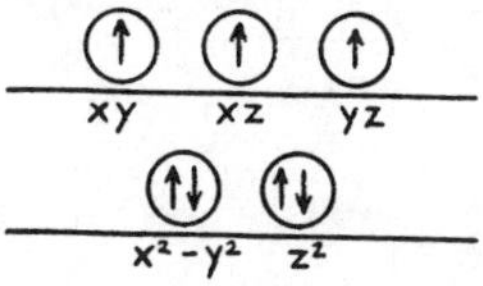

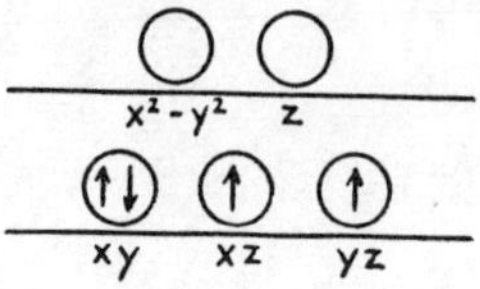

d) Palladium(II) is a d^8 ion. The d-electron configuration is

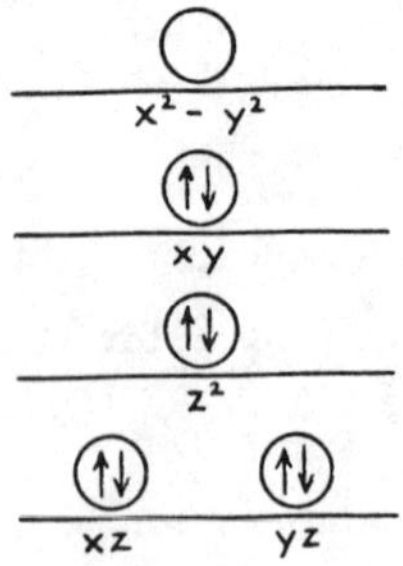

We predict that there are no unpaired electrons in $[PdCl_4]^{2-}$.

23-52. a) Copper(II) is a d^9 ion. The d-electron configuration is $t_{2g}^6 e_g^3$. There is one unpaired electron; thus $[Cu(NH_3)_6]^{2+}$ is paramagnetic.

b) Chromium(III) is a d^3 ion. The d-electron configuration is $t_{2g}^3 e_g^0$. There are three unpaired electrons, thus $[CrF_6]^{3-}$ is paramagnetic.

c) Cobalt(II) is a d^7 ion. The d-electron configuration is

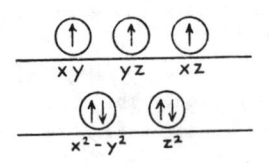

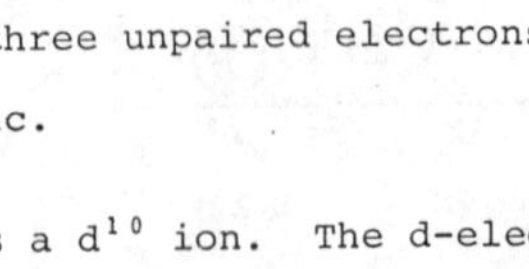

There are three unpaired electrons, thus $[CoCl_4]^{2-}$ is paramagnetic.

d) Zinc(II) is a d^{10} ion. The d-electron configuration is $t_{2g}^6 e_g^4$. There are no unpaired electrons, thus $[Zn(H_2O)_6]^{2+}$ is diamagnetic.

23-54. Iron(II) is a d^6 ion. Both complexes have an octahedral structure because each involves six ligands. The low-spin and high-spin d-electron configurations are

$$\begin{array}{cc} \underline{\bigcirc \quad \bigcirc} & \underline{\uparrow \quad \uparrow} \\ x^2-y^2 \quad z^2 & x^2-y^2 \quad z^2 \\ \underline{\uparrow\downarrow \; \uparrow\downarrow \; \uparrow\downarrow} & \underline{\uparrow\downarrow \; \uparrow \; \uparrow} \\ xy \quad xz \quad yz & xy \quad xz \quad yz \\ \text{low spin} & \text{high spin} \\ \text{diamagnetic} & \text{paramagnetic} \end{array}$$

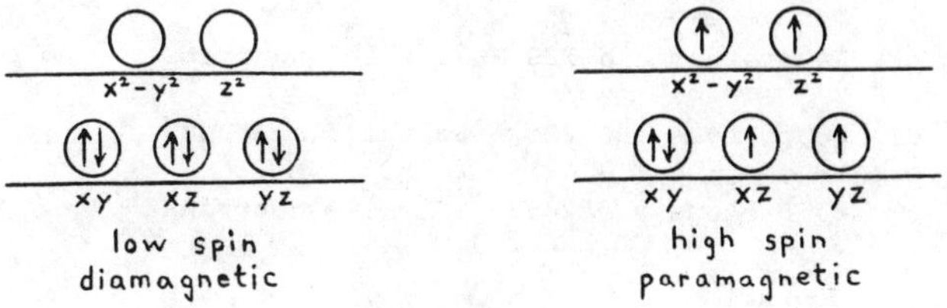

There are no unpaired electrons in the low-spin configuration, thus a complex in this configuration is diamagnetic. There are four unpaired electrons in the high-spin configuration, thus a complex in this configuration is paramagnetic. We see that the complex, $[Fe(H_2O)_6]^{2+}$ is a high-spin complex, whereas the complex $[Fe(CN)_6]^{4-}$ is a low-spin complex.

CHAPTER 24

SOLUTIONS TO THE EVEN-NUMBERED PROBLEMS

24-2. a) 34 protons and 82 - 34 = 48 neutrons

b) 30 protons and 70 - 30 = 40 neutrons

c) 74 protons and 180 - 74 = 106 neutrons

d) 88 protons and 223 - 88 = 135 neutrons

e) 97 protons and 245 - 97 = 148 neutrons

24-4. a) $^{43}_{19}K \longrightarrow ^{0}_{-1}e + ^{43}_{20}Ca$

b) $^{37}_{18}Ar + ^{0}_{-1}e \longrightarrow ^{37}_{17}Cl$

c) $^{208}_{87}Fr \longrightarrow ^{4}_{2}He + ^{204}_{85}At$

d) $^{83}_{38}Sr \longrightarrow ^{0}_{+1}e + ^{83}_{37}Rb$

24-6. a) $^{15}_{7}N + ^{1}_{1}H \longrightarrow ^{12}_{6}C + ^{4}_{2}He$

b) $^{3}_{2}He + ^{4}_{2}He \longrightarrow ^{7}_{4}Be + \gamma$

c) $^{9}_{4}Be + ^{2}_{1}H \longrightarrow ^{10}_{5}B + ^{1}_{0}n$

d) $^{81}_{37}Rb \longrightarrow ^{81}_{36}Kr + ^{0}_{+1}e$

e) $^{238}_{92}U + ^{16}_{8}O \longrightarrow ^{249}_{100}Fm + 5^{1}_{0}n$

24-8. a) $^{16}_{8}O + ^{1}_{1}H \longrightarrow ^{17}_{9}F + \gamma$

b) $^{20}_{10}Ne + ^{1}_{1}H \longrightarrow ^{21}_{11}Na + \gamma$

c) $^{17}_{8}O + ^{1}_{1}H \longrightarrow ^{14}_{7}N + ^{4}_{2}He$

d) $^{22}_{11}Na + ^{1}_{1}H \longrightarrow ^{19}_{10}Ne + ^{4}_{2}He$

e) $^{18}_{8}O + ^{4}_{2}He \longrightarrow ^{21}_{10}Ne + ^{1}_{0}n$

24-10. $^{27}_{13}Al + ^{4}_{2}He \longrightarrow ^{30}_{15}P + ^{1}_{0}n$

Phosphorus-30 decays by positron emission.

24-12. The proton to neutron ratio in potassium-37 is 19/18 = 1.13. The nucleus lies above the band of stability. We predict that potassium-42 decays by positron emission or electron capture.

24-14. The proton to neutron ratio in promethium-140 is 61/79 = 0.77. The nucleus lies above the band of stability, and so we predict that promethium-140 decays by positron emission or electron capture.

24-16. a) All nuclei with $Z \geqslant 84$ decay by α-emission.

b) The proton to neutron ratio in carbon-11 is 6/5 = 1.2. The nucleus lies above the band of stability. We predict that carbon-11 decays by positron emission.

c) The proton to neutron ratio in nitrogen-18 is 7/11 = 0.64. The nucleus lies below the band of stability. We predict that nitrogen-18 decays by β-emission.

d) The proton to neutron ratio in nitrogen-12 is 7/5 = 1.42. The nucleus lies above the band of stability. We predict that nitrogen-12 decays by positron emission.

e) The proton to neutron ratio in tritium is 1/3 = 0.33. The nucleus lies below the band of stability. We predict that hydrogen-3 (tritium) decays by β-emission.

24-18. a) Argon-36 has an even number of neutrons and protons and its proton to neutron ratio is 18/18 = 1.0. We predict that argon-36 is a stable isotope.

b) Potassium-40 has an odd number of neutrons and protons. We predict that potassium-40 is a radioactive isotope.

c) Nickel-58 has an even number of neutrons and protons and a magic number of protons. We predict that nickel-58 is a stable isotope.

d) Boron-8 has an odd number of neutrons and protons. We predict that boron-8 is a radioactive isotope.

e) Neon-20 has an even number of neutrons and protons and its proton to neutron ratio, 10/10, is equal to 1. We predict that neon-20 is stable.

24-20. a) Chlorine-38 has an odd number of neutrons and protons. We predict that chlorine-38 is radioactive.

b) Molybdenum-95 has an odd number of neutrons and an even number of protons; its proton to neutron ratio, 42/53 = 0.79, lies in the band of stability. We predict that molybdenum-95 is stable.

c) Calcium-40 has a magic number of neutrons and protons. We predict that calcium-40 is stable.

d) All isotopes with $Z \geqslant 84$ are radioactive. We predict that francium-225 is radioactive.

24-22. If we use Equation 24-1, then we have

$$\log \frac{N_0}{N} = \frac{0.301\,t}{t_{1/2}} = \frac{(0.301)(6.0\text{ hr})}{15.0\text{ hr}} = 0.1204$$

$$\frac{N_0}{N} = 1.39$$

or

$$\frac{N}{N_0} = 0.758$$

From Example 24-6, we have that

$$\frac{N}{N_0} = \frac{\text{present mass of Na-24}}{\text{initial mass of Na-24}}$$

Thus

$$\text{present mass of Na-24} = (0.758)(0.055\text{ mg})$$
$$= 0.042\text{ mg}$$

24-24. If we use Equation 24-1, then we have

$$\log\frac{N_0}{N} = \frac{0.30\,t}{t_{1/2}} = \frac{(0.301)(1\text{ d})\left(\frac{24\text{ hr}}{1\text{ d}}\right)}{36\text{ hr}}$$

$$= 0.2007$$

$$\frac{N_0}{N} = 1.59$$

or

$$\frac{N}{N_0} = 0.63$$

The fraction remaining is 0.63.

24-26. We have that

$$\log\frac{N_0}{N} = \frac{0.301t}{t_{1/2}}$$

When an isotope decays to 1% of its original value, then

$\frac{N}{N_0} = 0.01.$ If

$$\frac{N}{N_0} = 0.01$$

then

$$\frac{N_0}{N} = 100$$

and we have

$$\log 100 = \frac{0.30\ t}{30.2\text{yr}}$$

Solving for t, we have

$$t = \frac{(\log 100)(30.2\ \text{yr})}{0.301} = 201\ \text{yr}$$

24-28. We have

$$\log \frac{N_0}{N} = \frac{0.30\ t}{t_{1/2}} = \frac{(0.301)(100\ \text{yr})}{12.3\ \text{yr}} = 2.4472$$

$$\frac{N_0}{N} = 280$$

$$\frac{N}{N_0} = 0.00357$$

The fraction remaining is 3.57×10^{-3}.

24-30. Recall that the rate of radioactive decay is proportional to the number of nuclei present

$$\text{rate} \propto N$$

The ratio of N_0 to N is therefore equal to the ratio of the rates

$$\frac{N_0}{N} = \frac{\text{rate}_0}{\text{rate}}$$

Thus the equation

$$\log \frac{N_0}{N} = \frac{0.301t}{t_{1/2}}$$

becomes

$$\log\left(\frac{\text{rate}_0}{\text{rate}}\right) = \frac{0.30\ t}{t_{1/2}}$$

Substituting in the values of rate_0, rate, and t, we have

$$\log\left(\frac{12{,}000\ \text{DPM}}{1710\ \text{DPM}}\right) = \frac{(0.301)(8.0\ \text{hr})}{t_{1/2}}$$

Solving for $t_{1/2}$, we have

$$t_{1/2} = \frac{(0.301)(8.0\ \text{hr})}{0.8462} = 2.8\ \text{hr}$$

24-32. We have that

$$\log \frac{N_0}{N} = \frac{0.301t}{t_{1/2}} = \frac{(0.301)(6\ \text{hr})\left(\frac{60\ \text{min}}{\text{hr}}\right)}{25.0\ \text{min}}$$

$$= 4.3344$$

$$\frac{N_0}{N} = 21{,}600$$

or

$$\frac{N}{N_0} = 4.63 \times 10^{-5}$$

Recall that the rate of decay is proportional to the number of nuclei present

$$\text{rate} \propto N$$

thus

$$\frac{N}{N_0} = \frac{\text{rate}}{\text{rate}_0}$$

$$\text{rate at 2 P.M.} = (4.63 \times 10^{-5})(10{,}000 \text{ disintegrations}\cdot\text{min}^{-1})$$

$$= 0.46 \text{ disintegrations}\cdot\text{min}^{-1}$$

24-34. Using Equation (24-4), we have

$$t = (1.90 \times 10^4 \text{ years}) \log \frac{15.3}{R}$$

$$= (1.90 \times 10^4 \text{ years}) \log \frac{15.3}{6.24}$$

$$= 7.40 \times 10^3 \text{ years}$$

The charcoal is about 7400 years old (~5400 B.C.).

24-36. The rate of decay of living matter is 15.3 disintegrations per minute per gram carbon. The rate of decay in an 80 kg human is

$$(15.3 \text{ disintegrations}\cdot\text{min}^{-1}\cdot\text{g}^{-1})(15\text{x}10^3\text{g}) = 2.3 \times 10^5 \text{disintegrations}\cdot\text{min}^{-1}$$

The number of disintegrations per sec is

$$(2.3\text{x}10^5 \text{disintegrations}\cdot\text{min}^{-1})\left(\frac{1 \text{ min}}{60 \text{ s}}\right) = 3.8 \times 10^3 \text{ disintegrations}\cdot\text{s}^{-1}$$

There are about 3800 disintegrations of carbon-14 per second in the human body.

24-38. The ratio of carbon-14 contents is equal to the ratio of the rates of decay

$$\frac{\text{C-14 content}_{\text{living}}}{\text{C-14 content}_{\text{oak}}} = \frac{(\text{rate of decay})_{\text{living}}}{(\text{rate of decay})_{\text{oak}}} =$$

$$\frac{15.3 \text{ disintegrations per min per gram C}}{R}$$

$$\frac{\text{C-14 content}_{\text{oak}}}{\text{C-14 content}_{\text{living}}} = 0.65$$

thus

$$\frac{15.3}{R} = \frac{1}{0.65}$$

Using Equation (24-4), we have

$$t = (1.90 \times 10^4 \text{ years}) \log \frac{15.3}{R}$$

$$= (1.90 \times 10^4 \text{ years}) \log \frac{1}{0.65}$$

$$= 3.55 \times 10^3 \text{ years} = 3600 \text{ years old} \quad (1600 \text{ B.C.})$$

24-40. The age of the sample is given by

$$t = \frac{t_{\frac{1}{2}}}{0.301\,t} \log \frac{N_0}{N}$$

where

$$\frac{N_0}{N} = \frac{\text{initial mass of } {}^{238}\text{U}}{\text{present mass of } {}^{238}\text{U}}$$

The initial mass of ^{238}U is the sum of the present mass and the mass that has decayed. The mass of uranium that has decayed to lead is

$$\text{mass of } {}^{238}\text{U that decayed} = \left(\frac{238}{206}\right) \text{mass of } {}^{206}\text{Pb}$$

$$= \left(\frac{238}{206}\right)(0.460 \text{ mg}) = 0.531 \text{ mg}$$

Therefore

$$\text{initial mass of } {}^{238}\text{U} = 1.50 \text{ mg} + 0.531 \text{ mg} = 2.03 \text{ mg}$$

Thus

$$\frac{N_0}{N} = \frac{2.03 \text{ mg}}{1.50 \text{ mg}} = 1.353$$

The age of the sample is

$$t = \frac{4.51 \times 10^9 \text{ yr}}{0.301} \log 1.353$$

$$= 1.97 \times 10^9 \text{ yr}$$

24-42. The age of the rock is given by

$$t = \frac{t_{\frac{1}{2}}}{0.301} \log \frac{N_0}{N}$$

where

$$\frac{N_0}{N} = \frac{\text{initial mass of } ^{87}\text{Rb}}{\text{present mass of } ^{87}\text{Rb}}$$

The initial mass of ^{87}Rb is the sum of the present mass and the mass that has decayed

$$\frac{N_0}{N} = \frac{\text{mass of } ^{87}\text{Rb that has decayed + present mass of } ^{87}\text{Rb}}{\text{present mass of } ^{87}\text{Rb}}$$

The mass of ^{87}Rb that has decayed is equal to the mass of ^{87}Sr because the strontium-87 comes from decay of rubidium-87. Thus

$$\text{mass ratio } ^{87}\text{Sr}/^{87}\text{Rb} = \frac{\text{mass of } ^{87}\text{Sr}}{\text{present mass of } ^{87}\text{Rb}} = \frac{\text{mass of } ^{87}\text{Rb that decayed}}{\text{present mass of } ^{87}\text{Rb}}$$

$$= 0.056$$

Therefore

$$\frac{N_0}{N} = 0.056 + 1 = 1.056$$

The age of the rocks is

$$t = \frac{4.9 \times 10^{10}\ \text{yr}}{0.301} \log 1.056$$

$$= 3.9 \times 10^{9}\ \text{yr}$$

24-44. The age of the rock is given by

$$t = \frac{t_{1/2}}{0.301} \log \frac{N_0}{N}$$

$$\frac{N_0}{N} = \frac{\text{initial mass of } {}^{40}\text{K}}{\text{present mass of } {}^{40}\text{K}}$$

The initial mass of ^{40}K is the sum of the present mass and the mass that has decayed

$$\frac{N_0}{N} = \frac{\text{mass of } {}^{40}\text{K that has decayed + present mass of } {}^{40}\text{K}}{\text{present mass of } {}^{40}\text{K}}$$

$$= \frac{\text{mass of } {}^{40}\text{K that has decayed}}{\text{present mass of } {}^{40}\text{K}} + 1$$

We shall use the measured mass ratio ^{40}Ar/^{40}K to determine the ratio of mass of ^{40}K that has decayed to the present mass. Only 0.107 (10.7%) of the ^{40}K decays to ^{40}Ar. The mass of ^{40}Ar produced from the decay of ^{40}K is

$$\text{mass } {}^{40}\text{Ar} = (0.107)\left(\frac{40}{40}\right)(\text{mass of } {}^{40}\text{K that decayed})$$

$$\text{the mass ratio} = \frac{{}^{40}\text{Ar}}{{}^{40}\text{K}} = \frac{(0.107)\left(\frac{40}{40}\right)(\text{mass of } {}^{40}\text{K that decayed})}{\text{present mass of } {}^{40}\text{K}}$$

Thus

$$\frac{\text{mass of } {}^{40}\text{K that decayed}}{\text{present mass of } {}^{40}\text{K}} = \frac{6.0 \times 10^{-5}}{(0.107)(1)} = 5.61 \times 10^{-4}$$

and

$$\frac{N_0}{N} = 5.61 \times 10^{-4} + 1 = 1.00056$$

The age of the rock is

$$t = \frac{1.3 \times 10^9 \text{ yr}}{0.301} \log(1.00056)$$

$$= 1.1 \times 10^6 \text{ yr}$$

24-46. We have the following masses

$${}^{1}_{1}H \qquad 1.0078 \text{ amu}$$
$${}^{1}_{0}n \qquad 1.0087 \text{ amu}$$

The mass difference between ${}^{56}_{26}Fe$ and its constituent particles is

$$\Delta m = (26 \times 1.0078 \text{ amu}) + (30 \times 1.0087 \text{ amu}) - 55.9346 \text{ amu}$$

$$= 0.5292 \text{ amu}$$

which correspond to an energy of

$$\Delta E = c^2 \Delta m = (9.00 \times 10^{16} m^2 \cdot s^{-2})(0.5292 \text{ amu})(1.66 \times 10^{-27} kg \cdot amu^{-1})$$

$$= 7.91 \times 10^{-11} \text{ J}$$

The binding energy per nucleon is

$$\frac{7.91 \times 10^{-11} \text{ J}}{56 \text{ nucleons}} = 1.41 \times 10^{-12} \text{ J} \cdot \text{nucleon}^{-1}$$

24-48. We have the masses

$${}^{1}_{1}H \qquad 1.0078 \text{ amu}$$
$${}^{1}_{0}n \qquad 1.0087 \text{ amu}$$

The mass difference between ${}^{20}_{10}Ne$ and its constituent particles is

$$\Delta m = (10 \times 1.0078\ \text{amu}) + (10 \times 1.0087\ \text{amu}) - 19.9924\ \text{amu}$$

$$= 0.1726\ \text{amu}$$

which corresponds to an energy of

$$\Delta E = c^2\Delta m = (9.00 \times 10^{16}\text{m}^2\cdot\text{s}^{-2})(0.1726\ \text{amu})(1.66 \times 10^{-27}\text{kg}\cdot\text{amu}^{-1})$$

$$= 2.58 \times 10^{-11}\ \text{J}$$

The binding energy per nucleon is

$$\frac{2.58 \times 10^{-11}\ \text{J}}{20\ \text{nucleons}} = 1.29 \times 10^{-12}\ \text{J}\cdot\text{nucleon}^{-1}$$

24-50. The difference in masses between the products and reactants is

$$\Delta m = (2 \times 4.0026\ \text{amu}) - 1.0078\ \text{amu} - 7.0160\ \text{amu}$$

$$= -0.0186\ \text{amu}$$

The energy released by the loss of this mass is

$$\Delta E = c^2\Delta m = (9.00 \times 10^{16}\text{m}^2\cdot\text{s}^{-2})(0.0186\ \text{amu})(1.66 \times 10^{-27}\text{kg}\cdot\text{amu}^{-1})$$

$$= 2.78 \times 10^{-12}\ \text{J per atom of lithium consumed}$$

The energy released per mole of lithium consumed is

$$\Delta E = (2.78 \times 10^{-12}\text{J}\cdot\text{atom}^{-1})(6.022 \times 10^{23}\text{atom}\cdot\text{mol}^{-1})$$

$$= 1.67 \times 10^{12}\text{J}\cdot\text{mol}^{-1}$$

The energy released when 1.0 gram of lithium is consumed is

$$E = (1.67 \times 10^{12}\text{J}\cdot\text{mol}^{-1})\left(\frac{1\ \text{mol}}{7.0160\ \text{g}}\right) = 2.386 \times 10^{11}\text{J}\cdot\text{g}^{-1}$$

The number of moles of octane needed to produce this amount of energy is

$$\text{moles of octane} = \frac{2.386 \times 10^{11}\ \text{J}}{5.45 \times 10^{6}\ \text{J}\cdot\text{mol}^{-1}} = 4.378 \times 10^{4}\ \text{mol}$$

$$\text{mass of octane} = (4.378 \times 10^{4}\ \text{mol})\left(\frac{114.22\ \text{g}}{1\ \text{mol}}\right)$$

$$= 5.00 \times 10^{6}\ \text{g} = 5.00 \times 10^{3}\ \text{kg}$$

24-52. The difference in mass between products and reactants is

$$\Delta m = 140.9137\ \text{amu} + 87.9142\ \text{amu} + (7 \times 1.0087\ \text{amu}) - 235.0439\ \text{amu} - 1.0087\ \text{amu}$$

$$= -0.1638\ \text{amu}$$

which correspond to an energy of

$$\Delta E = c^2\Delta m = (9.00 \times 10^{16}\text{m}^2\cdot\text{s}^{-2})(0.1638\ \text{amu})(1.66 \times 10^{-27}\text{kg}\cdot\text{amu}^{-1})$$

$$= 2.447 \times 10^{-11}\text{J}\cdot\text{atom}^{-1} \text{ of } {}^{235}\text{U}$$

The energy released per gram of ${}^{235}\text{U}$ is

$$E = (2.447 \times 10^{-11}\text{J}\cdot\text{atom}^{-1})(6.022 \times 10^{23}\text{atom}\cdot\text{mol}^{-1})\left(\frac{1\ \text{mol}}{235.0439\ \text{g}}\right)$$

$$= 6.27 \times 10^{10}\text{J}\cdot\text{g}^{-1}$$

The energy released in a 50 kiloton bomb is

$$E = (50\ \text{kton})\left(\frac{10^{3}\text{ton}}{1\ \text{kton}}\right)\left(\frac{10^{3}\text{kg}}{1\ \text{ton}}\right)(2.5 \times 10^{6}\ \text{J}\cdot\text{kg}^{-1})$$

$$= 1.25 \times 10^{14}\ \text{J}$$

$$\text{mass of } {}^{235}\text{U consumed} = \frac{1.25 \times 10^{14}\ \text{J}}{6.27 \times 10^{10}\ \text{J}\cdot\text{g}^{-1}} = 1.99 \times 10^{3}\ \text{g}$$

$$= 1.99\ \text{kg}$$

24-54. The mass of an electron and of a positron is 5.4858×10^{-4} amu. The loss of mass is

$$\Delta m = 2 \times 5.4858 \times 10^{-4} \text{ amu} = 1.0972 \times 10^{-3} \text{ amu}$$

The energy produced by the reaction is

$$\Delta E = c^2\Delta m = (9.00 \times 10^{16} m^2\cdot s^{-2})(1.0972 \times 10^{-3} \text{ amu})(1.66 \times 10^{-27} kg\cdot amu^{-1})$$

$$= 1.64 \times 10^{-13} \text{ J}$$

The energy of each gamma ray is

$$E = \left(\frac{1}{2}\right)(1.64 \times 10^{-13} \text{ J}) = 8.196 \times 10^{-14}$$

Recall that

$$E = h\nu$$

and so the frequency of each gamma ray is

$$\nu = \frac{E}{h} = \frac{8.196 \times 10^{-14}}{6.626 \times 10^{-34} J\cdot s} = 1.24 \times 10^{20} \text{ s}^{-1}$$

24-56. The energy released per mole of sulfur that reacts is

$$E = 2.97 \times 10^5 \text{ J}\cdot\text{mol}^{-1}$$

The mass loss that corresponds to this energy is

$$\Delta m = \frac{E}{c^2} = \frac{2.97 \times 10^5 \text{ J}\cdot\text{mol}^{-1}}{9.00 \times 10^{16} \text{ m}^2\cdot\text{s}^{-2}}$$

$$= 3.30 \times 10^{-12} \text{kg per mole of sulfur that reacts}$$

24-58. The number of moles of uranium-235 required to produce 3×10^{17}kJ is

$$\text{moles of U-235} = \frac{3 \times 10^{20} \text{ J}}{2 \times 10^{13} \text{ J}\cdot\text{mol}^{-1}} = 1.5 \times 10^7 \text{ mol}$$

$$\text{mass of U-235} = (1.5 \times 10^7 \text{ mol})\left(\frac{235 \text{ g}}{1 \text{ mol}}\right)\left(\frac{1 \text{ kg}}{1000 \text{ g}}\right)\left(\frac{1 \text{ ton}}{1000 \text{ kg}}\right)$$

$$= 3.5 \times 10^3 \text{ metric tons}$$

The mass of naturally-occurring uranium needed to produce 3.5×10^3 metric tons of U-235 is

$$\text{mass of uranium} = \frac{3.5 \times 10^3 \text{ metric tons}}{0.007} = 5 \times 10^5 \text{ metric tons}$$

The time that the world supply will last is

$$\text{time} = \frac{10^6 \text{ metric tons}}{5 \times 10^5 \text{ metric tons}\cdot\text{yr}^{-1}} = 2 \text{ yr}$$

24-60. The mass of U-235 in 1 kg of uranium is

$$\text{mass of U-235} = (0.03)(1 \text{ kg}) = 30 \text{ g}$$

The mass of U-235 that can be used is

$$\text{mass of U-235 that reacts} = \left(\frac{1}{3}\right)(30 \text{ g}) = 10 \text{ g}$$

The number of U-235 atoms in 10 g is

$$\text{number of U-235} = (10 \text{ g})\left(\frac{1 \text{ mol}}{235 \text{ g}}\right)(6.022 \times 10^{23} \text{ atom}\cdot\text{mol}^{-1})$$

$$= 2.56 \times 10^{22} \text{ atoms}$$

The energy released by the fission of 2.56×10^{22} atoms of U-235 is

$$E = (2.9 \times 10^{-14} \text{ kJ}\cdot\text{atom}^{-1})(2.56 \times 10^{22} \text{ atom})$$

$$= 7.4 \times 10^8 \text{ kJ} = 7.4 \times 10^{11} \text{ J}$$

The available energy per kilogram of naturally occurring uranium is

$$E = (0.30)(7.4 \times 10^{11} \text{ J}) = 2.2 \times 10^{11} \text{ J}$$

The energy produced in one year is

$$E = (1000 \text{ megawatt})\left(\frac{10^6\text{watt}}{\text{megawatt}}\right)\left(\frac{1\text{J}\cdot\text{s}^{-1}}{1 \text{ watt}}\right)\left(\frac{60 \text{ s}}{1 \text{ min}}\right)\left(\frac{60 \text{ min}}{1 \text{ hr}}\right)\left(\frac{24 \text{ hr}}{1 \text{ d}}\right)\left(\frac{365 \text{ d}}{1 \text{ yr}}\right)(1 \text{ yr})$$

$$= 3.15 \times 10^{16} \text{ J}$$

The amount of uranium fuel is

$$\text{mass of uranium} = \frac{3.15 \times 10^{16} \text{ J}}{2.2 \times 10^{11}\text{J}\cdot\text{kg}^{-1} \text{ uranium}}$$

$$= 1.4 \times 10^5 \text{ kg}$$

24-62. The specific activity is given by

$$\text{specific activity} = \left(\frac{4.2 \times 10^{23}\text{disintegrations}\cdot\text{g}^{-1}}{Mt_{1/2}}\right)$$

We first must convert the half-life to seconds

$$t_{1/2} = (284 \text{ d})\left(\frac{24 \text{ hr}}{1 \text{ d}}\right)\left(\frac{60 \text{ min}}{1 \text{ hr}}\right)\left(\frac{60 \text{ min}}{1 \text{ min}}\right) = 2.45 \times 10^7 \text{ s}$$

The specific activity of cerium-144 is given by

$$\text{specific activity} = \frac{4.2 \times 10^{23} \text{ disintegrations}\cdot\text{g}^{-1}}{(144)(2.45 \times 10^7 \text{ s})}$$

$$= 1.2 \times 10^{14} \text{ disintegrations}\cdot\text{s}^{-1}\cdot\text{g}^{-1}$$

In terms of curies, we have

$$\text{specific activity} = \frac{1.2 \times 10^{14} \text{ disintegrations}\cdot\text{s}^{-1}\cdot\text{g}^{-1}}{3.7 \times 10^{10} \text{ disintegrations}\cdot\text{s}^{-1}\cdot\text{Ci}^{-1}}$$

$$= 3.2 \times 10^3 \text{ Ci}\cdot\text{g}^{-1}$$

24-64. We first must convert the half-life to seconds

$$t_{1/2} = (7.4 \times 10^5 \text{ yr})\left(\frac{365 \text{ d}}{1 \text{ yr}}\right)\left(\frac{24 \text{ hr}}{1 \text{ d}}\right)\left(\frac{60 \text{ min}}{1 \text{ hr}}\right)\left(\frac{60 \text{ s}}{1 \text{ min}}\right)$$

$$= 2.33\times 10^{13} \text{ s}$$

The specific activity of aluminum-26 is given by

$$\text{specific activity} = \frac{4.2 \times 10^{23} \text{ disintegrations}\cdot\text{g}^{-1}}{(26)(2.33 \times 10^{13} \text{ s})}$$

$$= 6.9 \times 10^{8} \text{ disintegrations}\cdot\text{s}^{-1}\cdot\text{g}^{-1}$$

$$= \frac{\times 10^{8} \text{ disintegrations}\cdot\text{s}^{-1}\cdot\text{g}^{-1}}{3.7 \times 10^{10} \text{ disintegrations}\cdot\text{s}^{-1}\cdot\text{Ci}^{-1}}$$

$$= 0.019 \text{ Ci}\cdot\text{g}^{-1}$$

24-66. We first must convert the half-life to seconds

$$t_{1/2} = (25 \text{ min})\left(\frac{60 \text{ s}}{1 \text{ min}}\right) = 1500 \text{ s}$$

The specific activity of iodine-128 is

$$\text{specific activity} = \frac{4.2 \times 10^{23} \text{ disintegrations}\cdot\text{g}^{-1}}{(128)(1500 \text{ s})}$$

$$= 2.19 \times 10^{18} \text{ disintegrations}\cdot\text{s}^{-1}\cdot\text{g}^{-1}$$

The activity of a 100 μCi dose is

$$\text{activity} = (100\ \mu\text{Ci})\left(\frac{1 \text{ Ci}}{10^{6}\ \mu\text{Ci}}\right)(3.7 \times 10^{10} \text{ disintegrations}\cdot\text{s}^{-1}\cdot\text{Ci}^{-1})$$

$$= 3.7 \times 10^{6} \text{ disintegrations}\cdot\text{s}^{-1}$$

The mass of iodine-128 that produces this activity is

$$\text{mass of I-128} = \frac{3.7 \times 10^{6} \text{ disintegrations}\cdot\text{s}^{-1}}{2.19 \times 10^{18} \text{ disintegrations}\cdot\text{s}^{-1}\cdot\text{g}^{-1}}$$

$$= 1.69 \times 10^{-12} \text{ g}$$

The mass of $Na^{128}I$ is

$$(1.69 \times 10^{-12} \text{g } ^{128}\text{I})\left(\frac{1 \text{ mol}^{128}\text{I}}{128 \text{ g}^{128}\text{I}}\right)\left(\frac{1 \text{ mol Na}^{128}\text{I}}{1 \text{ mol}^{128}\text{I}}\right)\left(\frac{151 \text{ g Na}^{128}\text{I}}{1 \text{ mol Na}^{128}\text{I}}\right)\left(\frac{10^{3}\text{mg}}{1 \text{ g}}\right)$$

$$= 2.0 \times 10^{-9} \text{ mg}$$

24-68. Assuming that the activity due to phosphorus-32 remains constant during the time of the experiment, we write

$$(\text{activity})_{\text{start}} = (\text{activity})_{\text{later}}$$

or

$$50{,}000 \text{ disintegrations}\cdot\text{min}^{-1} = (10.0 \text{ disintegrations}\cdot\text{min}^{-1}\cdot\text{mL}^{-1})(\text{volume of blood})$$

Solving for the volume of blood, we find that

$$\text{volume of blood} = \frac{50{,}000 \text{ disintegrations}\cdot\text{min}^{-1}}{10.0 \text{ disintegrations}\cdot\text{min}^{-1}\cdot\text{mL}^{-1}}$$

$$= 5000 \text{ mL} = 5.0 \text{ L}$$

24-70. The number of grams of barium in the precipitate is

$$\text{mass of Ba}^{2+} = \frac{3270 \text{ disintegrations}\cdot\text{min}^{-1}}{7.6 \times 10^{7} \text{ disintegration}\cdot\text{min}^{-1}\cdot\text{g}^{-1}}$$

$$= 4.30 \times 10^{-5} \text{ g}$$

The number of moles of barium-131 is

$$\text{moles of Ba-131} = (4.30 \times 10^{-5} \text{ g})\left(\frac{1 \text{ mol}}{131 \text{ g}}\right) = 3.28 \times 10^{-7} \text{ mol}$$

The number of moles of $SO_4^{2-}(aq)$ is

$$\text{moles of } SO_4^{2-}(aq) = \text{moles of } BaSO_4(s) = \text{moles of Ba-131}$$

$$= 3.28 \times 10^{-7} \text{ mol}$$

Assuming that essentially all of the sulfate is precipitated, the concentration of sulfate ion is

$$[SO_4^{2-}] = \frac{3.28 \times 10^{-7} \text{ mol}}{0.010 \text{ L}} = 3.3 \times 10^{-5}$$

24-72. The total activity of sulfur-35 is

$$\text{activity} = (14{,}000 \text{ disintegrations}\cdot\text{min}^{-1}\cdot\text{mL}^{-1})(75 \text{ mL})$$

$$= 1.05 \times 10^{6} \text{ disintegrations}\cdot\text{min}^{-1}$$

The total number of moles of $SO_4^{2-}(aq)$ is

$$\text{moles of } SO_4^{2-}(aq) = (0.010\text{ M})(0.075\text{ L}) = 7.50 \times 10^{-4}\text{ mol}$$

The molar activity of sulfur-35 is

$$\text{molar activity} = \frac{1.05 \times 10^{6}\text{ disintegrations}\cdot\text{min}^{-1}}{7.50 \times 10^{-4}\text{ mol}}$$

$$= 1.40 \times 10^{9}\text{ disintegrations}\cdot\text{min}^{-1}\cdot\text{mol}^{-1}$$

The total activity of sulfur-35 after the two solutions are mixed is

$$\text{activity} = (183\text{ disintegrations}\cdot\text{min}^{-1}\cdot\text{mL}^{-1})(150\text{ mL})$$

$$= 2.745 \times 10^{4}\text{ disintegrations}\cdot\text{min}^{-1}$$

The number of moles of $SO_4^{2-}(aq)$ in the solution is

$$\text{moles of } SO_4^{2-}(aq) = \frac{2.745 \times 10^{4}\text{ disintegrations}\cdot\text{min}^{-1}}{1.40 \times 10^{9}\text{ disintegrations}\cdot\text{min}^{-1}\cdot\text{mol}^{-1}}$$

$$= 1.96 \times 10^{-5}\text{ mol}$$

The concentration of sulfate ion is

$$[SO_4^{2-}] = \frac{1.96 \times 10^{-5}\text{ mol}}{0.150\text{ L}} = 1.3 \times 10^{-4}\text{ M}$$

The concentration of $Pb^{2+}(aq)$ is

$$[Pb^{2+}] = 1.31 \times 10^{-4}\text{ M}$$

The solubility product of $PbSO_4$ is given by

$$K_{sp} = [Pb^{2+}][SO_4^{2-}]$$

$$= (1.31 \times 10^{-4}\text{ M})(1.31 \times 10^{-4}\text{ M})$$

$$= 1.7 \times 10^{-8}\text{ M}^2$$

CHAPTER 25

SOLUTIONS TO THE EVEN-NUMBERED PROBLEMS

25-2. a) $C_3H_8(g) + H_2SO_4(aq) \longrightarrow$ N.R.

b) This is the reaction for the combustion of C_5H_{12}

$$C_5H_{12}(g) + 8O_2(g) \longrightarrow 5CO_2(g) + 6H_2O(g)$$

c) $C_2H_6(g) + HCl(aq) \longrightarrow$ N.R.

d) $C_5H_{12}(l) + Cl_2(g) \xrightarrow{UV} C_5H_{11}Cl(l) + HCl(g)$

Because this is a free-radical reaction, dichloropentanes and polychloropentanes are also formed.

25-4. a) The pair is identical. One molecule can be rotated 180^0 to superimpose upon the other molecule.

b) The pair is identical. The chlorine atom is attached to the first carbon atom in each molecule.

c) The pair is different. The pair are structural isomers because the methyl groups are attached to different carbon atoms.

d) The pair is identical. The groups in one molecule can be rotated around a carbon-carbon bond so that it is superimposable on the other molecule.

25-6. The structural formulas for isomers of dichloropentane are

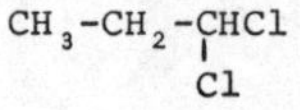

1,1-dichloropropane

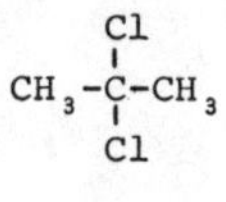

2,2-dichloropropane

$$CH_3\text{-}\underset{\displaystyle Cl}{\underset{|}{C}H}\text{-}CH_2Cl \qquad \text{1,2-dichloropropane}$$

$$ClCH_2\text{-}CH_2\text{-}CH_2Cl \qquad \text{1,3-dichloropropane}$$

25-8. a) The longest consecutive chain of carbon atoms is five

$$^1CH_3\text{-}^2CH_2\text{-}^3\overset{|}{C}H\text{ -}^4CH_2\text{-}^5CH_3$$

and so we shall name this molecule as a derivative of pentane. The IUPAC name is 3-chloropentane.

b) The longest consecutive chain of carbon atoms is four

$$^4CH_3\text{-}^3\overset{|}{C}H\text{ - }^2\underset{|}{\overset{|}{C}}H\text{ -}^1CH_3$$

and so we shall name this molecule as a derivative of butane. The IUPAC name is 2,2-dimethyl-3-nitrobutane.

c) The longest consecutive chain of carbon atoms is four

$$^4CH_3\text{-}^3CH_2\text{- }\overset{|2}{C}H\text{ - }\overset{1}{C}H_3$$

The IUPAC name is 2-chlorobutane.

d) The longest consecutive chain of carbon atoms is four (not three)

$$^1CH_3\text{-}^2\overset{|}{C}H\text{ -}^3CH_2\text{-}^4CH_3$$

The IUPAC name is 2-methylbutane.

25-10. a) The name violates Rule 3. The chloro groups have not been assigned the lowest numbers. The correct IUPAC name is 1,2-dichloropropane.

b) The formula for the compound is

$$CH_3-\underset{\underset{CH_3}{\overset{\displaystyle CH_2}{|}}}{\overset{\displaystyle |}{CH}}-CH_2Br$$

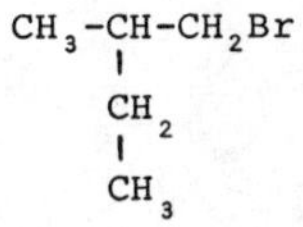

This name violates both Rule 2 and Rule 3. The correct IUPAC name is 1-bromo-2-methylbutane.

c) The formula for the compound is

$$CH_3-\underset{\underset{}{\overset{\displaystyle CH_3}{\overset{\displaystyle |}{}}}}{CH}-CH_2-\underset{\underset{}{\overset{\displaystyle CH_3}{\overset{\displaystyle |}{}}}}{CH_2}$$

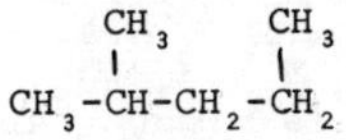

The longest chain has not been used (Rule 2). The correct IUPAC name is 2-methylpentane.

d) One of the methyl groups has not been numbered (Rule 6). The correct IUPAC name is either 2,2-dimethylbutane or 2,3-dimethylbutane.

25-12. The formulas are

$$\underset{\underset{Cl}{|}}{CH_3CHCH_2CH_3} \qquad \underset{\underset{Cl}{|}}{CH_3CH_2CHCH_3}$$

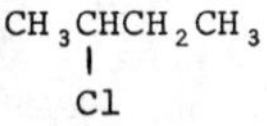

2-chlorobutane 3-chlorobutane

They do not differ.

The formulas are

$$\underset{\underset{Cl}{|}}{CH_3CHCH_2CH_2CH_3} \qquad \underset{\underset{Cl}{|}}{CH_3CH_2CHCH_2CH_3}$$

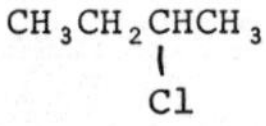

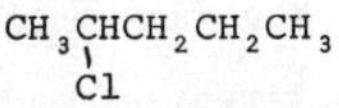

2-chloropentane 3-chloropentane

They are structural isomers and hence they differ. 2-chloropentane and 4-chloropentane are the same compound; 4-chloropentane is an incorrect name.

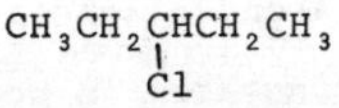

25-14. a) CH₃-CH₂-CH₂-CH₂-CH₂-CH₃   hexane

   b) CH₃-CH-CH₂-CH₂-CH-CH₃   2,5-dimethylhexane
           | |
           CH₃   CH₃

           CH₃
           |
   c) CH₃-C-CH₃   2,2-dimethylpropane
           |
           CH₃

              Cl Cl
              | |
   d) CH₃-CH-C-C-Cl   1,1,1,2,2,3-hexachlorobutane
           | | |
           Cl Cl Cl

25-16. a) The parent alkane is butane. The name indicates that two
            chlorine atoms are bonded to the first carbon atom and one
            chlorine atom is bonded to the second carbon atom. The
            structural formula is

                        Cl Cl
                        | |
                     Cl-CH-CH-CH₂-CH₃

   b) The parent alkane is ethane. The name indicates that three
       chlorine atoms are bonded to the first carbon atom. The
       structural formula is

                          Cl
                          |
                       Cl-C-CH₃
                          |
                          Cl

   c) The parent alkane is pentane. The name indicates that a
       chlorine atom is bonded to the first, second and third carbon
       atoms. The structural formula is

                    Cl-CH₂-CH-CH-CH₂-CH₃
                          | |
                          Cl Cl

   d) The parent alkane is hexane. The name indicates that two
       chlorine atoms are bonded to the second carbon atom and one
       chlorine atom is bonded to the fourth carbon atom. The

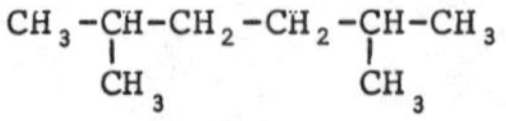

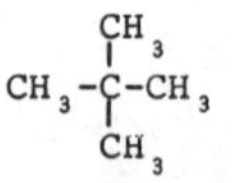

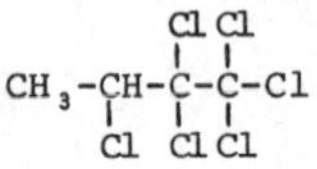

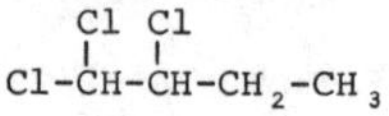

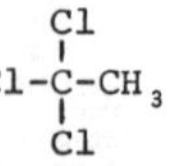

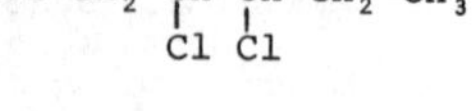

structural formula is

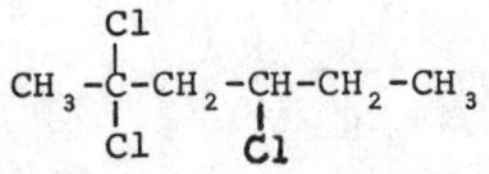

$$CH_3-\underset{\underset{Cl}{|}}{\overset{\overset{Cl}{|}}{C}}-CH_2-\underset{\underset{Cl}{|}}{CH}-CH_2-CH_3$$

25-18. a) $CH_3\underset{\underset{}{}}{\overset{\overset{CH_3}{|}}{C}}HCH=CHCH_2CH_3$

b) $CH_2=\overset{\overset{CH_3}{|}}{C}CH_2CH_2CH_3$

c) $CH_3\underset{\underset{CH_3}{|}}{\overset{\overset{CH_3}{|}}{C}}=CCH_3$

d) $CH_3CH=\overset{\overset{CH_3}{|}}{C}CH_2CH_3$

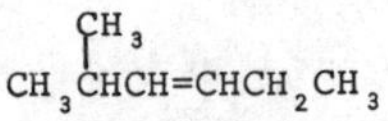

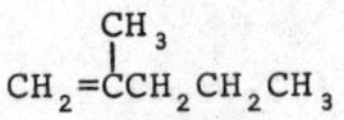

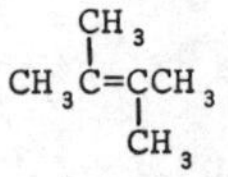

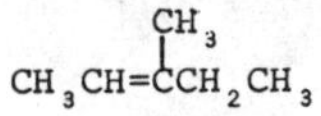

25-20. a) The reaction is

$$CH_2=CHCH_2CH_3 + Br_2 \longrightarrow \underset{Br\ Br}{\underset{|\ \ |}{CH_2CHCH_2CH_3}}$$

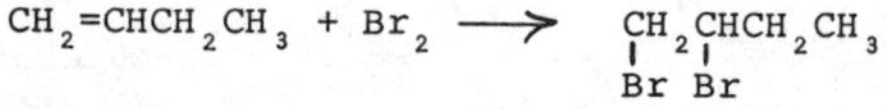

1,2-dibromobutane

b) The reaction is

$$CH_3CH=CHCH_3 + Br_2 \longrightarrow \underset{BrBr}{\underset{|\ |}{CH_3CHCHCH_3}}$$

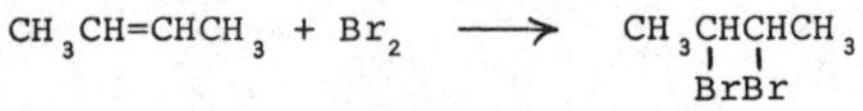

2,3-dibromobutane

c) The reaction is

$$CH_2=CHCH=CHCH_2CH_3 + Br_2 \longrightarrow \underset{Br\ Br}{\underset{|\ \ |}{CH_2CHCH=CHCH_2CH_3}}$$

1,2-dibromo-3-hexene
and

$$\underset{BrBr}{\underset{|\ |}{CH_2=CHCHCHCH_2CH_3}}$$

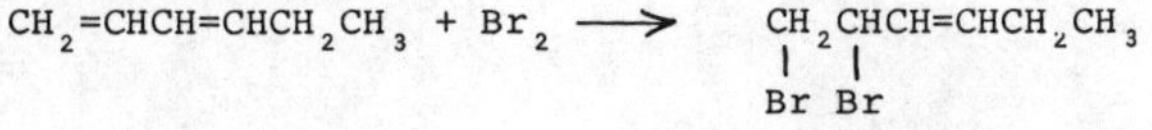

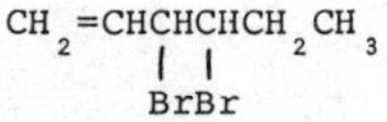

3,4-dibromo-1-hexene

25-22. The Lewis formula for 2-pentene is

$$\begin{array}{c} H\ H\ H\ H\ H \\ |\ \ |\ \ \ \ \ |\ \ |\ \ | \\ H-C-C=C-C-C-H \\ |\ \ \ \ \ \ \ \ \ \ \ \ |\ \ | \\ H\ \ \ \ \ \ \ \ \ H\ H \end{array}$$

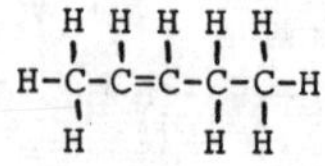

a) 2-pentene + $Cl_2(g) \longrightarrow CH_3-\underset{Cl}{CH}-\underset{Cl}{CH}-CH_2-CH_3$

2,3-dichloropentane

b) We must use Markovnikov's rule in this case

2-pentene + HCl $\longrightarrow CH_3-CH_2-\underset{Cl}{CH}-CH_2-CH_3$

3-chloropentane

and

$CH_3-\underset{Cl}{CH}-CH_2-CH_2-CH_3$

2-chloropentane

c) 2-pentene + $H_2O(l) \xrightarrow{\text{acid}} CH_3-CH_2-\underset{OH}{CH}-CH_2-CH_3$

3-pentanol

and

$CH_3-\underset{OH}{CH}-CH_2-CH_2-CH_3$

2-pentanol

d) 2-pentene + $H_2(g) \xrightarrow{Pt} CH_3-CH_2-CH_2-CH_2-CH_3$

pentane

25-24. a) $\underset{H_3C}{\overset{H_3C}{>}}C=C\underset{CH_3}{\overset{H}{<}}$ (l) + $Br_2(l) \longrightarrow CH_3-\underset{Br}{\overset{CH_3}{\underset{|}{C}}}-\underset{Br}{CHCH_3}$ (l)

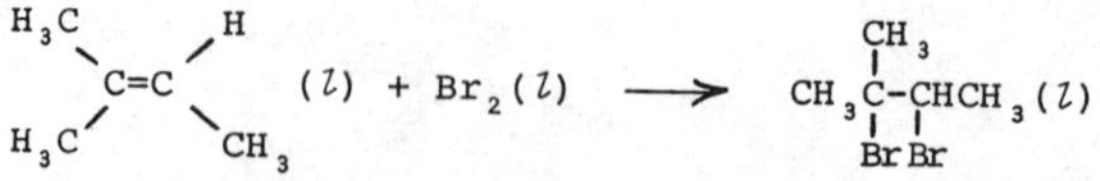

b) We must use Markovnikov's rule

$$\underset{H_3C}{\overset{H_3C}{>}}C=C\underset{CH_3}{\overset{H}{<}} \;(l) + HCl(g) \longrightarrow CH_3\underset{Cl}{\overset{CH_3}{\underset{|}{C}}}CH_2CH_3\,(l)$$

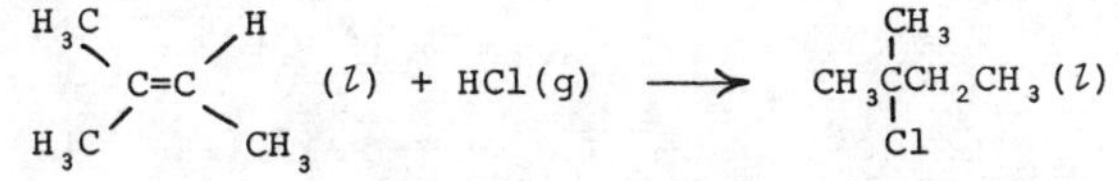

c) We must use Markovnikov's rule

$$\underset{H_3C}{\overset{H_3C}{>}}C=C\underset{CH_3}{\overset{H}{<}} \;(l) + H_2O(l) \xrightarrow{\text{acid}} CH_3\underset{OH}{\overset{CH_3}{\underset{|}{C}}}CH_2CH_3\,(l)$$

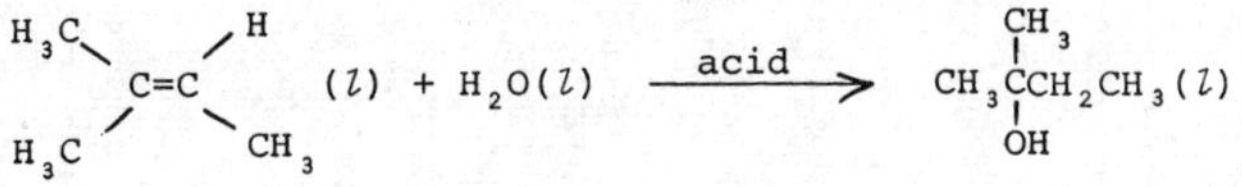

25-26. a) We shall react Br_2 with the alkene $CH_3CH=CHCH_3$.

b) We shall react two moles of Cl_2 with the alkyne $CH_3C\equiv CCH_3$.

c) We shall react H_2O in the presence of an acid with an alkene in accord with Markovnikov's rule to obtain the desired alcohol. We may use $(CH_3)_2C=CH_2$.

d) We shall react Br_2 with $CH_3\underset{\underset{CH_3}{|}}{CH}CH=CH_2$

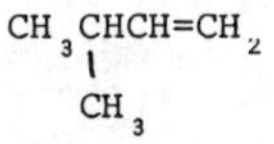

25-28. We shall add HCl to each alkene according to Markovnikov's rule

a) $CH_2=CHCHCH_3 + HCl \longrightarrow CH_3\underset{Cl\;Cl}{\underset{|\;\;|}{CH}CHCH_3}$
$\quad\quad\;\;\underset{Cl}{\overset{|}{}}$

b) $BrCH_2CH=CHCH_3 + HCl \longrightarrow BrCH_2\underset{Cl}{\underset{|}{CH}}CH_2CH_3$

and

$\quad\quad\quad\quad\quad\quad\quad\quad\quad\quad BrCH_2CH_2\underset{Cl}{\underset{|}{CH}}CH_3$

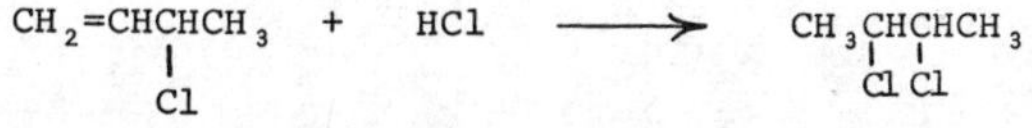

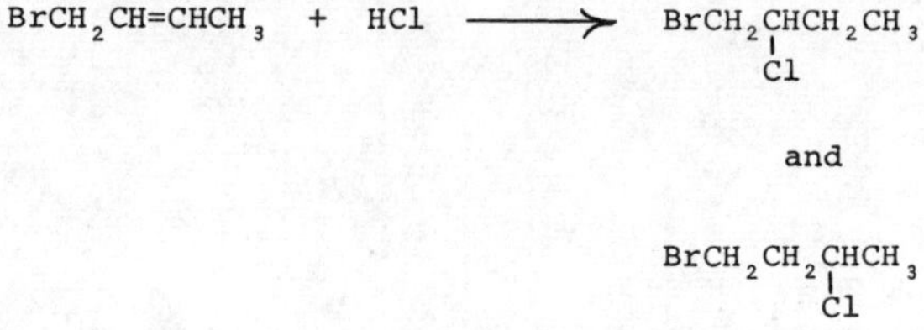

c) $CH_2=CCH_3 + HCl \longrightarrow CH_3\underset{CH_3}{\overset{Cl}{\underset{|}{C}}}CH_3$
        $|$
     CH_3

d) $ClCH=CHCH_3 + HCl \longrightarrow ClCH_2\underset{Cl}{\overset{|}{C}H}CH_3$ and

$Cl_2CHCH_2CH_3$

25-30. a) $\underset{H}{\overset{H}{}}C=C\underset{CH_3}{\overset{CH_2CH_2CH_3}{}}$ does not show cis-trans isomerism

b) $\underset{H}{\overset{CH_3\text{-}CH\text{-}CH_3}{}}C=C\underset{H}{\overset{CH_3}{}}$ $\underset{H}{\overset{CH_3\text{-}CH\text{-}CH_3}{}}C=C\underset{CH_3}{\overset{H}{}}$

     cis   trans

c) $\underset{CH_3}{\overset{CH_3}{}}C=C\underset{CH_2CH_3}{\overset{H}{}}$ does not show cis-trans isomerism

25-32. a) $CH_3CH_2\underset{CH_3}{\overset{CH_3}{\underset{|}{\overset{|}{C}}}}CH_2OH$

b) $CH_3\underset{OH}{\overset{|}{C}H} CH_2CH_2\underset{CH_3}{\overset{|}{C}H}CH_3$

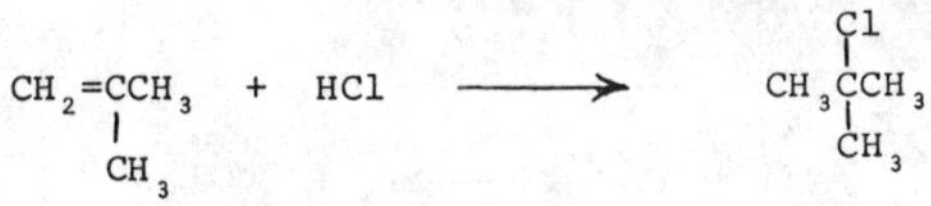

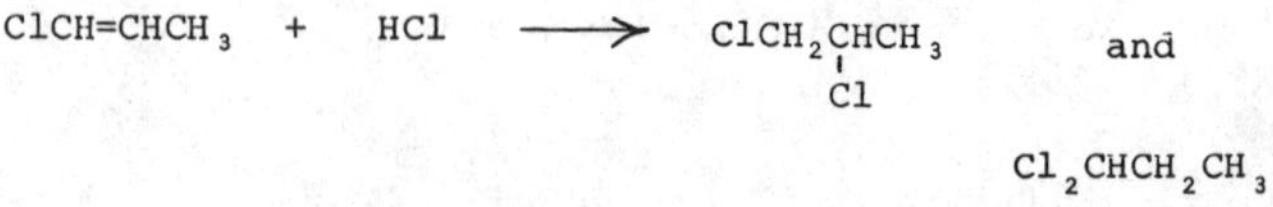

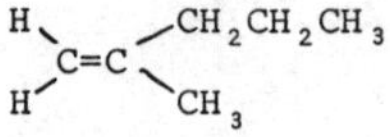

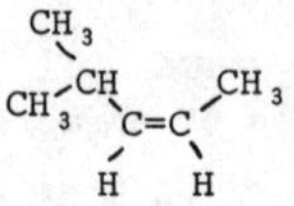

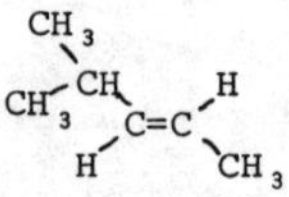

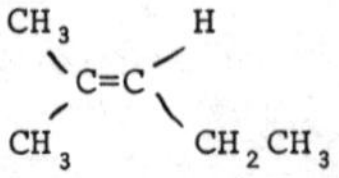

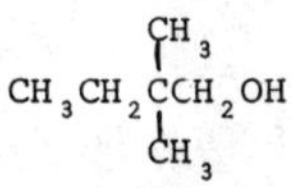

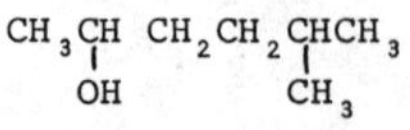

c) $ClCH_2CHCH_2CH_2CH_2CH_3$
$\quad\quad\quad\quad\quad|$
$\quad\quad\quad\quad OH$

d) $ClCH_2CHCHCH_2CH_3$
$\quad\quad\quad\quad\quad| \;\; |$
$\quad\quad\quad\quad Cl\,OH$

25-34. a) a tertiary alcohol

b) a secondary alcohol

c) a primary alcohol

d) a primary alcohol

25-36. $CH_3O^-(aq) + H_2O(l) \longrightarrow CH_3OH(aq) + OH^-(aq)$

25-38. Each amino group will react with HBr, thus

$$H_2N\text{-}CH_2CH_2\text{-}NH_2(aq) + 2HBr(aq) \longrightarrow {}^-Br\,{}^+H_3NCH_2CH_2NH_3^+Br^-(aq)$$

25-40. a) $CH_3C{\equiv}CH$

b) $CH_3CH_2C{\equiv}CCH_2CH_3$

c) $CH_3CH_2C{\equiv}CCHCH_2CH_2CH_3$
$\quad\quad\quad\quad\quad\quad\quad\quad\quad|$
$\quad\quad\quad\quad\quad\quad\quad\quad CH_2CH_3$

$\quad\quad\quad CH_3$
$\quad\quad\quad\;|$
d) $CH_3CC{\equiv}CCH_2CH_3$
$\quad\quad\quad\;|$
$\quad\quad\quad CH_3$

25-42. We can break the reaction down into two steps. We shall use Markovnikov's rule to predict the product of each step. The first step is

$$CH_3CH_2C{\equiv}CH(g) + HBr(g) \longrightarrow CH_3CH_2\underset{\displaystyle Br}{\underset{|}{C}}{=}CH_2(g)$$

The second step is

$$CH_3CH_2\underset{\displaystyle Br}{\underset{|}{C}}{=}CH_2(g) + HBr(g) \longrightarrow CH_3CH_2\overset{\displaystyle Br}{\overset{|}{\underset{\displaystyle Br}{\underset{|}{C}}}}CH_3(l)$$

The product is 2,2-dibromobutane.

25-44. a) This reaction can be broken down into two steps. We shall use Markovnikov's rule to predict the product in each step. The first step is

$$CH_3C{\equiv}CCH_3(g) + HCl(g) \longrightarrow CH_3\underset{\displaystyle Cl}{\underset{|}{C}}{=}CHCH_3(g)$$

The second step is

$$CH_3\underset{\displaystyle Cl}{\underset{|}{C}}{=}CHCH_3(g) + HCl(g) \longrightarrow CH_3\overset{\displaystyle Cl}{\overset{|}{\underset{\displaystyle Cl}{\underset{|}{C}}}}CH_2CH_3(l)$$

b) $$CH_3C{\equiv}CCH_3(g) + 2H_2(g) \xrightarrow{Ni(s)} CH_3CH_2CH_2CH_3(g)$$

c) $$CH_3C{\equiv}CCH_3(g) + 2Cl_2(g) \longrightarrow CH_3\overset{\displaystyle Cl}{\overset{|}{\underset{\displaystyle Cl}{\underset{|}{C}}}}{-}\overset{\displaystyle Cl}{\overset{|}{\underset{\displaystyle Cl}{\underset{|}{C}}}}CH_3(l)$$

25-46. a) 1-nitro-3-chlorobenzene

b) hexamethylbenzene

c) 1,4-dichlorobenzene or p-dichlorobenzene

d) 1,2-dichlorobenzene or o-dichlorobenzene

25-48. a) CH_3CH_2CHO

c) 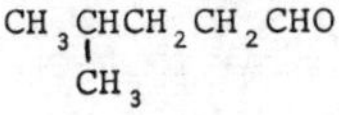

b) 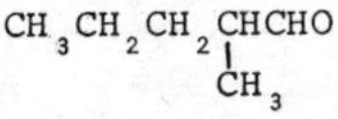

d) 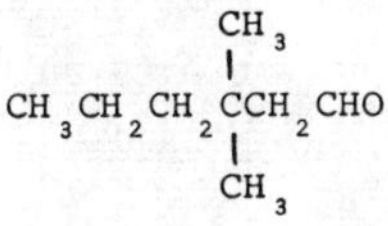

25-50. a) The alcohol is $CH_3CH_2CH_2OH$. We would use propanal

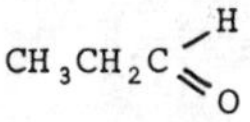

b) The alcohol is 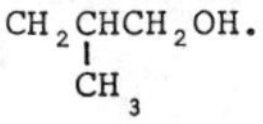 We would use 2-methylpropanal

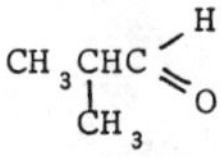

c) The alcohol is 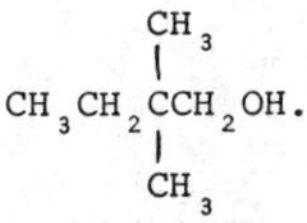 We would use 2,2-dimethylbutanal

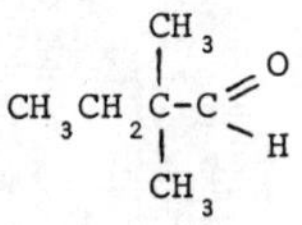

25-52. a) 3-chlorobutanoic acid

b) 2,2-dimethylpropanoic acid

c) 4-chloro-3-methylpentanoic acid

d) 2,2,3,3,3-pentachloropropanoic acid

25-54. $CH_3CH_2CH_2COOH(aq)$ + $CH_3CH_2OH(aq)$ ⟶ 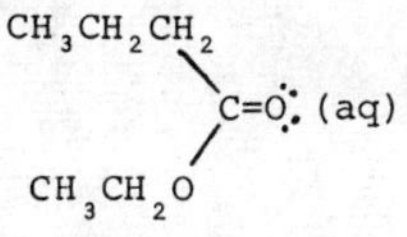 (aq) + $H_2O(l)$

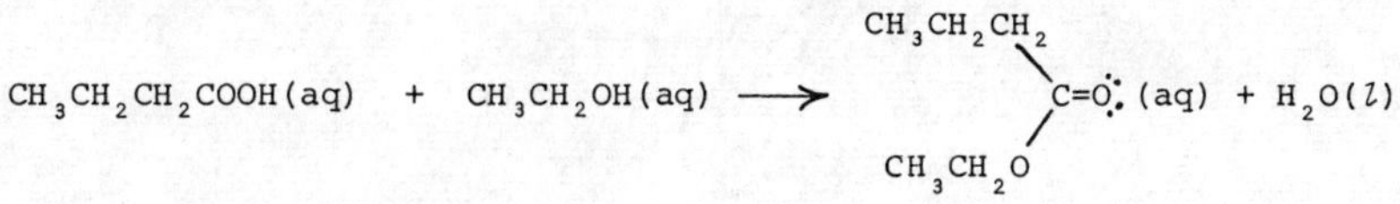

butyric acid ethanol ethyl butyrate

25-56. a) This is a neutralization reaction. The balanced equation is

$$CH_3CH_2COOH(aq) + NH_3(aq) \longrightarrow NH_4CH_3CH_2COO(aq)$$

b) The reaction between an acid and an alcohol yields an ester. The balanced equation is

$$CH_3CH_2COOH(aq) + CH_3OH(aq) \longrightarrow \underset{CH_3O}{\overset{CH_3CH_2}{>}}C=\ddot{O}: \,(aq) + H_2O(l)$$

c) $$CH_3CH_2COOH(aq) + CH_3CH_2OH(aq) \longrightarrow \underset{CH_3CH_2O}{\overset{CH_3CH_2}{>}}C=\ddot{O}: \,(aq) + H_2O(l)$$

25-58. a) $Na^+ \left[CH_3\underset{Cl}{CH}-C\overset{\ddot{O}:}{\underset{\ddot{O}:}{\lessgtr}} \right]^-$

b) $Rb^+ \left[H-C\overset{\ddot{O}:}{\underset{\ddot{O}:}{\lessgtr}} \right]^-$

c) $Sr^{2+} \left[\underset{CH_3}{\overset{CH_3}{\mid}}CH_3-C-C\overset{\ddot{O}:}{\underset{\ddot{O}:}{\lessgtr}} \right]^-_2$

d) $La^{3+} \left[CH_3-C\overset{\ddot{O}:}{\underset{\ddot{O}:}{\lessgtr}} \right]^-_3$

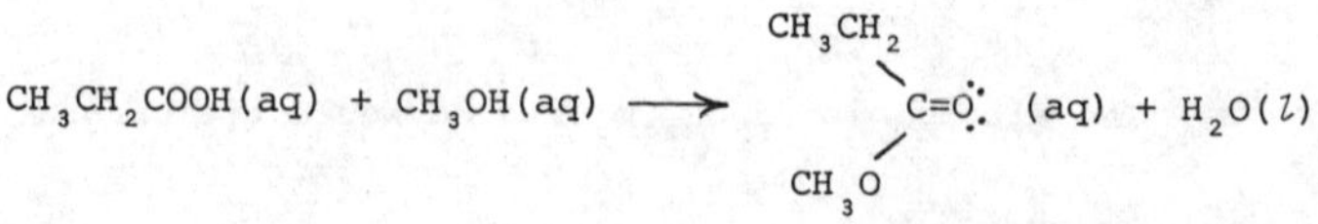

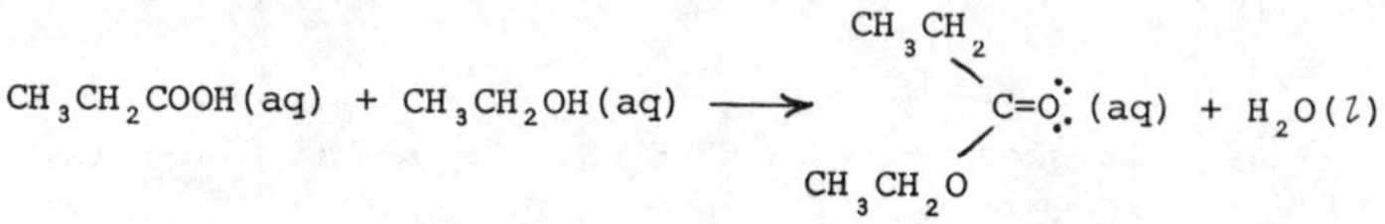

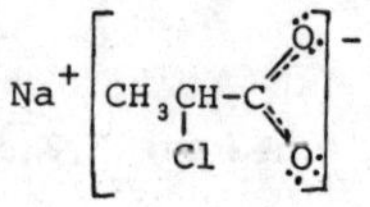

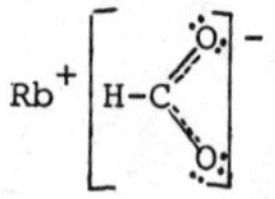

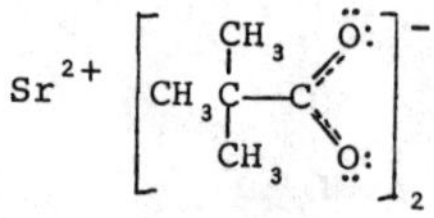

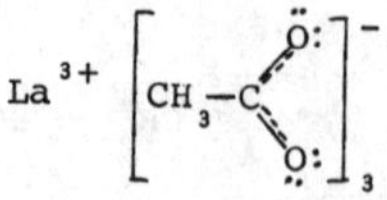

CHAPTER 26

SOLUTIONS TO THE EVEN-NUMBERED PROBLEMS

26-2. a)

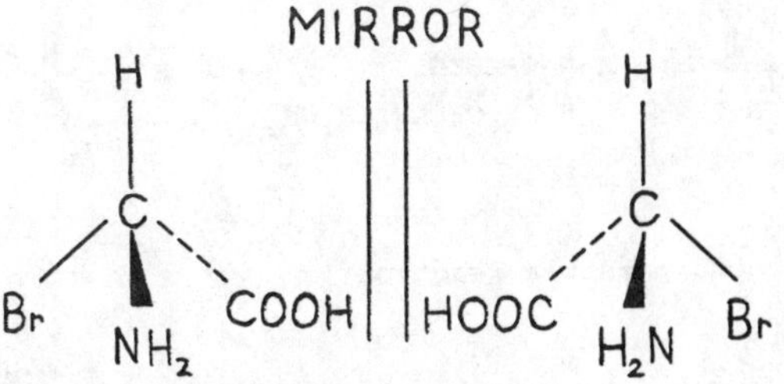

b) No.

c)

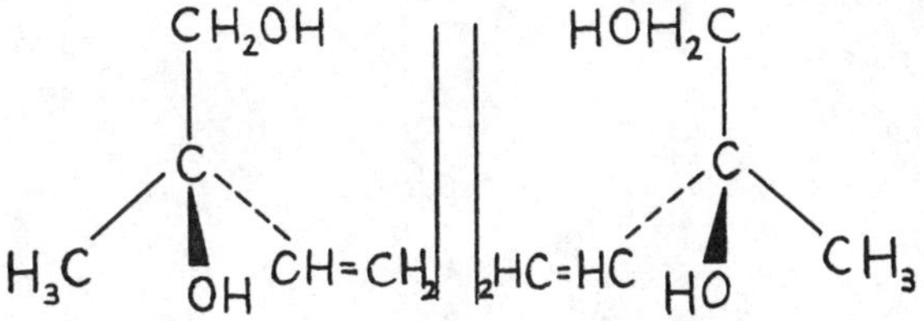

d) No.

e)

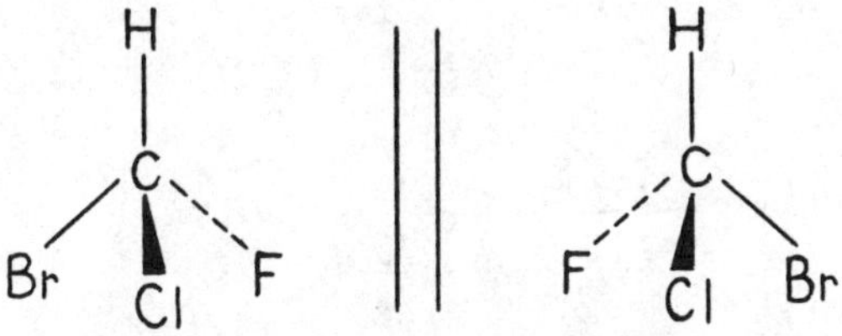

26-4. From Table 26-1, we find that the formula for the side group lysine is

$$-CH_2CH_2CH_2CH_2NH_2$$

We learned in Chapter 17 that $-NH_2$ behaves as a base in water The amino side group in lysine reacts with water according to

$$-CH_2CH_2CH_2CH_2NH_2(aq) + H_2O(l) \rightleftharpoons -CH_2CH_2CH_2CH_2NH_3^+(aq) + OH^-(aq)$$

26-6. The formulas for threonine and lysine are given in Table 26-1. One possible reaction between them is

$$H_2N-\underset{\underset{OH}{CHCH_3}}{\overset{H}{\underset{|}{C}}}-\overset{\overset{..}{O}}{\underset{||}{C}}-OH + H-\overset{H}{\underset{|}{\overset{..}{N}}}-\underset{CH_2CH_2CH_2NH_2}{\overset{H}{\underset{|}{C}}}-COOH \longrightarrow H_2N-\underset{\underset{OH}{CHCH_3}}{\overset{H}{\underset{|}{C}}}-\overset{\overset{..}{O}}{\underset{||}{C}}-\overset{H}{\underset{|}{\overset{..}{N}}}-\underset{CH_2CH_2CH_2NH_2}{\overset{H}{\underset{|}{C}}}-COOH + H_2O$$

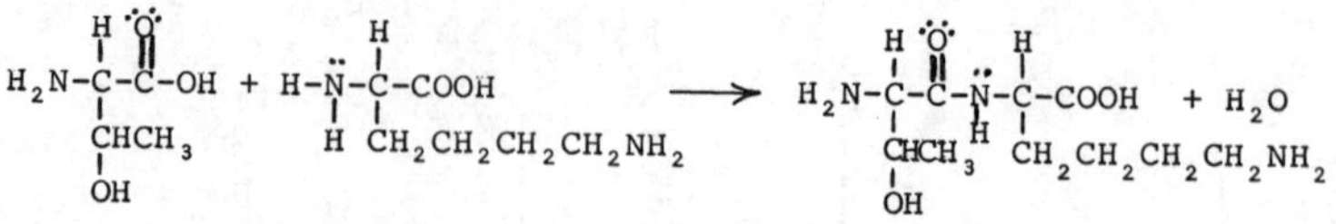

The second possible reaction is

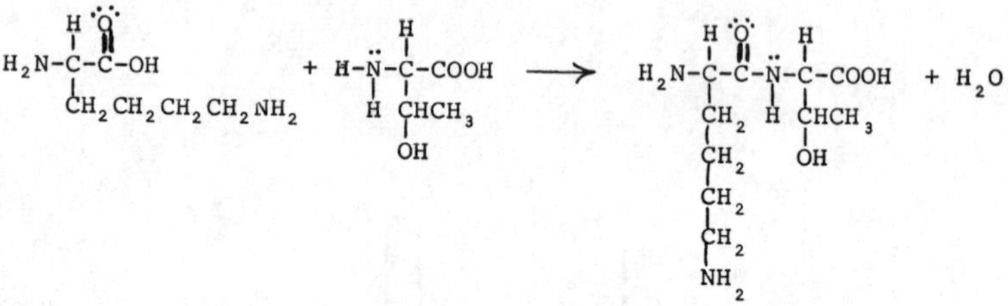

$$H_2N-\underset{\underset{}{CH_2CH_2CH_2CH_2NH_2}}{\overset{H}{\underset{|}{C}}}-\overset{\overset{..}{O}}{\underset{||}{C}}-OH + H-\overset{H}{\underset{|}{\overset{..}{N}}}-\underset{\underset{OH}{CHCH_3}}{\overset{H}{\underset{|}{C}}}-COOH \longrightarrow H_2N-\underset{\underset{\underset{NH_2}{CH_2}}{\underset{CH_2}{\underset{CH_2}{CH_2}}}}{\overset{H}{\underset{|}{C}}}-\overset{\overset{..}{O}}{\underset{||}{C}}-\overset{H}{\underset{|}{\overset{..}{N}}}-\underset{\underset{OH}{CHCH_3}}{\overset{H}{\underset{|}{C}}}-COOH + H_2O$$

26-8.

$$H_2N-\underset{\underset{CH_3}{CHCH_3}}{\overset{H}{\underset{|}{C}}}-\overset{\overset{..}{O}}{\underset{||}{C}}-\overset{H}{\underset{|}{\overset{..}{N}}}-\underset{\underset{O}{CH_2\underset{||}{C}-NH_2}}{\overset{H}{\underset{|}{C}}}-COOH$$

            val   asn

$$H_2N-\underset{\underset{O}{CH_2\underset{||}{C}-NH_2}}{\overset{H}{\underset{|}{C}}}-\overset{\overset{..}{O}}{\underset{||}{C}}-\overset{H}{\underset{|}{\overset{..}{N}}}-\underset{\underset{CH_3}{CHCH_3}}{\overset{H}{\underset{|}{C}}}-COOH$$

            asn   val

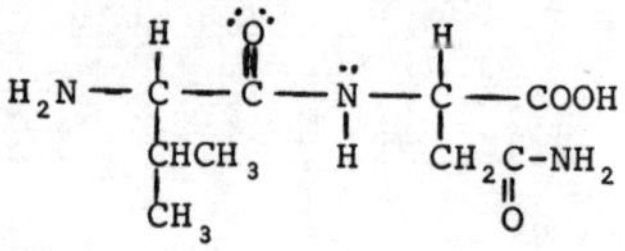

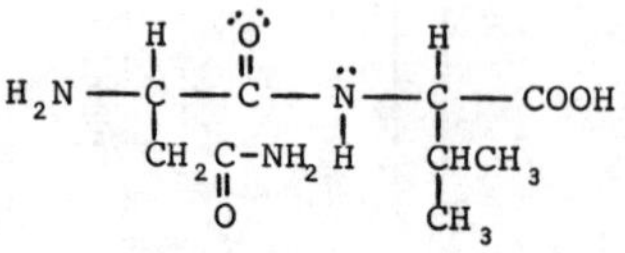

26-10. We can form six tripeptides from three different amino acids. If we represent the side groups of the three amino acids by G_1, G_2 and G_3, then the tripeptides are

$$H_2N-\underset{\underset{G_1}{|}}{\overset{\overset{H}{|}}{C}}-\overset{\overset{\cdot\cdot}{\overset{\cdot\cdot}{O}}}{\overset{||}{C}}-\underset{\underset{H}{|}}{\overset{\overset{H}{|}}{N}}-\underset{\underset{G_2}{|}}{\overset{\overset{H}{|}}{C}}-\overset{\overset{\cdot\cdot}{\overset{\cdot\cdot}{O}}}{\overset{||}{C}}-\underset{\underset{H}{|}}{\overset{\overset{H}{|}}{N}}-\underset{\underset{G_3}{|}}{\overset{\overset{H}{|}}{C}}-COOH$$

$$H_2N-\underset{G_1}{C}-\underset{||}{\overset{O}{C}}-\underset{H}{N}-\underset{G_3}{C}-\underset{||}{\overset{O}{C}}-\underset{H}{N}-\underset{G_2}{C}-COOH$$

$$H_2N-\underset{G_2}{C}-\underset{||}{\overset{O}{C}}-\underset{H}{N}-\underset{G_1}{C}-\underset{||}{\overset{O}{C}}-\underset{H}{N}-\underset{G_3}{C}-COOH$$

$$H_2N-\underset{G_2}{C}-\underset{||}{\overset{O}{C}}-\underset{H}{N}-\underset{G_3}{C}-\underset{||}{\overset{O}{C}}-\underset{H}{N}-\underset{G_1}{C}-COOH$$

$$H_2N-\underset{G_3}{C}-\underset{||}{\overset{O}{C}}-\underset{H}{N}-\underset{G_1}{C}-\underset{||}{\overset{O}{C}}-\underset{H}{N}-\underset{G_2}{C}-COOH$$

$$H_2N-\underset{G_3}{C}-\underset{||}{\overset{O}{C}}-\underset{H}{N}-\underset{G_2}{C}-\underset{||}{\overset{O}{C}}-\underset{H}{N}-\underset{G_1}{C}-COOH$$

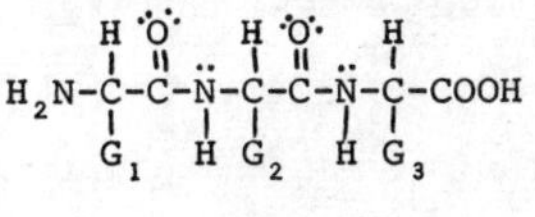

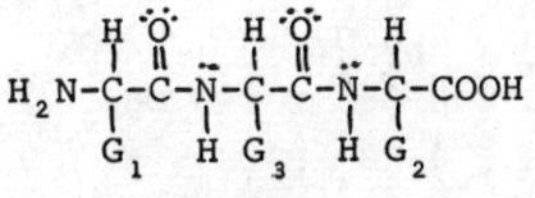

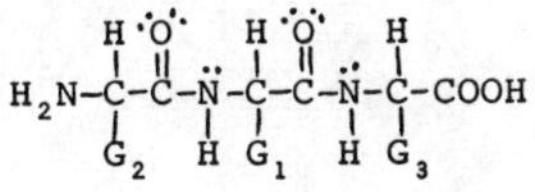

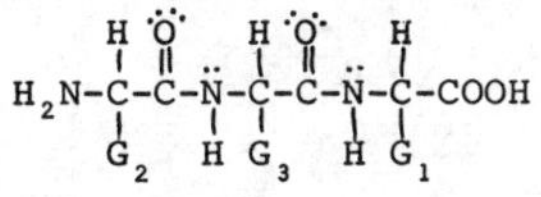

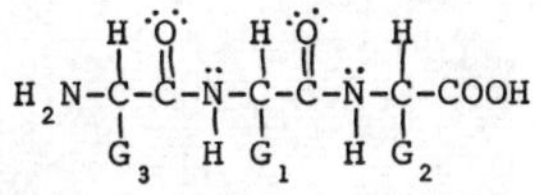

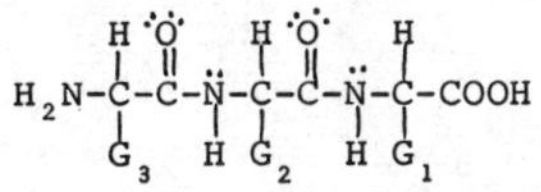

26-12. Referring to Table 26-1, we find that asn, val, and cys are the amino acids asparagine, valine, and cysteine. The structural formula for the tripeptide is

$$H_2N-\underset{\underset{\underset{\underset{NH_2}{|}}{\underset{C=O}{|}}}{\underset{CH_2}{|}}}{C}-\underset{||}{\overset{O}{C}}-\underset{H}{N}-\underset{\underset{\underset{CH_3}{|}}{\underset{CHCH_3}{|}}}{C}-\underset{||}{\overset{O}{C}}-\underset{H}{N}-\underset{\underset{CH_2SH}{|}}{C}-COOH$$

       asn   val   cys

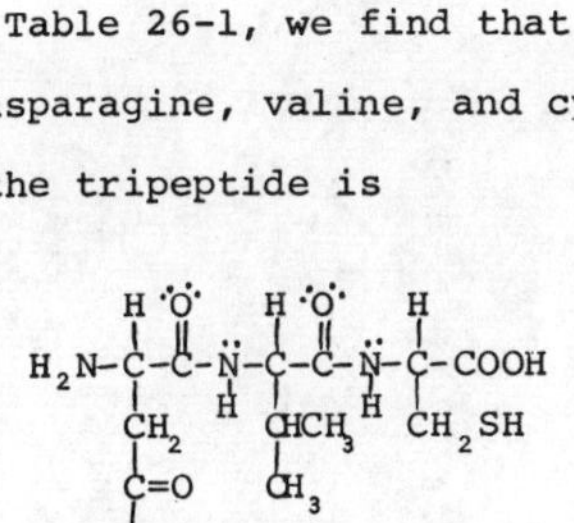

26-14. Referring to Table 26-1, we find that tyr, gly, phe, and met are the amino acids tyrosine, glycine, phenylalanine, and methionine. The structural formula for met-enkephalin is

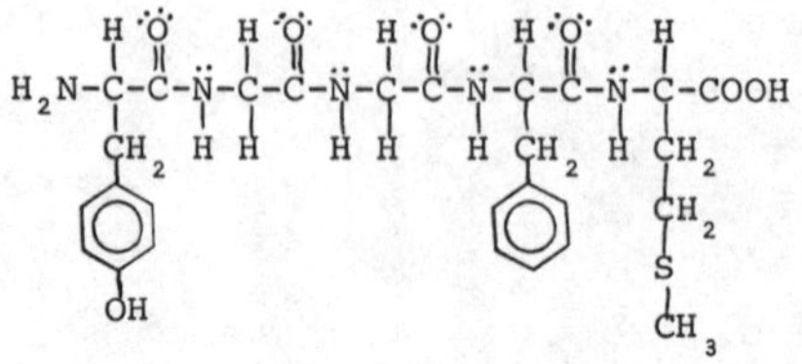

26-16.

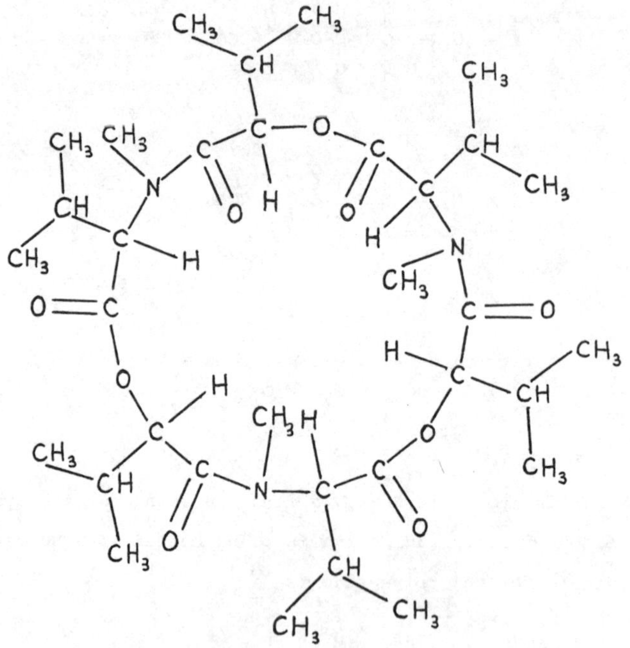

26-18. Amino acids with polar or charged side groups occur on the surface of a protein. Thus the amino acids lysine, lys, and asparagine, asn, will occur on the surface of a protein.

26-20. Amino acids with nonpolar side groups cluster in the interior of a protein. Thus the amino acids valine, val, and proline, pro, will cluster in the interior of a protein.

26-22. a) The oxygen atom can form a hydrogen bond to a hydrogen atom in H_2O.

b) The oxygen atom can form a hydrogen bond to a hydrogen atom in H_2O; H in –OH can bond to O in water.

c) The oxygen atom and the nitrogen atom can form a hydrogen bond to a hydrogen atom in H_2O.

d) The two oxygen atoms can form a hydrogen bond to a hydrogen atom in H_2O.

e) There are none.

26-24. Some factors that govern the secondary and tertiary structure of proteins are

1) The presence of disulfide bonds.
2) Hydrogen bonds within the peptide backbone.
3) Interactions between the amino acid side groups and the solvent, water.

26-26. The disaccharide, cellobiose, is composed of two β-glucose molecules. The reaction between them that produces cellobiose is

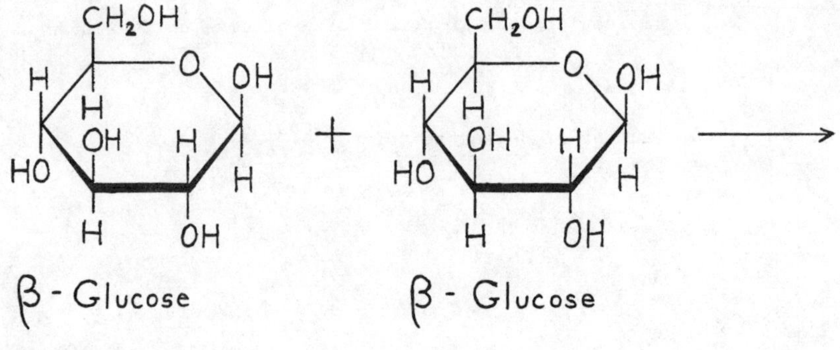

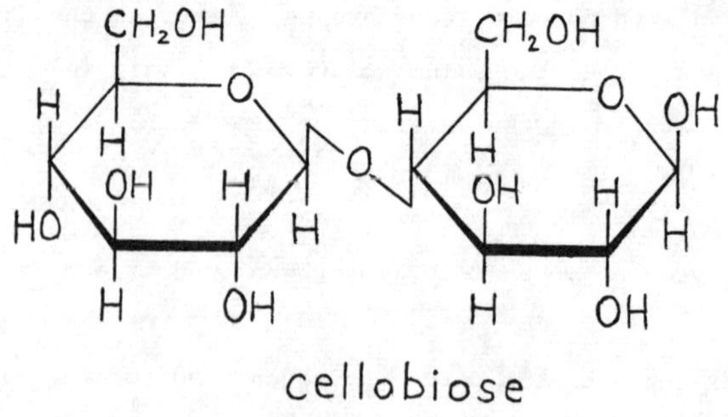

26-28. The Lewis formula for raffinose is

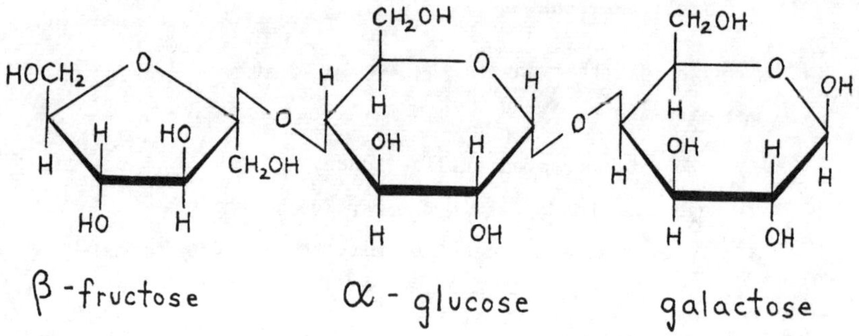

26-30. The molecular mass of β-glucose ($C_6H_{12}O_6$) minus one H_2O that is lost is 162. The molecular mass of the cellulose polymer is estimated by

$$\text{molecular mass} \approx \text{number of } \beta\text{-glucose units} \times 162$$

Thus

$$\text{number of } \beta\text{-glucose units} = \frac{500{,}000}{162} = 3100$$

26-32.

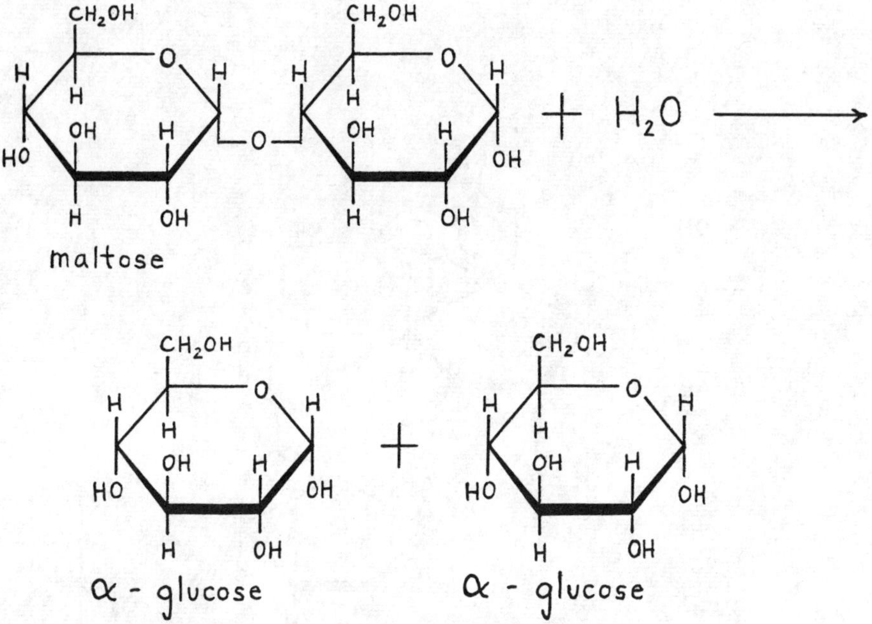

maltose
α - glucose
α - glucose

26-34. The sugar in DNA polynucleotides is deoxyribose. The DNA triplet ATC is deoxyadenosine-deoxythymidine-deoxycytidine. The Lewis formula for the DNA triplet ATC is

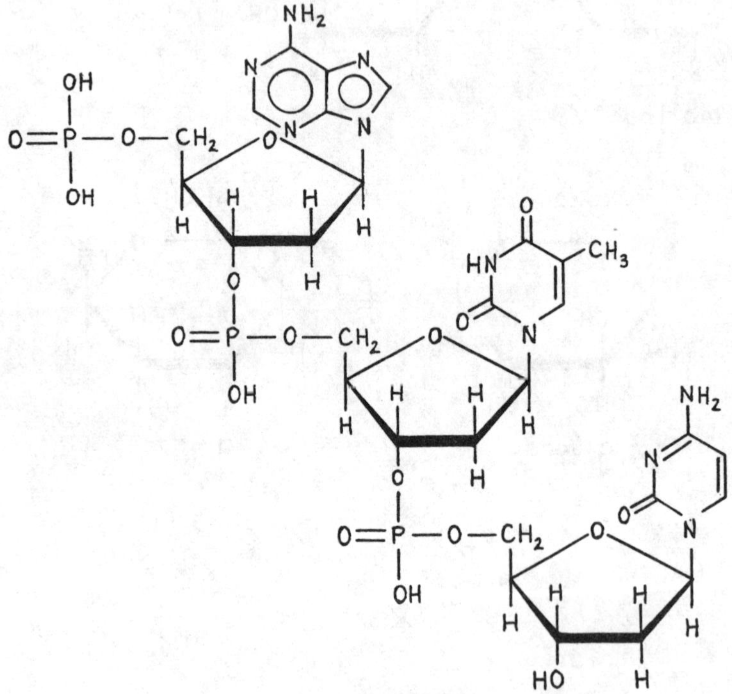

26-36. The sugar in RNA is ribose. The RNA triplet CUG is cytidine-uridine-guanosine. The Lewis formula for the RNA triplet is

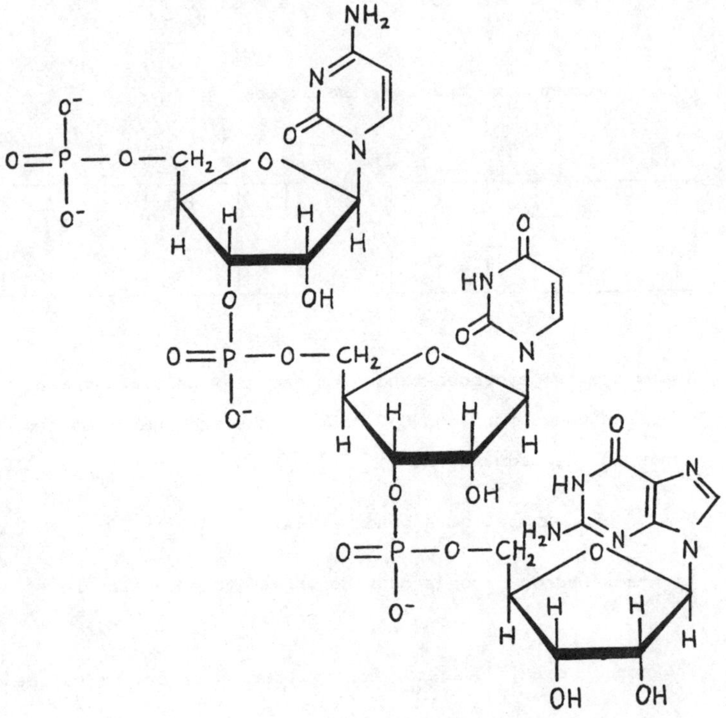

26-38. The two sequences must be complementary to each other; A and T must be opposite to each other and G and C must be opposite to each other. The other sequence must have the base sequence GTACCGATT.

26-40. We must have T and A opposite each other and G and C opposite each other. The complementary base sequence is

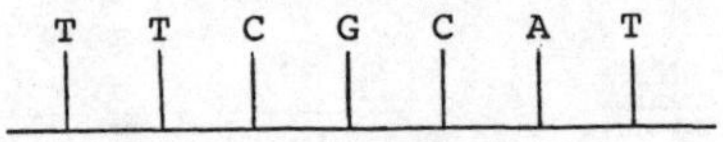

26-42. The two strands come apart to give

```
| | | | | | | ①      | | | | | | | ②
T C G T A C G       A G C A T G C
```

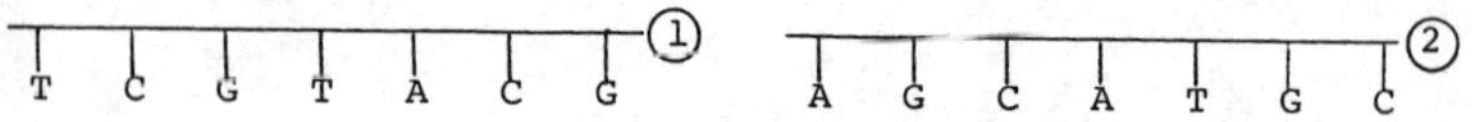

The complements to the two strands are

```
| | | | | | | ①      | | | | | | | ②
T C G T A C G       A G C A T G C
A G C A T G C       T C G T A C G
| | | | | | |       | | | | | | |
```

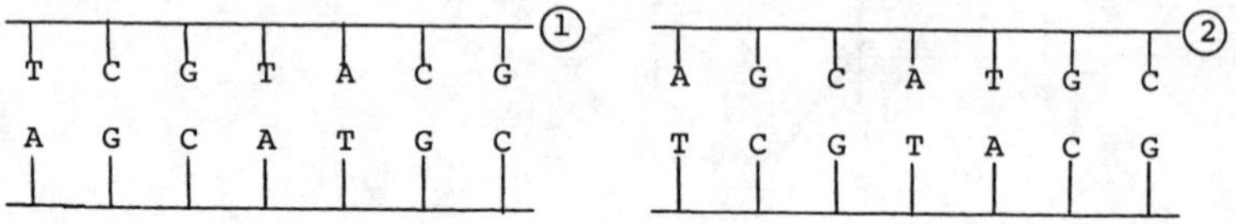

26-44. There are two hydrogen bonds for each A-T pair and three hydrogen bonds for each G-C pair. Thus the number of hydrogen bonds in the sequence is

number of H-bonds = (3x2) + (4x3) = 18

Eighteen hydrogen bonds must be broken to separate the strands.

26-46. The structural formula for the zwitterionic form of valine is

$$H_3\overset{+}{N}-\underset{|}{\overset{H}{\underset{CHCH_3}{\overset{|}{C}}}}-COO^-$$
$$\hspace{2em}|$$
$$\hspace{2em}CH_3$$

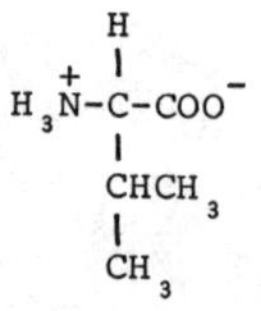

26-48. The rate law in terms of 1/rate is given by

$$\frac{1}{\text{rate}} = \frac{K + [S]}{k[E_o][S]}$$

Thus

$$\frac{1}{\text{rate}} = \frac{K}{k[E_o][S]} + \frac{1}{k[E_o]}$$

If we let 1/rate = y and $1/[S]$ = x, then we have

$$y = \frac{K}{k[E_o]} x + \frac{1}{k[E_o]}$$

which is the equation for a straight line. The equation for a straight line is

$$y = ax + b$$

where a is the slope and b is the y intercept. Thus a plot of 1/rate versus $1/[S]$ is a straight line with a slope of

$a = \dfrac{K}{k[E_o]}$ and a y intercept of $b = \dfrac{1}{k[E_o]}$.

26-50. a) The limiting reaction rate is given by

$$\text{rate} = k[E_o]$$

The value of k for acetylcholine esterase is 25,000 s^{-1}. Thus

$$\text{rate} = (25{,}000\ s^{-1})(1.0 \times 10^{-5}\ M)$$

$$= 0.25\ M{\cdot}s^{-1}$$

b) The turnover number, k_2, is the maximum number of substrate molecules that react per second per enzyme molecule, thus $1/k_2$ is the time for reaction of one substrate molecule

$$\frac{1}{k_2} = \frac{1}{25{,}000\ s^{-1}} = 4.0 \times 10^{-5}\ s$$

26-52. The rate of hydrolysis is given by

$$\text{rate} = k[E_o]$$

The rate is proportional to the concentration of the enzyme; the greater the value of $[E_o]$, the faster the rate. Thus the kidney solution contains more urease than the liver solution. The ratio of the rates of hydrolysis is equal to the ratio of the concentrations of the enzyme.

$$\frac{\text{rate}_{\text{kidney}}}{\text{rate}_{\text{liver}}} = \frac{k[E_o]_{\text{kidney}}}{k[E_o]_{\text{liver}}} = \frac{[E_o]_{\text{kidney}}}{[E_o]_{\text{liver}}}$$

$$= \frac{0.67\ \text{M}\cdot\text{min}^{-1}}{0.032\ \text{M}\cdot\text{min}^{-1}} = 21$$

The kidney contains 21 times more urease than the liver.

26-54. For the reaction

$$C_6H_{12}O_6(aq) + 6O_2(g) \longrightarrow 6CO_2(g) + 6H_2O(l)$$

$$\Delta G^0_{rxn} = -2.87 \times 10^3\ \text{kJ}\cdot\text{mol}^{-1}$$

Thus for the reaction

$$6CO_2(g) + 6H_2O(l) \longrightarrow C_6H_{12}O_6(aq) + 6O_2(g)$$

$$\Delta G^0_{rxn} = -(-2.87 \times 10^3\ \text{kJ}\cdot\text{mol}^{-1}) = 2.87 \times 10^3\ \text{kJ}\cdot\text{mol}^{-1}$$

The reaction is not spontaneous. Energy is required to drive the reaction toward the formation of glucose.

26-56. If we add the equations

$$\text{glucose}(aq) + 6O_2(g) \longrightarrow 6CO_2(g) + 6H_2O(l)$$

$$\Delta G^0_{rxn} = -2.87 \times 10^3 \text{ kJ}$$

$$38ADP(aq) + 38HPO_4^{2-}(aq) \longrightarrow 38ATP(aq) + 38H_2O(l)$$

$$\Delta G^0_{rxn} = (38 \text{ mol})(29 \text{ kJ}\cdot\text{mol}^{-1}) = 1.10 \times 10^3 \text{ kJ}$$

then we obtain

$$\text{glucose}(aq) + 38ADP(aq) + 38HPO_4^{2-}(aq) + 6O_2(g) \longrightarrow$$

$$6CO_2(g) + 44H_2O(l) + 38ATP(aq)$$

$$\Delta G^0_{rxn} = -2.87 \times 10^3 \text{ kJ} + 1.10 \times 10^3 \text{ kJ}$$

$$= -1.77 \times 10^3 \text{ kJ}$$

The reaction is spontaneous under standard conditions.

26-58. For the reaction

$$ATP(aq) + H_2O(l) \longrightarrow ADP(aq) + HPO_4^{2-}(aq)$$

we have

$$\Delta G_{rxn} = \Delta G^0_{rxn} + RT\ln \frac{[ADP][HPO_4^{2-}]}{[ATP]}$$

Thus at 37°C

$$\Delta G_{rxn} = -31 \times 10^3 \frac{J}{mol}$$

$$+(8.31 \text{ J}\cdot\text{K}^{-1}\cdot\text{mol}^{-1})(310 \text{ K})\ln \frac{(0.5\text{x}10^{-3}\text{M})(1.0\text{x}10^{-3}\text{M})}{(50\text{x}10^{-3}\text{M})}$$

$$\Delta G_{rxn} = -61 \times 10^3 \frac{J}{mol}$$

26-60. The value of ΔG_{rxn} is given by

$$\Delta G_{rxn} = \Delta G^0_{rxn} + RT\ln \frac{[\text{lactic acid}]^2}{[\text{glucose}]}$$

At 25°C with the stated concentrations we have

$$\Delta G_{rxn} = -200 \times 10^3 \frac{J}{mol} + (8.31\ J\cdot K^{-1}\cdot mol^{-1})(298\ K)\ln\frac{(2.9\times10^{-3}M)^2}{(5.0\times10^{-3}M)}$$

$$\Delta G_{rxn} = -216 \times 10^3\ J\cdot mol^{-1}$$

Interchapter A. Oxygen

A-1. Sand is predominantly silicon dioxide, SiO_2.

A-2. Oxygen is produced commercially by the fractional distillation of air.

A-3. Most of the oxygen in the Earth's atmosphere is due to photosynthesis.

A-4. The overall reaction of photosynthesis is

$$CO_2(g) + H_2O(g) \xrightarrow{\text{visible light}} \text{carbohydrate} + O_2(g)$$

A-5. Oxygen can be prepared in the laboratory by

(a) the thermal decomposition of $KClO_3(s)$

$$2KClO_3(s) \longrightarrow 2KCl(s) + 3O_2(g).$$

(b) the addition of $Na_2O_2(s)$ to water

$$2Na_2O_2(s) + 2H_2O(\ell) \longrightarrow 4NaOH(aq) + O_2(g)$$

(c) the electrolysis of water

$$2H_2O(\ell) \xrightarrow{\text{electrolysis}} 2H_2(g) + O_2(g)$$

A-6. A catalyst is a substance that facilitates a chemical reaction and yet is not consumed in the reaction.

A-7. Combustion is a reaction with oxygen ("burning") in which a large amount of heat is evolved.

A-8. The reaction for the combustion of methane, the principal component of natural gas, is

$$CH_4(g) + 2O_2(g) \longrightarrow CO_2(g) + 2H_2O(g)$$

A-9. The heat-producing reaction in an oxy-acetylene torch is

$$2C_2H_2(g) + 5O_2(g) \longrightarrow 4CO_2(g) + 2H_2O(g)$$

A-10. The burning of a candle can be represented by the reactions

$$C_{20}H_{42}(s) \longrightarrow C_{20}H_{42}(\ell)$$

$$2C_{20}H_{42}(\ell) + 61O_2(g) \longrightarrow 40CO_2(g) + 42H_2O(g)$$

A-11. Allotropy occurs if two different forms of an element have different numbers of arrangements of the atoms in the molecules.

A-12. The two allotropes of oxygen are O_2 and O_3 (ozone).

Interchapter B. Separation and Purification

B-1. Filtration-a solution-plus-solid mixture is separated by passing the mixture through a porous medium such as filter paper. The solid particles are trapped by the filter paper and the solution passes through. (see Figure B-4).

B-2. The contrasting property of gold and sand, on which panning for gold depends, is that gold is much more dense than sand.

B-3. Distillation-a solution in which there is dissolved a non-volatile solid can be separated into its components by vaporizing the solvent and then condensing and collecting it. (see Figure B-6).

B-4. The role of a condensor in a distillation apparatus is to condense the vaporized solvent by lowering its temperature.

B-5. A liquid that readily vaporizes is said to be volatile.

B-6. Fractional distillation is used to separate a mixture of volatile liquids.

B-7. Fractional distillation-a solution of two volatile liquids is separated into its components by repeatedly vaporizing and condensing the solution in a fractional distillation column. (see Figure B-7).

B-8. In a gas-chromatographic separation, a mixture of two or more vapors is swept through a narrow column packed with a solid that is coated with a nonvolatile liquid. The vapors that dissolve more readily in the liquid coating lag behind those that dissolve less readily and hence exit the column later. (see Figures B-8 and B-9).

Interchapter C. Nitrogen.

C-1. Nitrogen fixation is the conversion of elemental atmospheric nitrogen into soluble nitrogen-containing compound(s).

C-2. The Haber process can be described by

$$N_2(g) + 3H_2(g) \xrightarrow[500°C]{300atm} 2NH_3(g)$$

C-3. The Ostwald process can be described by the reactions

(1) $4NH_3(g) + 5O_2(g) \longrightarrow 4NO(g) + 6H_2O(g)$

(2) $2NO(g) + O_2(g) \longrightarrow 2NO_2(g)$

(3) $3NO_2(g) + H_2O(\ell) \longrightarrow 2HNO_3(aq) + NO(g)$

C-4. Azides are compounds containing the N_3^- ion. They can be made from hydrazoic acid, HN_3, which is produced in the reaction

$$N_2H_4(aq) + HNO_2(aq) \longrightarrow HN_3(aq) + 2H_2O(\ell)$$

C-5. Nitrides are compounds containing the N^{3-} ion. A few metals react directly with nitrogen to produce azides.

C-6. (a) NH_3 (f) NO_2

(b) HNO_3 (g) NaN_3

(c) $NaNO_2$ (h) Li_3N

(d) N_2O (i) NH_4NO_3

(e) NO (j) N_2H_4

C-7. The Raschig synthesis of hydrazine utilizes the reaction

$$2NH_3(aq) + ClO^-(aq) \xrightarrow{OH^-(aq)} N_2H_4(aq) + H_2O(\ell) + Cl^-(aq)$$

C-8. You should never mix household ammonia and bleach because toxic chloramines, such as H_2NCl and $HNCl_2$, are produced, as well as hydrazine which is a carcinogen.

C-9. Air is 78% (by volume) N_2.

C-10. $3NO_2(g) + D_2O(\ell) \longrightarrow 2DNO_3(\text{in } D_2O) + NO(g)$

Interchapter D. Sulfur.

D-1. In the Frasch process for sulfur extraction, three concentric pipes are sunk into sulfur-bearing calcite rock. Water at 180°C is forced down the outermost pipe to melt the sulfur. Hot compressed air is forced down the innermost pipe and mixes with the molten sulfur, forming a foam of water, air, and sulfur. The mixture rises to the surface through the center pipe. The resulting dried sulfur has a purity of 99.5 percent. (see Figure D-2).

D-2. Zinc can be obtained from zinc blende (ZnS) by the reactions

$$\underset{\text{(zinc blende)}}{2ZnS(s)} + 3O_2(g) \longrightarrow 2ZnO(s) + 2SO_2(g)$$

$$ZnO(s) + C(s) \longrightarrow Zn(\ell) + CO(g)$$

D-3. (a) SO_2 (f) FeS_2

(b) SO_3 (g) $CaSO_4 \cdot 2H_2O$

(c) H_2SO_4 (h) HgS

(d) H_2SO_3 (i) $Na_2S_2O_3 \cdot 5H_2O$

(e) H_2S

D-4. The changes that sulfur undergoes upon being heated from 90°C to 450°C can be represented by the scheme

$$S(\text{rhombic}) \xrightarrow{96°C} S(\text{monoclinic}) \xrightarrow{119°C} S_8(\text{thin, pale yellow liquid})$$

$$\downarrow \approx 150°C$$

$$S_8(\text{vapor}) \xleftarrow{445°C} S_n(\text{thick, reddish-brown liquid})$$

(see Figure D-9).

D-5. The contact process for the production of sulfuric acid can be represented by the reactions

$$S(s) + O_2(g) \longrightarrow SO_2(g)$$

$$2SO_2(g) + O_2(g) \xrightarrow{V_2O_5(s)} 2SO_3(g)$$

$$H_2SO_4(\ell) + SO_3(g) \longrightarrow \underset{\text{oleum}}{H_2S_2O_7(35\% \text{ in } H_2SO_4)}$$

$$H_2S_2O_7(\text{in } H_2SO_4) + H_2O(\ell) \longrightarrow 2H_2SO_4(aq)$$

D-6. A balanced equation for the conversion of gypsum to Plaster of Paris is

$$2CaSO_4 \cdot 2H_2O(s) \longrightarrow 2CaSO_4 \cdot \tfrac{1}{2}H_2O(s) + 3H_2O(g)$$

D-7. Salts of sulfuric acid are called sulfates. Salts of sulfurous acid are called sulfites.

D-8. The formation of the fertilizer ammonium sulfate from ammonia and sulfuric acid is described by

$$2NH_3(g) + H_2SO_4(\ell) \longrightarrow (NH_4)_2SO_4(s)$$

D-9. Do not attempt to increase the acidity of soil by adding concentrated sulfuric acid because concentrated sulfuric acid is too strong an acid. It is also a strong dehydrating agent, and destroys plant roots.

D-10. The formula for Epsom salt is $MgSO_4 \cdot 7H_2O(s)$.

Interchapter E. The Chemistry of the Atmosphere.

E-1. The four regions into which the Earth's atmosphere can be divided are the troposphere, the stratosphere, the mesosphere, and the ionosphere. (see Figure E-2).

E-2. All our weather takes place in the troposphere.

E-3. The origin of nitrogen in our atmosphere is due to volcanic activity.

E-4. An increase in the concentration of carbon dioxide in the atmosphere may lead to an increase in the average surface temperature of the Earth and a subsequent increase in the levels of the oceans due to the melting of the polar ice caps, possibly flooding coastal areas.

E-5. Clear nights are colder than cloudy nights because the infrared radiation emitted by the Earth is not trapped by the atmosphere and so there is a heat loss.

E-6. The noble gases were discovered by a careful analysis of the density of nitrogen obtained from the thermal decomposition of $NH_4NO_2(s)$

$$NH_4NO_2(s) \longrightarrow N_2(g) + 2H_2O(g)$$

and nitrogen obtained from the atmosphere after having removed all the known other gases.

E-7. Photochemical smog is a mixture of irritating substances that are produced in the atmosphere by the action of sunlight. It is called photochemical smog because light is needed to sustain the reactions that produce the smog.

E-8. Acid rain is rain that is acidic due to the reactions of the atmospheric contaminants $SO_2(g)$ and $SO_3(g)$ with water vapor to produce $H_2SO_3(aq)$ and $H_2SO_4(aq)$.

E-9. A chemical reaction that describes the deterioration of limestone by acid rain is

$$CaCO_3(s) + H_2SO_4(aq) \longrightarrow CaSO_4(s) + H_2O(\ell) + CO_2(g)$$

E-10. We say that ozone is a life-saving constituent of the atmosphere because the atmosphere ozone absorbs the short wavelength, harmful, solar ultraviolet radiation.

E-11. The ozone layer is a region in the atmospheric (15 to 30 km) in which most atmospheric ozone occurs. (see Figure E-8).

E-12. Photoionization is ionization produced by radiation. Photoionization is responsible for the absorption of solar radiation with wavelengths less than 100 nm. (see Table E-3).

Interchapter F. Alkali Metals.

F-1. The alkali metals must be stored underwater because they react with the oxygen and the water vapor in air, but they do not react with kerosene.

F-2. Sodium metal is produced commercially by the electrolysis of NaCl(s).

F-3. Sodium is the least expensive metal per unit volume because sodium salts are abundant, easily converted to the metal, and also because sodium has a relatively low density for a metal.

F-4. The reactivity of the alkali metals increases with atomic number because the outer ns electron is more easily removed the larger is the atom.

F-5. Lithium is the only element that reacts directly with nitrogen at room temperature.

F-6. The raw materials in the Solvay process are NaCl(s), limestone ($CaCO_3$(s)) and water.

F-7. The chemical equations for the reactions in the Solvay process are

$$NH_3(aq) + CO_2(aq) + H_2O(\ell) \longrightarrow NH_4^+(aq) + HCO_3^-(aq)$$

$$NaCl(aq) + NH_4^+(aq) + HCO_3^-(aq) \xrightarrow{15°C} NaHCO_3(s) + NH_4Cl(aq)$$

$$2NaHCO_3(s) \xrightarrow{80°C} Na_2CO_3(s) + H_2O(\ell) + CO_2(g)$$

F-8. (a) $K(s) + O_2(g) \longrightarrow KO_2(s)$

(b) $2Na(s) + 2H_2O(\ell) \longrightarrow 2NaOH(aq) + H_2(g)$

(c) $K(s) + N_2(g) \longrightarrow$ no reaction

(d) $NaH(s) + H_2O(\ell) \longrightarrow NaOH(aq) + H_2(g)$

(e) $Li_3N(s) + 3H_2O(\ell) \longrightarrow 3LiOH(aq) + NH_3(g)$

(f) $2Na(s) + F_2(g) \longrightarrow 2NaF(s)$

Interchapter G. Spectroscopy.

G-1. Infrared spectrum-when a molecule is irradiated with infrared radiation of various wavelengths, the molecule absorbs radiation at certain wavelengths. A plot of the amount of infrared radiation absorbed versus wavelength of the radiation is called an infrared spectrum. (see Figures G-1 and G-2).

G-2. Infrared spectroscopy can be used to distinguish two substances from each other because the various vibrational motions of groups of atoms in the molecules lead to characteristic absorption of infrared radiation, which can be used to verify the presence of certain bonds or groups of atoms.

G-3. The motion that leads to the absorption of infrared radiation is vibrational motion.

G-4. The initials NMR stand for <u>n</u>uclear <u>m</u>agnetic <u>r</u>esonance.

G-5. The nuclear magnetic resonance discussed in Interchapter G is sometimes called proton magnetic resonance because the nuclear magnetic resonance discussed in the Interchapter deals only with protons.

G-6. Mass spectrum-if electrons of sufficiently high energy are directed at molecules in a gas phase, then the molecules can be fragmented into various ions that are characteristic of the molecule. The fragmentation pattern is called a mass spectrum. (see Figure G-6).

G-7. Mass spectrometry can be used to identify an unknown compound because the fragmentation pattern, or mass spectrum, of a molecule is characteristic of the molecule, much like a fingerprint.

G-8. A mass spectrometer works in the following manner: Gas molecules are ionized by electron bombardment, and the ions are accelerated by an electric field. The ion beam is passed through a magnetic field, where it is resolved into component beams of ions of equal mass. Light ions are deflected more strongly than heavy ions by the magnetic field. In a beam containing $C_5H_{12}^+$ and $C_4H_9^+$ ions, the lighter $C_4H_9^+$ ions are deflected more than the heavier $C_5H_{12}^+$

ions. The mass spectrometer depicted here is adjusted to detect the $C_5H_{12}^+$ ions. By changing the magnitude of the magnetic or electric field, the beam of $C_4H_9^+$ can be moved to strike the collector at the slit, where it would then pass through to the detector and be measured as a current.

G-9. The origin of the spectrum in X-ray fluorescence spectroscopy is the transition of electrons from outer shell to inner shell.

G-10. X-ray fluorescence spectroscopy can be used to identify unknown compounds because if a sample is subjected to an X-ray beam of sufficiently high energy, then the X-rays impinging on the sample eject electrons from the inner shells of the atoms in the sample. The ejection of these electrons is followed by the emission of X-radiation from the atom at a set of wavelengths that is characteristic of the particular element.

Interchapter H. Silicon, a Semimetal.

H-1. Silicon is produced by the following sequence of reactions

$$\underset{\text{(sand)}}{SiO_2(s)} + C(s) \longrightarrow \underset{\text{(impure)}}{Si(\ell)} + CO_2(g)$$

$$\underset{\text{(impure)}}{Si(s)} + 2Cl_2(g) \longrightarrow \underset{\text{(pure)}}{SiCl_4(\ell)}$$

$$SiCl_4(g) + 2Mg(s) \longrightarrow 2MgCl_2(s) + \underset{\text{(pure)}}{Si(\ell)}$$

H-2. Zone refining-an impure solid is packed tightly in a glass tube, and the tube is lowered slowly through a heating coil that melts the solid. Pure solid crystallizes out from the bottom of the melted zone, and the impurities concentrate in the moving molten zone. Silicon of purity up to 99.9999 percent can be obtained by zone refining.

H-3. The difference between a conductor, a semiconductor, and an insulator is as follows: Metals have no band gap, semiconductors have a small band gap and insulators have a large band gap. (see Figure H-3).

H-4. Comparison of normal, *n*-type, and *p*-type silicon--(a) Silicon has four valence electrons, and each silicon atom forms four 2-electron bonds to other silicon atoms. (b) Phosphorus has five valence electrons, and thus when a phosphorus atom substitutes for a silicon atom in a silicon crystal, there is an unused valence electron on each phosphorus atom that can become a conduction electron. (c) Boron has only three valence electrons, and thus when a boron atom substitutes for a silicon atom in a silicon crystal, there results an electron vacancy (a "hole"). Electrons from the silicon valence bond can move through the crystal by hopping from one vacancy site to another.

H-5. The Lewis formula of SiO_4^{4-} is

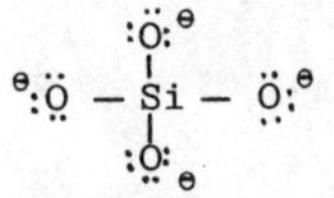

and we predict from VSEPR that SiO_4^{4-} is tetrahedral.

H-6. The silicate structure of asbestos is shown in Figure H-8, and that of mica is shown in Figure H-10.

H-7. The Lewis formula of a straight-chain silicate polyanion is

$$-\ddot{\underset{\underset{:\ddot{O}:^{\ominus}}{|}}{\overset{:\ddot{O}:^{\ominus}}{Si}}}-\ddot{O}-\underset{\underset{:\ddot{O}:^{\ominus}}{|}}{\overset{:\ddot{O}:^{\ominus}}{Si}}-\ddot{O}-\underset{\underset{:\ddot{O}:^{\ominus}}{|}}{\overset{:\ddot{O}:^{\ominus}}{Si}}-\ddot{O}-$$

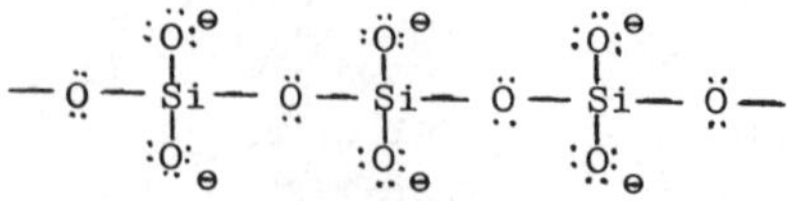

The SiO_4^{4-} units are linked together through oxygen atoms.

H-8. The Lewis formula of the cyclic polysilicate ion, $Si_6O_{18}^{12-}$, is

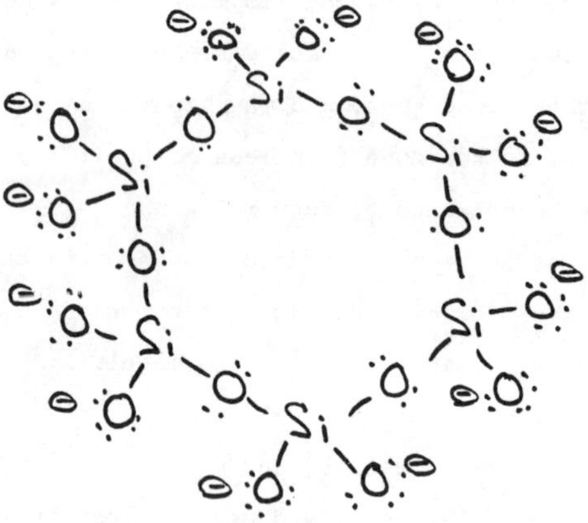

(see Figure H-7). This structure occurs in the mineral beryl.

H-9. Mica consists of two-dimensional, polymeric silicate sheets. (see Figure H-10).

H-10. The three principal components of glass are

SiO_2, Na_2O from Na_2CO_3, CaO from $CaCO_3$.

H-11. Photochromic eyeglasses have a small amount of added silver chloride dispersed throughout the glass. When sunlight strikes the glass, the tiny AgCl grains decompose into opaque clusters of silver atoms and chlorine atoms

$$AgCl \underset{dark}{\overset{sunlight}{\rightleftarrows}} Ag + Cl$$
   clear   opaque

H-12. The reaction by which HF attacks glass is

$$6HF(aq) + SiO_2(s) \longrightarrow H_2SiF_6(s) + 2H_2O(\ell)$$
              glass   etched glass

Interchapter I. Main-Group Metals.

I-1. The Group 2 metals are obtained by electrolysis.

I-2. You would not throw water on a magnesium fire because hot magnesium reacts violently with water.

I-3. The balanced equation for the reaction that occurs when a flashbulb flashes is

$$2Mg(s) + O_2(g) \longrightarrow 2MgO(s)$$

I-4. (a) $Mg(OH)_2$ (f) $CaSO_4 \cdot H_2O$

(b) $MgSO_4 \cdot 7H_2O$ (g) $CaSO_4 \cdot 2H_2O$

(c) CaO (h) $CaMg_3(SiO_3)_4$

(d) $CaCO_3$ (i) Al_2O_3

(e) $Ca(OH)_2$ (j) $Pb(CH_2CH_3)_4$

I-5. The substance with the greatest liquid range is gallium.

I-6. Balanced equations for the dissolution of gallium metal in HCl(aq) and NaOH(aq) are

$$2Ga(s) + 6H^+(aq) + 6Cl^-(aq) \longrightarrow 2Ga^{3+}(aq) + 6Cl^-(aq) + 3H_2(g)$$

$$2Ga(s) + 6H_2O(\ell) + 2Na^+(aq) + 2OH^-(aq) \longrightarrow 2Na^+(aq) + 2Ga(OH)_4^-(aq) + 3H_2(g$$

I-7. The metals of Groups 3, 4, and 5 are

Group 3 - aluminum, gallium, indium and thallium

Group 4 - lead and tin

Group 5 - bismuth

I-8. Tin and lead have ionic charges of +2 and +4 because their ground-state election configuration are $Sn([Kr]5s^2 4d^{10} 5p^2)$ and $Pb([Xe]6s^2 4f^{14} 5d^{10} 6p$). Each atom can lose two p electrons to become M(II) or two p electrons and two s electrons to become M(IV).

I-9. Tin disease is the result of the conversion of the metallic, lustrous white allotrope of tin to the brittle, gray allotrope.

I-10. Lead is produced from galena by the reaction

$$PbS(s) + C(s) + 2O_2(g) \longrightarrow Pb(\ell) + CO_2(g) + SO_2(g)$$

Interchapter J. Phosphorus.

J-1. White phosphorus is very reactive, igniting spontaneously in air at about 25°C. Red phosphorus is much less reactive than white phosphorus. For example, red phosphorus must be heated to 260°C before it burns in air.

J-2. P_4 is tetrahedral. (see Figure J-2).

J-3. Phosphate rock is $Ca_3(PO_4)_2$. Its most important use is in the production of fertilizer.

J-4. Phosphate rock cannot be used directly as fertilizer because it is insoluble in water.

J-5. The structures of P_4O_6 and P_4O_{10} are

P_4O_6

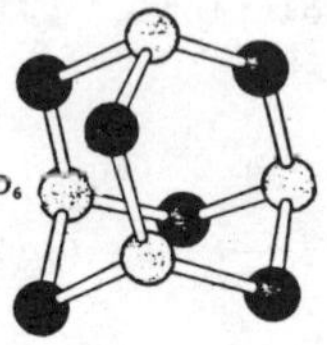

P_4O_{10}

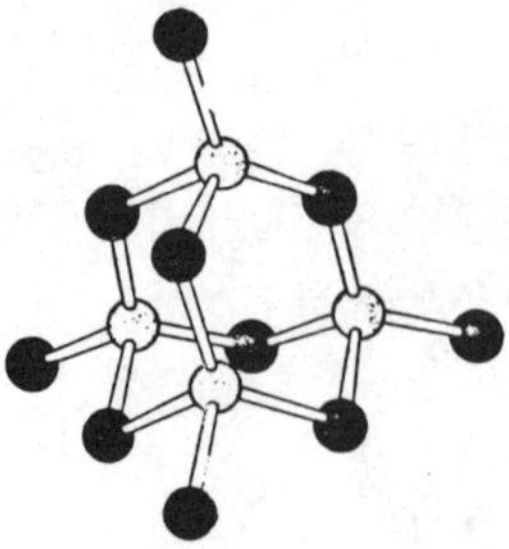

J-6. A desiccant is a drying agent. P_4O_{10} is a powerful desiccant.

J-7. The structures of H_2PO_4, HPO_4^{2-} and PO_4^{3-} are

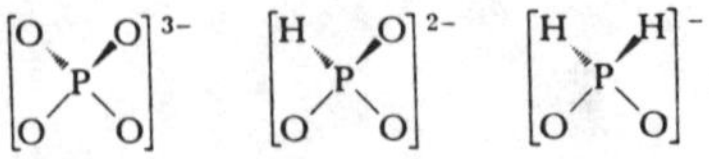

There are two moles of dissociable protons per mole of phosphorous acid and only one mole of dissociable protons per mole of hypophosphorous acid.

J-8. The action of baking powder can be described by

$$\underbrace{Ca(H_2PO_4)_2(s) + 2NaHCO_3(s)}_{\text{in baking powder}} \xrightarrow{300°C} \underbrace{2CO_2(g)}_{\text{rising agent}} + 2H_2O(g) + CaHPO_4(s) + Na_2HPO_4(s$$

J-9. Eutrophication-when aquatic organisms die, oxygen dissolved in the water is consumed in the decay of the organisms, thus depleting the oxygen supply.

J-10. Ammonia is a much stronger base than phosphine.

J-11. Two ways to prepare phosphine are

$$Ca_3P_2(s) + 6H_2O(\ell) \longrightarrow 2PH_3(g) + 3Ca(OH)_2(aq)$$

$$P_4(s) + 3OH^-(aq) + 3H_2O(\ell) \longrightarrow PH_3(g) + 3H_2PO_2^-(aq)$$

J-12. The difference between safety matches and strike-anywhere matches is as follows: Matches that can be ignited by striking on any rough surface contain a tip composed of the yellow P_4S_3 on top of a red portion that contains lead dioxide, PbO_2, together with antimony sulfide, Sb_2S_3. Friction causes the P_4S_3 to ignite in air, and the heat produced then initiates a reaction between antimony sulfide and lead dioxide that produces a flame. Safety matches consist of a mixture of potassium chlorate and antimony sulfide. The match is ignited by striking on a special rough surface composed of a mixture of red phosphorus, glue, and abrasive. The red phosphorus is ignited by friction and in turn ignites the reaction mixture in the matchhead.

Interchapter K. Natural Waters.

K-1. The principal ionic consitituents of seawater are

$$Cl^-,\ Na^+,\ SO_4^{2-},\ Mg^{2+},\ Ca^{2+},\ K^+,\ HCO_3^-$$

(see Table K-2).

K-2. The major source of dissolved salts in the oceans is vent holes in the ocean floor.

K-3. The four substances that are obtained from seawater on a commercial scale are pure water, NaCl, Br_2, and Mg.

K-4. Desalination is the separation of salts from seawater.

K-5. Bromine is obtained from seawater by the following process:

$$2Br^-(aq) + Cl_2(g) \longrightarrow Br_2(\ell) + 2Cl^-(aq)$$

The $Br_2(\ell)$ is swept out from the seawater by passing a stream of air through the solution.

K-6. The distinction between fresh water, brackish water, salty water and brine is

Table K-1 Classification scheme for water

Type of water	*Quantity of dissolved minerals/% by mass*
fresh	0– 0.1
brackish	0.1– 1
salty	1–10
brine	>10

K-7. The principal constituents of hard water are divalent cations such as $Ca^{2+}(aq)$, $Mg^{2+}(aq)$ and $Fe^{2+}(aq)$.

K-8. Soap molecules have a long hydrocarbon chain with one or more charged groups at one end of the chain. When soap molecules that are dissolved in water come into contact with grease, the hydrocarbon portions of the soap molecules stick into the grease, leaving the anion portions of the soap molecules at the grease-water interface. The penetration of the grease by the soap molecules is then followed by the formation of micelles, which encapsulate the grease particles and are subsequently rinsed away.

K-9. A micelle is a cluster of molecules that have a long hydrocarbon chain with one or more charged groups at one end of the chain. In aqueous solution, the hydrocarbon chains of the molecules "dissolve in each otner" to form a spherical region of hydrocarbon chains, and the charged groups occur at the surface of this spherical region at the hydrocarbon-water interface. (see Figure K-5).

K-10. Temporary hard water contains $HCO_3^-(aq)$ anions, along with $Ca^{2+}(aq)$ and/or $Mg^{2+}(aq)$. When temporary hard water is heated, $CaCO_3(s)$ or $MgCO_3(s)$ precipitates because of the reaction

$$M^{2+}(aq) + 2HCO_3^-(aq) \longrightarrow MCO_3(s) + H_2O(\ell) + CO_2(g)$$

In permanent hard water the primary anion is $SO_4^{2-}(aq)$. Both calcium sulfate and magnesium sulfate are soluble in hot water and are not precipitated by heating.

K-11. The formation of soap scum is described by the equation

$$\underset{\text{(from hard water)}}{Ca^{2+}(aq)} + \underset{\text{(from soap)}}{2C_{17}H_{35}COO^-(aq)} \longrightarrow \underset{\text{(soap scum)}}{Ca(C_{17}H_{35}COO)_2(s)}$$

K-12. To soften water is to remove divalent cations such as $Ca^{2+}(aq)$ and $Mg^{2+}(aq)$.

K-13. An ion-exchange resin works in the following manner: An ion-exchange resin contains acidic and basic groups. Metal cations such as Ca^{2+}, Mg^+ and Na^+ displace $H^+(aq)$ ions and become bound to the resin. Anions such as Cl^- and SO_4^{2-} displace $OH^-(aq)$ ions and also become bound to the resin. This two-stage ion-exchange process removes both cations and anions from the water. (see Figure K-4).

K-14. Primary treatment of sewage involves physical processes, such as the removal of large pieces of solid material by screening and settling in large sedimentation tanks, followed by the removal of fine solid particles by the addition of aluminum salts that produce a flocculent $Al(OH)_3$ precipitate. Secondary sewage treatment (Figure K-6b) involves biological processes. The water resulting from a primary treatment is run into aeration tanks where air in the presence of aerobic bacteria is bubbled through the water in order to remove organic wastes. The aeration is carried out over a bed of small stones to increase the surface area of the water and thus increase the rate of oxygen dissolution. The aeration step is followed by storage in a sedimentation tank and then chlorination to destroy microorganisms. Tertiary sewage treatment involves chemical processes. Tertiary sewage treatment is designed to remove specific chemical pollutants, such as nitrogen and phosphorus compounds.

Interchapter L. The Halogens.

L-1. Fluorine is able to stabilize unusually high oxidation states in many elements because fluorine has the highest electronegativity.

L-2. Elemental fluorine is obtained by the electrolysis of hydrogen fluoride dissolved in molten potassium fluoride

$$2HF(\text{in KF melt}) \xrightarrow{\text{electrolysis}} H_2(g) + F_2(g)$$

Chlorine is obtained by the electrolysis of molten sodium chloride

$$2NaCl(\ell) \xrightarrow{\text{electrolysis}} 2Na(\ell) + Cl_2(g)$$

Bromine and iodine are obtained by the oxidation of bromides and iodides with chlorine

$$2Br^-(aq) + Cl_2(g) \longrightarrow 2Cl^-(aq) + Br_2(\ell)$$

$$2I^-(aq) + Cl_2(g) \longrightarrow 2Cl^-(aq) + I_2(s)$$

L-3. Glass is etched or "frosted" via the reaction

$$\underset{\text{glass}}{SiO_2(s)} + 6HF(aq) \longrightarrow \underset{\text{etched glass}}{H_2SiF_6(s)} + 2H_2O(\ell)$$

L-4. (a) NaCl (d) CaF_2

(b) KCl (e) Na_3AlF_6

(c) $KNgCl_3 \cdot 6H_2O$ (f) $Ca_{10}F_2(PO_4)_6$

L-5. The chemical formulas and major use of ethyl chloride and sodium hypochlorite are

(a) CH_3CH_2Cl - externally applied local anesthetic

(b) NaClO - bleaching agent

L-6. The chemical formulas and names of the oxyacids of chlorine are

HClO - hypochlorous acid

$HClO_2$ - chlorous acid

$HClO_3$ - chloric acid

$HClO_4$ - perchloric acid.

L-7. Household bleach is a dilute aqueous solution of NaClO.

L-8. KIO_3 can be prepared from $I_2(s)$ by the following process:

$$I_2(s) + 5H_2O_2(aq) \longrightarrow 2IO_3^-(aq) + 4H_2O(\ell) + 2H^+(aq)$$

$$HIO_3(aq) + KOH(aq) \longrightarrow KIO_3(aq) + H_2O(\ell)$$

L-9. (a) bromous acid (c) perbromic acid

(b) hypoiodous acid (d) iodic acid

L-10. (a) nitrous acid (c) hypophosphorous acid

(b) sulfurous acid (d) phosphorous acid

(e) hyponitrous acid

(a) potassium sulfite (c) potassium iodite

(b) calcium nitrite (d) magnesium hypobromite

Interchapter M. The Transition Metals.

M-1. The d transition metals are Sc-Zn, Y-Cd, and Lu-Hg.

M-2. Tungsten is the metal with the highest melting point.

M-3. The two most dense metals are iridium and osmium.

M-4. Iron is the most abundant transition metal.

M-5. Titanium metal is produced by the following reactions:

$$\underset{\text{rutile}}{TiO_2(s)} + 2C(s) + 2Cl_2(g) \longrightarrow TiCl_4(g) + 2CO(g)$$

$$TiCl_4(g) + 2Mg(s) \longrightarrow 2MgCl_2(s) + Ti(s)$$

M-6. The percentage of gold in 14-karat gold is

$$\frac{14}{24} \times 100 = 58\%$$

M-7. Iron is produced in a blast furnace in the following manner:

A mixture of iron ore, coke, and limestone ($CaCO_3$) is loaded into the top, and preheated compressed air and oxygen are blown in near the bottom. The reaction of the coke and the oxygen to produce carbon dioxide gives off a great deal of heat, and the temperature in the lower region of a blast furnace is around 1900°C. As the CO_2 rises, it reacts with more coke to produce hot carbon monoxide, which reduces the iron ore to iron. The molten iron metal is denser than the other substances and drops to the bottom, where it can be drained off to form ingots of what is called *pig iron*. The function of the limestone is to remove the sand and gravel that normally occur with iron ore.

The principal reactions that occur in a blast furnace are:

$$C(s) + O_2(g) \longrightarrow CO_2(g)$$

$$C(s) + CO_2(g) \longrightarrow 2CO(g)$$

$$3CO(g) + Fe_2O_3(s) \longrightarrow 2Fe(\ell) + 3CO_2(g)$$

M-8. Pig iron, the product of a blast furnace, contains about 4 or 5 percent carbon together with lesser amounts of silicon, manganese, phosphorus and sulfur. Pig iron is brittle, difficult to weld, and is not strong enough for structural applications.

M-9. Slag, which is primarily molten calcium silicate, is produced in a blast furnace by the reaction:

$$\underset{\text{lime}}{CaO(s)} + \underset{\text{sand, gravel}}{SiO_2(s)} \longrightarrow \underset{\text{slag}}{CaSiO_3(\ell)}$$

Slag is used in building materials, such as cement and concrete aggregate, rock-wool insulation, and cinder block.

M-10. The basic oxygen process is a process used in the conversion of pig iron to steel in which hot pure oxygen gas is blown through the molten pig iron to oxidize carbon and phosphorus impurities.

M-11. The differences between mild steel, medium steel and high-carbon steel is as follows:

Carbon steel that contains less than 0.2 percent carbon is called *mild steel*. Mild steels are malleable and ductile and are used where load-bearing ability is not a consideration. *Medium steels*, which contain 0.2 to 0.6 percent carbon, are used for such structural materials as beams and girders and for railroad equipment. *High-carbon steels* contain 0.8 to 1.5 percent carbon and are used to make drill bits, knives, and other tools in which hardness is important.

M-12. The reactions are described on page 904; The overall process is:

$$2Fe(s) + \frac{3}{2} O_2(g) + 3H_2O(\ell) \longrightarrow Fe_2O_3 \cdot 3H_2O(s)$$

M-13. The corrosion of aluminum is not as serious a problem as the corrosion of iron because aluminum forms an aluminum oxide coating that is impervious to oxygen.

M-14. A sacrificial anode works in the following manner:

Consider protecting an iron pipe with a sacrificial zinc anode. Zinc is a stronger reducing agent than iron and thus is preferentially oxidized. The electrons produced in the oxidation flow to the iron pipe, on the surface of which O_2 is reduced to hydroxide ion. The net process is $2Zn(s) + O_2(aq) + 2H_2O(\ell) \longrightarrow 2Zn(OH)_2(s)$, and the iron remains intact.

Interchapter N. Synthetic Polymers.

N-1. The polymerization of ethylene to polyethylene is an example of an addition polymerization reaction.

N-2. The formation of dacron from para-terephthalic acid and ethylene glycol is an example of a condensation polymerization reaction.

N-3. Teflon, $\{CF_2CF_2\}_n$, is formed in an addition polymerization reaction. Dacron, $\{OCH_2CH_2\text{-}O\text{-}\underset{\underset{O}{\|}}{C}\text{-}\underset{}{\bigcirc}\text{-}\underset{\underset{O}{\|}}{C}\}_n$, is formed in a condensation polymerization reaction.

N-4. Nylon, $\{\underset{\underset{H}{|}}{N}\text{-}CH_2CH_2CH_2CH_2CH_2CH_2\text{-}\underset{\underset{H}{|}}{\underset{\|}{N}}\underset{\underset{O}{\|}}{\text{-}C}\text{-}CH_2CH_2CH_2CH_2\text{-}\underset{\underset{O}{\|}}{C}\}_n$

is formed in a condensation polymerization reaction.

Polystyrene, $\{CH_2\underset{\underset{\bigcirc}{|}}{CH}\}_n$, is formed in an addition polymerization reaction.

N-5. An equation for the condensation reaction that is shown in the frontispiece to Interchapter N is

$$n\ H_2N(CH_2)_6NH_2 + n\ ClC(CH_2)_4CCl \longrightarrow$$
$$\underset{\underset{O}{\|}}{}\underset{\underset{O}{\|}}{}$$

$$\{\underset{\underset{H}{|}}{N}(CH_2)_4\underset{\underset{H}{|}}{N}\text{-}\underset{\underset{O}{\|}}{C}(CH_2)_4\underset{\underset{O}{\|}}{C}\}_n + n\ HCl$$

N-6. The chemical formula for the polymer unit in natural rubber is

$$\begin{array}{c}-CH_2CH_2-\\ \diagdown\diagup\\ C=C\\ \diagup\diagdown\\ H_3CH\end{array}$$

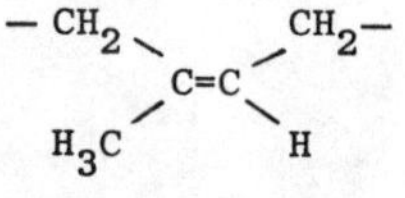

N-7. The possible basic polymer units in a one-to-one copolymer formed from ethylene and propylene are

$$-CH_2CH_2\underset{\underset{CH_3}{|}}{CH}CH_2-$$

or

$$\{CH_2CH_2CH_2\underset{\underset{CH_3}{|}}{CH}\}$$

There will also be runs of ethylene units $-\!(CH_2CH_2)\!-$ and propylene units $-\!(CH_2\underset{\displaystyle CH_3}{\underset{|}{C}}H)\!-$ or $-\!(\underset{\displaystyle CH_3}{\underset{|}{C}}HCH_2)\!-$.

N-8. The vuclanization of natural rubber involves the formation of -S-S- cross-links between the polyisoprene chains. See the marginal comment on page 1028.